Foundation of Engineering Mechanics

V. Palmov, Vibrations of Elasto-Plastic Bodies

Springer-Verlag Berlin Heidelberg GmbH

V. Palmov

Vibrations of Elasto-Plastic Bodies

Translated by A. Belyaev

With 57 Figures

Springer

Series Editors:
Prof. Dr. V. I. Babitsky
Loughborough University
Department of Mechanical Engineering
LE11 3TU Loughborough, Leicestershire
Great Britain

Prof. Dr. J. Wittenburg
Universität Karlsruhe (TH)
Institut für Mechanik
Kaiserstraße 12
D-76128 Karlsruhe / Germany

Author:
Prof. Dr. Vladimir Palmov
State Technical University of St. Petersburg
Department of Mechanics and Control Processes
Dean of the Faculty of Physics and Mechanics
Polytekhnicheskaya str. 29
195251 St. Petersburg / Russia

Translator:
Dr. Alexander Belyaev
Johannes Kepler Universität Linz
Institut für Mechanik und Maschinenlehre
Altenbergstraße 69
4040 Linz / Österreich

ISBN 978-3-642-08352-5

Cataloging-in-Publication Data applied for

Die Deutsche Bibliothek - CIP-Einheitsaufnahme
Palmov, Vladimir A.: Vibrations of elasto-plastic bodies / V. Palmov. Transl. by A. Belyaev.
(Foundations of engineering mechanics)
ISBN 978-3-642-08352-5 ISBN 978-3-540-69636-0 (eBook)
DOI 10.1007/978-3-540-69636-0

Originally published by Springer-Verlag Berlin Heidelberg New York in 1998

Typesetting: Camera ready by author
Cover: de'blik, Berlin
SPIN: 10650856 62/3020 - 5 4 3 2 1 0 - Printed on acid -free paper

Preface

Undeservedly little attention is paid in the vast literature on the theories of vibration and plasticity to the problem of steady-state vibrations in elastoplastic bodies. This problem, however, is of considerable interest and has many important applications.

The problem of low-cyclic fatigue of metals, which is now in a well developed state is one such application. The investigations within this area are actually directed to collecting experimental facts about repeated cyclic loadings, cf. [47]. Theoretical investigations within this area usually consider the hysteretic loops and the construction of models of plasticity theory which are applicable to the analysis of repeated loadings and the study of the simplest dynamic problems.

Another area of application of the theory of the vibration of elastoplastic bodies is the applied theory of amplitude-dependent internal damping. Another name for this theory is the theory of energy dissipation in vibrating bodies. In accordance with the point of view of Davidenkov "internal damping" in many metals, alloys and structural materials under considerable stress presents exactly the effect of microplastic deformations. Therefore, it may be described by the methods of plasticity theory. This point of view is no doubt fruitful for the theory of energy dissipation in vibrating bodies, as it allows one to write down the constitutive equations appropriate both for vibrational analysis of three-dimensional stress states and an investigation of nonharmonic deformation. These problems are known to be important for the theory of internal damping.

The theory of the vibration of elastoplastic bodies contains two principal questions. The first question is the choice of an appropriate version of the plasticity theory, while the second one is a search for the mathematical tools for investigating dynamic problems.

The microplasticity theories developed in the papers by Ishlinsky, Besseling, Novozhilov, Kadashevich, Iwan *et al.* turn out to be the most suitable approaches to the dynamic problems. The most appropriate mathematical tools are the method of harmonic linearisation and the method of statistical linearisation which can be specially adapted to nonlinear tensorial equations. They allow harmonic, polyharmonic and random vibrations to be studied in the framework of the same model.

The two first Chapters are of introductory character. A brief overview of the basic concepts and methods of continuum mechanics is given there. Considerable attention is paid to the thermodynamic principles, the rheological models and the construction of microplasticity theories.

The second Chapter is devoted to the substantiation of the applicability and expediency of utilizing plasticity theory in the problem of "amplitude-dependent" internal damping. An analysis of a one-dimensional variant of a simple microplasticity theory, namely the theory of elastoplastic materials with the Ishlinsky rheological model, is given, and the cases of loading, unloading and cyclic deformation are considered. It is shown that the well-known Davidenkov's formula which is widely used in the theory of "amplitude-dependent" internal friction can be derived from the general equations by means of plausible physical arguments.

A comparison of the behaviour of elastoplastic materials with some known effects of amplitude-dependent internal friction is also given. For an example problem, the present theory is compared with the theories by Pisarenko, Panovko and Sorokin.

The equations of elastoplastic materials in the case of a three-dimensional state of stress are studied in the third Chapter. The properties of the curves of loading and unloading, as well as the properties of the cyclic deformation diagrams, are investigated. The analysis of the cyclic deformation is performed by the method of harmonic linearisation. A complex shear modulus is introduced and it is shown that the complex shear modulus does not depend upon the deformation frequency, but does depend on the amplitude value of the shear stress intensity.

The fourth Chapter is concerned with single-frequency vibrations of elastoplastic bodies of arbitrary form. The Galerkin method in the form of a one-term-approximation with consequent use of the method of harmonic linearisation is applied to the analysis of resonance and near-resonance vibrations. The normal modes of free elastic vibration of a body under consideration are taken as the coordinate functions for the Galerkin method. Closed form expressions for the coefficients of the equation for the amplitude-frequency characteristic and for the decrement of free decaying vibrations due to the vibration mode are derived. These formulae are of a general character and applicable for analysis of vibrations in homogeneous and heterogeneous bodies having an arbitrary state of stress. The formulae take an especially simple form provided that a homogeneous stress state is realised in the volume under deformation. This shows that the vibration decrement depends upon the type of stress state and the amplitude of the shear stress intensity. A special Section is concerned with the consideration of some specific states of stress. The vibration decrements due to the derived equations are compared with those obtained in experimental tests, and satisfactory agreement between the theory and test data, for the energy dissipation, is ascertained.

The fifth Chapter is devoted to the analysis of random deformations of elastoplastic materials. An analysis is performed by means of the method of statistical linearisation, and in this case the problem is also proved to reduce to the concept of complex shear modulus. The value of the modulus is shown to depend on the normalised spectral density of stresses, the root-mean-square of the shear stress intensity and frequency. A few calculations for the complex shear modulus in the cases of polyharmonic deformations and a simple type of broad-band random deformations are given. A special Section is devoted to the evaluation of the reaction of a material subject to random loading, provided that its reaction under harmonic deformation is known from tests.

Random vibrations of elastoplastic bodies are studied in the sixth Chapter. A solution of the boundary-value problem is obtained by Galerkin's method by utilizing a series expansion in terms of normal modes of free elastic body. Closed form expressions for the decrements of free decaying vibration due to the normal modes are given in the case of asymptotically small plastic strains. A cross-influence of the frequency components of vibration is observed. The equations for a random broad-band loading and a particular case of random loading with a line spectrum are derived.

A Section of the sixth Chapter is devoted to a theoretical analysis of biharmonic vibrations for those particular situations which are described in the experimental investigations of the energy dissipation performed in Kiev. A comparison of the theoretical prediction and the tests data is provided, and satisfactory agreement is established.

Vibration of an infinite plate subject to broad-band loading is studied at the end of the Chapter. This problem is remarkable in that regard that its result is essentially dependent on the character of the energy dissipation over the whole frequency range, and not on near-resonance domains as is the case for finite bodies. The spectral theory of homogeneous random fields is used to construct the solution. Further, the theory of asymptotic evaluations of integrals are applied for estimation of the spectral density of vibrations. A simple closed form solution is obtained in the case of linear dissipation of energy. While modelling the energy dissipation by the methods of plasticity theory the problem reduces to an implicit nonlinear integral equation, its solution being easily obtained by the method of successive approximations for any particular spectral density.

The approach applied to the analysis of plate vibrations is generalised to the vibrations in shallow shells. The solution of this elastoplastic problem turns out to be very complicated. For this reason, some simplifications of the problem are considered.

The last two Chapters are devoted to the problem of vibration propagation in dissipative media with ordinary and complex structure. The simplest statement of the problem is as follows. Given a semi-infinite rod subject to a harmonic or non-harmonic loading at its end, the vibration decay with distance from the vibrational source is sought.

A rod of ordinary structure with various rheological constitutive equations is considered in the seventh Chapter. An elastoplastic material is also studied among other materials. The vibration turns out to occupy either the whole rod or a finite domain near the loaded end, the result depending upon the constitutive equation of material. For example, vibration propagates down a rod for a finite distance in the case of a rigid-plastic material with a linear hardening. In the case of so-called amplitude-dependent "internal friction" vibration occupies the entire rod and possesses a saturation property. This property implies that there exists a limit of the attainable level of vibration which depends on the material properties, the distance from the vibration source and frequency, and does not depend on the loading intensity. Considerable difficulties arise when analysing vibrations of rods of finite length. A compact solution turns out to be obtained by means of a special modification of the method of statistical linearisation. This allows one to obtain the solution of the nonlinear problem by means of a single term approximation without the need to use the conventional many term expansion in terms of the normal modes.

Vibrations in media with complex structure are analysed in the eighth and final Chapter. A specific variant of a medium with complex structure is considered. The medium is postulated to consist of a certain carrier medium described by the classical equations of the dynamic theory of elasticity, and an infinite set of non-interacting oscillators with a continuous spectrum of eigenfrequencies attached to the carrier structure. Some physical arguments on the applicability of the model of this medium with complex structure for a phenomenological description of the behaviour of complex dynamical structures are given. The one-dimensional theory is analysed in detail. The solution of the problem of vibration propagation indicates that the character of the vibration decay depends weakly on the oscillators' damping and is mainly determined by the spectral properties of the set of oscillators. A spatial decay of vibration in this medium turns out to be finite even for oscillators with vanishingly damping. This fact, which may seem to be a paradox, is explained in such a way that the oscillators act as dynamic absorbers. An application of the theory for some practical problems is reported. It is mentioned that the theory provides a convenient way of describing and analysing complex modern dynamic structures. The theory of vibroconductivity is proposed as a description of high frequency vibration in complex structures. The theory is based on an evident analogy between high frequency vibration and thermal motion. Its boundary-value problem is identical to that of thermal conductivity with a distributed heat sink, the latter modelling the vibration absorption by the secondary systems of the complex structures.

L.M. Zubov, P.A. Zhilin and O.A. Zverev looked through the manuscript of the Russian version of this book. The author expresses his sincerely gratitude for their discussions and proposals. The author is also thankful to A.I. Lurie for his thoughts on the problems considered in the Russian version of the book and useful advises.

The author is thankful to A.K. Belyaev who translated the book into English and identified several improvements. The first person who read the manuscript in English was S. McWilliam. The author expresses his sincere gratitude for the valuable advice he received.

Translator's Preface

Professor V.A. Palmov has been my teacher since 1972, and translating this book has been a great pleasure and honour. The present book is actually a revised version of the original, which was published in Russian in 1976.

Although I tried to do my best while translating the book, some typing and other mistakes may have occurred in the translation, for which I would like to apologize. Also, I did not always succeed in finding the original references in English, and had to inversely translate them from Russian into English. I apologise to the authors for possible inaccuracies in the titles of some references.

Firstly, I would like to express my sincere gratitude to the author with whom I had the pleasure and privilege to communicate with permanently during the translation. He was always ready to discuss the smallest details of the translation and was tolerant of my mistakes.

Secondly, I appreciate the constant attention and support of Professor Hans Irschik, from the University of Linz, Austria. I carried out the translation during my stay at his Institute, and his vivid interest and useful advice helped me to find the correct terminology and expressions.

And finally, I am greatly obliged to Dr. Stewart McWilliam, from the University of Nottingham, UK who read the manuscript very thoroughly and not only converted my Russian English into British English but also made a series of profound remarks on the manuscript.

Contents

1
Foundations and equations of continuum mechanics

1.1 Kinematics of a continuous medium

Real solids have discrete structure. Descriptions of the mechanics of such solids usually ignore this important physical property and adopt the so-called continuous medium approach, i.e. the medium is considered to be a continuous aggregate of material points in motion. The properties of the medium are assumed not to change even if an infinitesimally small piece is considered. Continuous medium is a useful abstraction which allows one to apply differential and integral calculus while studying the motion of deformable bodies.

The motion of the continuous medium is considered in a certain fixed system of coordinates, e.g. a system of rectangular Cartesian coordinates $x_1x_2x_3$ depicted in Fig. 1.1. The position of a generic material point M of the continuum at time t is given by the position vector $\mathbf{r}$ which has the components x_1, x_2 and x_3

$$\mathbf{r} = \mathbf{i}_k x_k, \tag{1.1}$$

where $\mathbf{i}_k$ are referred to unit base vectors of the coordinate axes.

Use is made throughout of the traditional convention in tensor analysis of summation over the repeated subscript (in this particular case over k).

We suppose that each material point of the continuous medium corresponds the only position $\mathbf{r}$ at any instant t of time and vise versa, i.e. each value of $\mathbf{r}$ at any instant t of time corresponds the only material point. It

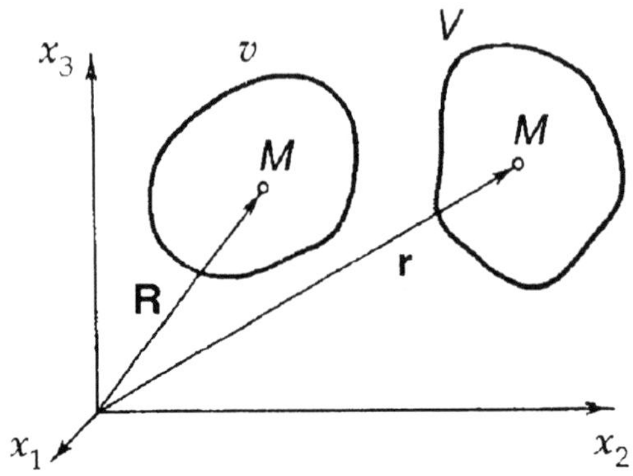

FIGURE 1.1. Initial (v) and current (V) positions of a material volume.

allows one to individualize the material points of the continuum by means of their position at some particular instant of time.

The individualization of the material points in terms of their position at some fixed instant $t = t_0$ of time (e.g. at the initial instant of time) introduces the so-called Lagrangian approach to continuum mechanics. Individualization of the material points in terms of their position at the actual instant of time t introduces the Eulerian approach.

The position of a generic material point M at the initial instant $t = t_0$ of time is given by the vector

$$\mathbf{R} = \mathbf{i}_k X_k, \tag{1.2}$$

where X_k are the components of the vector $\mathbf{R}$.

In the framework of the Lagrangian approach the material point is completely specified by the given value of $\mathbf{R}$. For this reason the posterior motion of the point should be given in the form of a single-valued function of the arguments $\mathbf{R}$ and t

$$\mathbf{r} = \mathbf{r}(\mathbf{R}, t). \tag{1.3}$$

Provided that the dependence (1.3) is given for each point $\mathbf{R}$ of the initial volume v, this dependence is then said to determine the motion of the material volume. The latter is a volume which contains the same material points at any instant of time. Initial (v) and actual (V) positions of the material volume are shown in Fig. 1.1. The Lagrangian coordinate $\mathbf{R}$ of any material point remains constant during the motion of the point. By virtue of (1.3) one can easily find the velocity and the acceleration of the material points

$$\mathbf{v} = \dot{\mathbf{r}} = \frac{\partial \mathbf{r}}{\partial t} = \mathbf{v}(\mathbf{R}, t), \tag{1.4}$$

$$\mathbf{w} = \ddot{\mathbf{r}} = \frac{\partial^2 \mathbf{r}}{\partial t^2} = \mathbf{w}(\mathbf{R}, t). \tag{1.5}$$

In the framework of the Eulerian approach the material points are specified by their position $\mathbf{r}$ at the current instant of time t. This means that a single-valued dependence

$$\mathbf{R} = \mathbf{R}(\mathbf{r}, t) \tag{1.6}$$

is assumed to be known. The dependence (1.6) is the inverse of (1.3).

Substituting (1.6) into (1.4) one obtains the field of velocities in terms of the Eulerian coordinate $\mathbf{r}$ and time t

$$\mathbf{v} = \mathbf{v}(\mathbf{r}, t). \tag{1.7}$$

Let us obtain an equation for the material derivative of some quantity, i.e. the rate of change of this quantity for a given material point using the Eulerian approach. Consider a tensor, vector or scalar field

$$\mathbf{\Lambda} = \mathbf{\Lambda}(\mathbf{r}, t). \tag{1.8}$$

While deriving the equation for the material derivative of $\mathbf{\Lambda}$ one should take into account that $\mathbf{\Lambda}$ depends on time both explicitly and implicitly in terms of $\mathbf{r}$ due to (1.3) wherein the Lagrangian coordinate $\mathbf{R}$ remains constant. Using (1.4) we obtain

$$\frac{d\mathbf{\Lambda}}{dt} = \dot{\mathbf{\Lambda}} = \frac{\partial \mathbf{\Lambda}}{\partial t} + \mathbf{v} \cdot (\nabla \mathbf{\Lambda}) = \frac{\partial \mathbf{\Lambda}}{\partial t} + (\mathbf{\Lambda} \nabla) \cdot \mathbf{v}, \tag{1.9}$$

where ∇ is the Hamilton operator

$$\nabla = \mathbf{i}_k \frac{\partial}{\partial x_k}, \tag{1.10}$$

and the velocity $\mathbf{v}$ should be taken in the form of (1.7).

Here and throughout the rest of this book we use the so-called direct tensor calculus which is conventional in recent papers on mechanics of deformable bodies, cf. [99], [46], [162], [163], and [180]. The reader is assumed to be acquainted with the direct tensor calculus which can be found, for example, in the appendices of the treatises [99] and [100]. Particularly, the dot in eq. (1.9) denotes the scalar product. Vectors and tensors which are not linked by an operation sign form an indefinite product. For instance, such an indefinite product is seen in the parentheses in eq. (1.9) and is understood as an indefinite product of the Hamilton operator and the tensor $\mathbf{\Lambda}$. The result of this product is a tensor whose rank is the rank of tensor $\mathbf{\Lambda}$ plus unit.

The displacement of a generic material point is determined by the difference

$$\mathbf{u} = \mathbf{r} - \mathbf{R}. \tag{1.11}$$

By means of (1.2) and (1.3) the displacement $\mathbf{u}$ can be expressed either in terms of the Lagrangian variables $\mathbf{R}$ and t or in terms of the Eulerian variables $\mathbf{r}$ and t. This leads to two different representations for the displacement

$$\mathbf{u} = \mathbf{u}(\mathbf{R}, t), \quad \mathbf{u} = \mathbf{u}(\mathbf{r}, t). \tag{1.12}$$

Finally, please notice that the introduced functions $\mathbf{u}$, $\mathbf{r}$, $\mathbf{v}$ and $\mathbf{w}$ are assumed to be continuous functions of time and spatial variables $\mathbf{R}$ and $\mathbf{r}$. They are also supposed to have continuous derivatives of the order which is necessary for further analysis.

1.2 Law of conservation of mass

Inertial properties of continuum are characterized by the mass density ρ which is usually given in Eulerian variables, i.e.

$$\rho = \rho(\mathbf{r}, t). \tag{1.13}$$

The mass of the material volume V is given by

$$m = \int_V \rho dV. \tag{1.14}$$

The law of conservation of mass says that the mass contained in a certain material volume remains constant in time, i.e.

$$\dot{m} = 0. \tag{1.15}$$

This equation is an integral form of the mass conservation law. One obtains another form, i.e. a differential form of the law, by substituting (1.14) into (1.15). While estimating the time derivative of the volume integral (1.14) one should bear in mind that change of the mass m occurs both because of time which is explicitly present in ρ (cf. eq. (1.13)) and because of the volume change caused by the motion of the surface O of the material volume V. A surface point moves in space with the velocity of the corresponding material point $\mathbf{v}$, the volume change depending upon the normal component of the velocity only

$$v_n = \mathbf{n} \cdot \mathbf{v},$$

where $\mathbf{n}$ is the exterior unit normal to the surface O.

Taking into account the aforementioned property we can write the following

$$\dot{m} = \int_V \frac{\partial \rho}{\partial t} dV + \int_O \mathbf{n} \cdot \mathbf{v} \rho dO = 0. \tag{1.16}$$

Applying the Ostrogradsky-Gauss integral formula to the surface integral yields

$$\int_V \left[\frac{\partial \rho}{\partial t} + \nabla \cdot (\mathbf{v}\rho) \right] dV = 0. \tag{1.17}$$

Since the material volume V is arbitrary, the function in the square brackets must vanish at any point of the body. i.e.

$$\frac{\partial \rho}{\partial t} + \nabla \cdot (\mathbf{v}\rho) = 0. \tag{1.18}$$

By means of the material derivative of ρ we obtain another equation

$$\dot{\rho} + \rho (\nabla \cdot \mathbf{v}) = 0. \tag{1.19}$$

Equations (1.18) and (1.19) are the differential forms of the mass conservation law.

Finally, let us remind the rule of differentiation of an integral over the material volume with respect to time. Let us consider an integral

$$\mathbf{M} = \int_V \mathbf{\Lambda} \rho dV, \tag{1.20}$$

where $\mathbf{\Lambda}$ is a tensor, a vector or a scalar.

We obtain by analogy with the above

$$\frac{d}{dt} \int_V \mathbf{\Lambda} \rho dV = \int_V \left[\frac{\partial (\mathbf{\Lambda} \rho)}{\partial t} + \nabla \cdot (\rho \mathbf{v} \mathbf{\Lambda}) \right] dV.$$

Simplifying the integrand of this equation by means of (1.18) and (1.9) yields

$$\frac{d}{dt} \int_V \mathbf{\Lambda} \rho dV = \int_V \frac{d\mathbf{\Lambda}}{dt} \rho dV. \tag{1.21}$$

1.3 Strain and rate of strain

The instantaneous motion of an infinitesimally small vicinity of a material point $\mathbf{r}$ is characterized by the tensor $\nabla \mathbf{v}$ where $\mathbf{v}$ is the velocity due to

(1.7). The first invariant of this tensor, i.e. $\nabla \cdot \mathbf{v}$, determines the rate of change of the mass density as follows from the mass conservation law (1.19). The deviator of this tensor is denoted as follows

$$\mathbf{D} = \nabla\mathbf{v} - \frac{1}{3}\nabla \cdot \mathbf{v}\mathbf{E}, \tag{1.22}$$

where $\mathbf{E}$ is the unity tensor.

The symmetric part of the deviator

$$\mathbf{d} = \mathbf{D}^s \tag{1.23}$$

determines the rate of distortion of a small vicinity of a material point whereas the skew symmetric part

$$\boldsymbol{\Omega} = \mathbf{D}^a \tag{1.24}$$

represents the velocity of the rigid-body rotation of the above vicinity. The spin tensor $\boldsymbol{\Omega}$ is expressed in terms of the vorticity vector $\boldsymbol{\omega}$

$$\boldsymbol{\omega} = \frac{1}{2}\nabla \times \mathbf{v} \tag{1.25}$$

as follows

$$\boldsymbol{\Omega} = -\mathbf{E} \times \boldsymbol{\omega} = -\boldsymbol{\omega} \times \mathbf{E}. \tag{1.26}$$

In continuum mechanics, finite deformation is characterized by various strain measures and strain tensors. As we prefer the Eulerian approach we will use the following strain measure

$$\mathbf{g} = (\nabla\mathbf{R}) \cdot (\mathbf{R}\nabla), \tag{1.27}$$

which is referred to as the Almansi strain measure.

Let us find the relation between the material time derivative of $\mathbf{g}$ and the motion characteristics. To this end we take an expression for the velocity $\mathbf{v}$ as the material time derivative of $\mathbf{u}(\mathbf{r}, t)$, i.e.

$$\mathbf{v} = \dot{\mathbf{u}} = \frac{\partial \mathbf{u}}{\partial t} + \mathbf{v} \cdot (\nabla\mathbf{u}), \tag{1.28}$$

and obtain the product $\nabla\mathbf{v}$ such that

$$\nabla\mathbf{v} = \frac{\partial \nabla\mathbf{u}}{\partial t} + (\nabla\mathbf{v}) \cdot (\nabla\mathbf{u}) + \mathbf{v} \cdot (\nabla\nabla\mathbf{u}). \tag{1.29}$$

This leads to the following relation

$$(\nabla\mathbf{v}) \cdot (\mathbf{E} - \nabla\mathbf{u}) = (\nabla\mathbf{u})^{\bullet},$$

which due to the definition of displacement (1.11) may be written in the following form

$$(\nabla \mathbf{R})^{\bullet} = -(\nabla \mathbf{v}) \cdot (\nabla \mathbf{R}). \tag{1.30}$$

Transposition of the tensors in both sides of this equations yields the following auxiliary equation

$$(\mathbf{R}\nabla)^{\bullet} = -(\mathbf{R}\nabla) \cdot (\mathbf{v}\nabla). \tag{1.31}$$

By means of (1.30) and (1.31) we easily obtain the material derivative

$$\dot{\mathbf{g}} = -(\nabla \mathbf{v}) \cdot \mathbf{g} - \mathbf{g} \cdot (\mathbf{v}\nabla). \tag{1.32}$$

Materials are known to behave essentially differently at the volume change than at the shape change. For this reason it is reasonable to extract the dilatation from the tensor $\mathbf{g}$. The dilatation is determined by the mass density and is bound up with the third invariant of $\mathbf{g}$ as follows, cf. [99], [100], [46] and [37]

$$\rho = \rho_0 \sqrt{I_3(\mathbf{g})}, \tag{1.33}$$

where ρ_0 is the initial mass density and $I_3(\mathbf{g})$ is the third invariant of tensor $\mathbf{g}$.

Equation (1.33) can be proven by substituting it into (1.19). By means of the following relation

$$\frac{\partial I_3}{\partial \mathbf{g}} = I_3(\mathbf{g})\, \mathbf{g}^{-1}, \tag{1.34}$$

cf. [99] and [100], and by virtue of (1.32) we obtain consequently

$$\dot{\rho} = \rho_0 \frac{\partial \sqrt{I_3}}{\partial \mathbf{g}} : \dot{\mathbf{g}} = -\frac{\rho}{2} \mathbf{g}^{-1} : [(\nabla \mathbf{v}) \cdot \mathbf{g} + \mathbf{g} \cdot (\mathbf{v}\nabla)] = -\rho \nabla \cdot \mathbf{v}, \tag{1.35}$$

which shows that (1.19) is satisfied. In (1.35) : denotes the double scalar product.

The third invariant of $\mathbf{g}$ has been shown to be responsible for the mass density.

In order to describe a shape change at deformation we introduce the following tensor

$$\mathbf{\Lambda} = \left(\frac{\rho}{\rho_0}\right)^{-2/3} \mathbf{g}. \tag{1.36}$$

It is easy to prove that its third invariant is equal to unity. Indeed, we have

$$I_3(\mathbf{\Lambda}) = \left(\frac{\rho}{\rho_0}\right)^{-2} I_3(\mathbf{g}) = 1, \tag{1.37}$$

the latter equality being a direct consequence of (1.33).

Therefore, tensor $\mathbf{\Lambda}$ characterises deformation without change of the mass density, i.e. distortion of the shape of small elements of the medium. For this reason, this tensor is referred to as the *strain of shape change.*

Let us link now the strain of shape change $\mathbf{\Lambda}$ to the motion characteristics. To this aim, consider the material derivative

$$\dot{\mathbf{\Lambda}} = \left(\frac{\rho}{\rho_0}\right)^{-2/3} \dot{\mathbf{g}} - \frac{2}{3\rho_0}\left(\frac{\rho}{\rho_0}\right)^{-2/3-1} \dot{\rho}\mathbf{g}.$$

Substituting the expressions for the derivatives of $\mathbf{g}$ and ρ given by (1.32) and (1.19) yields

$$\dot{\mathbf{\Lambda}} = -\mathbf{D} \cdot \mathbf{\Lambda} - \mathbf{\Lambda} \cdot \mathbf{D}^T \tag{1.38}$$

or

$$\dot{\mathbf{\Lambda}} = -\mathbf{d} \cdot \mathbf{\Lambda} - \mathbf{\Lambda} \cdot \mathbf{d} - \mathbf{\Omega} \cdot \mathbf{\Lambda} + \mathbf{\Lambda} \cdot \mathbf{\Omega}, \tag{1.39}$$

transpose being denoted by T.

Presence of the tensor $\mathbf{D}$, eq. (1.22), in eq. (1.38) emphasizes the mechanical meaning and validity of the name of the tensor $\mathbf{\Lambda}$ because $\mathbf{D}$ characterises the motion of a small neighbourhood of the point without volume change.

Remark: Consider now equation (1.38), not as a consequence of the definition (1.36) but as an equation for the symmetric tensor $\mathbf{\Lambda}$, and complete it with the initial condition

$$t = t_0, \quad \mathbf{\Lambda} = \mathbf{E}. \tag{1.40}$$

Let us prove that the third invariant of the tensor $\mathbf{\Lambda}$ is always equal to unity. We make use of (1.34) which is valid for any symmetric tensor and express the material derivative of $I_3(\mathbf{\Lambda})$ using (1.38). The following chain of equalities

$$\begin{aligned}\frac{dI_3}{dt} &= \frac{\partial I_3(\mathbf{\Lambda})}{\partial \mathbf{\Lambda}} : \dot{\mathbf{\Lambda}} = -I_3(\mathbf{\Lambda})\left[\mathbf{\Lambda}^{-1} : (\mathbf{D} \cdot \mathbf{\Lambda}) + \mathbf{\Lambda}^{-1} : (\mathbf{\Lambda} \cdot \mathbf{D}^T)\right] = \\ &= -I_3(\mathbf{\Lambda})\left[\mathbf{\Lambda} \cdot \mathbf{\Lambda}^{-1} : \mathbf{D} + (\mathbf{\Lambda}^{-1} \cdot \mathbf{\Lambda}) : \mathbf{D}^T\right]\end{aligned}$$

holds in a fixed coordinate system.

The following equalities

$$\mathbf{\Lambda} \cdot \mathbf{\Lambda}^{-1} = \mathbf{\Lambda}^{-1} \cdot \mathbf{\Lambda} = \mathbf{E}, \ \mathbf{E} : \mathbf{D} = \mathbf{E} : \mathbf{D}^T = 0,$$

are valid due to the definitions of inverse tensor and deviator. Using them we obtain

$$\frac{dI_3}{dt} = 0.$$

Hence, the invariant $I_3(\mathbf{\Lambda})$ remains constant during the motion of the medium and is equal to unity in any material point due to the initial condition (1.40).

1.4 Rotation tensor

In addition to the orthogonal trihedron of unit base vectors $\mathbf{i}_k$ of a fixed coordinate system x_1, x_2 and x_3, we also introduce an orthogonal trihedron of unit vectors $\mathbf{e}_k$. The trihedron $\mathbf{e}_k$ is assumed to rotate with the rotational velocity $\boldsymbol{\omega}$ of the medium at this point, so according to the theory of rigid body kinematics

$$\dot{\mathbf{e}}_k = \boldsymbol{\omega} \times \mathbf{e}_k$$

or, using (1.26),

$$\dot{\mathbf{e}}_k = -\boldsymbol{\Omega} \cdot \mathbf{e}_k. \tag{1.41}$$

Here we assume that at the initial instant of time the following initial conditions

$$t = t_0, \quad \mathbf{e}_k = \mathbf{i}_k \tag{1.42}$$

are prescribed. The tensor of finite rotation of an infinitesimally small vicinity of a material point is defined as follows

$$\mathbf{B} = \mathbf{e}_k \mathbf{i}_k. \tag{1.43}$$

Note some evident properties of the definition (1.43)

$$\mathbf{B}^T \cdot \mathbf{B} = \mathbf{E}, \quad \mathbf{B} \cdot \mathbf{B}^T = \mathbf{E}, \tag{1.44}$$

which expresses the characteristic properties of any rotation tensor. A consequence of the definitions (1.43), (1.41) and (1.42) is that

$$\dot{\mathbf{B}} = -\boldsymbol{\Omega} \cdot \mathbf{B} \tag{1.45}$$

and the initial condition is such that when

$$t = t_0, \quad \mathbf{B} = \mathbf{E}. \tag{1.46}$$

Let us prove an inverse statement: tensor $\mathbf{B}$ determined by differential equation (1.45) and initial condition (1.46) is a rotation tensor, i.e. it satisfies conditions (1.44).

To this end we take the derivative of the product $\mathbf{B}^T \cdot \mathbf{B}$ due to (1.45). From this we obtain the following chain of equalities

$$\frac{d}{dt}\left(\mathbf{B}^T \cdot \mathbf{B}\right) = \left(\dot{\mathbf{B}}\right)^T \cdot \mathbf{B} + (\mathbf{B})^T \cdot \dot{\mathbf{B}} = -(\boldsymbol{\Omega} \cdot \mathbf{B})^T \cdot \mathbf{B} - \mathbf{B}^T \cdot \boldsymbol{\Omega} \cdot \mathbf{B} =$$
$$= \mathbf{B}^T \cdot \boldsymbol{\Omega} \cdot \mathbf{B} - \mathbf{B}^T \cdot \boldsymbol{\Omega} \cdot \mathbf{B} = \mathbf{0}.$$

Hence, at any instant of time the product $\mathbf{B}^T \cdot \mathbf{B}$ remains a constant tensor and according to the initial conditions is equal to $\mathbf{E}$. Therefore, the first condition (1.44) is satisfied.

Consider now the product

$$\mathbf{A} = \mathbf{B} \cdot \mathbf{B}^T.$$

Due to (1.45) the time derivative of this equation is given by

$$\dot{\mathbf{A}} = -\boldsymbol{\Omega} \cdot \mathbf{A} + \mathbf{A} \cdot \boldsymbol{\Omega}, \tag{1.47}$$

which is an ordinary differential equation provided that the Lagrangian approach is used. Due to (1.46) the initial condition is given by

$$t = t_0, \quad \mathbf{A} = \mathbf{E}. \tag{1.48}$$

For a given function $\boldsymbol{\Omega} = \boldsymbol{\Omega}(t)$ a solution of the Cauchy problem, eqs. (1.47) and (1.48), may be obtained by the method of successive approximations [6], i.e.

$$\dot{\mathbf{A}}_{n+1} = -\boldsymbol{\Omega} \cdot \mathbf{A}_n + \mathbf{A}_n \cdot \boldsymbol{\Omega}, \quad \mathbf{A}_0 = \mathbf{E},$$

$$t = t_0, \quad \mathbf{A}_{n+1} = \mathbf{E}.$$

The iteration process converges at the first step and yields

$$\mathbf{A}_n = \mathbf{E}.$$

Hence, for $n \to \infty$ we have

$$\mathbf{A} = \mathbf{E},$$

and therefore the second equality (1.45) is also satisfied.

We proceed now to the evaluation of the material derivative in a rotating coordinate system. Consider a tensor of the second rank $\mathbf{M}$ given in a rotating base $\mathbf{e}_k$ such that

$$\mathbf{M} = \mathbf{e}_k \mathbf{e}_l M_{kl}. \tag{1.49}$$

The material derivative of this, in a rotating coordinate system, is given by

$$\mathbf{M}^{\nabla} = \mathbf{e}_k \mathbf{e}_l \dot{M}_{kl}. \tag{1.50}$$

The superscript ∇ is usually used for the Jaumann derivative. Let us prove that definition (1.50) does correspond to the Jaumann derivative. To this end, we shall evaluate the material derivative of $\mathbf{M}$ (1.49) by means of equation (1.41). The result is

$$\dot{\mathbf{M}} = \mathbf{M}^{\nabla} - \boldsymbol{\Omega} \cdot \mathbf{M} + \mathbf{M} \cdot \boldsymbol{\Omega},$$

which allows us to get an invariant expression

$$\mathbf{M}^{\nabla} = \dot{\mathbf{M}} + \boldsymbol{\Omega} \cdot \mathbf{M} - \mathbf{M} \cdot \boldsymbol{\Omega}. \tag{1.51}$$

The tensor on the right hand side is the Jaumann derivative of $\mathbf{M}$, cf. [152] and [153], that completes the proof. Hence the derivative in a rotating coordinate system coincides with the Jaumann derivative and justifies the above denotation.

Below we will use the Jaumann derivative of tensor $\boldsymbol{\Lambda}$. Its explicit expression is found by means of (1.39) to be

$$\boldsymbol{\Lambda}^{\nabla} = -\mathbf{d} \cdot \boldsymbol{\Lambda} - \boldsymbol{\Lambda} \cdot \mathbf{d}, \tag{1.52}$$

with tensor $\mathbf{d}$ being given by formula (1.23). Note that the latter equation does not contain tensor $\boldsymbol{\Omega}$ which determines the angular velocity of rotation of the vicinity of the material point. This fact demonstrates the convenience of the rotational base.

1.5 Equations of dynamics for a continuous medium

We will describe the state of stress in a continuum by the Cauchy true stress tensor $\boldsymbol{\tau}$. The traction vector on an arbitrary surface with exterior unit normal $\mathbf{n}$ is given by the Cauchy formula, cf. [99], [192] and [100]

$$\boldsymbol{\tau}_n = \mathbf{n} \cdot \boldsymbol{\tau}. \tag{1.53}$$

Further, the external load acting on the material points is given by a body force vector $\mathbf{K}$.

The equations governing the dynamics of an arbitrary material volume V have the following form, cf. [99], [192] and [100]

$$\frac{d}{dt} \int_V \mathbf{v} \rho dV = \int_V \mathbf{K} \rho dV + \int_O \mathbf{n} \cdot \boldsymbol{\tau} dO, \tag{1.54}$$

$$\frac{d}{dt}\int_V \mathbf{r}\times\mathbf{v}\rho dV = \int_V \mathbf{r}\times\mathbf{K}\rho dV - \int_O \mathbf{n}\cdot\boldsymbol{\tau}\times\mathbf{r}dO, \tag{1.55}$$

where O is the boundary surface of volume V. The first equation is concerned with the law of change of linear momentum, while the second is concerned with the law of change of moment of momentum. Equations (1.54) and (1.55) represent an integral form of the equations of dynamics. Another form, i.e. a differential form, is obtained by applying the Ostrogradsky-Gauss theorem. The result is well known

$$\nabla\cdot\boldsymbol{\tau} + \rho(\mathbf{K} - \dot{\mathbf{v}}) = 0, \tag{1.56}$$

$$\boldsymbol{\tau}^T = \boldsymbol{\tau}. \tag{1.57}$$

The latter equation expresses the symmetry nature of the stress tensor.

In what follows we will decompose the stress tensor into a spherical part and a deviator, such that

$$\boldsymbol{\tau} = \sigma\mathbf{E} + \mathbf{s}. \tag{1.58}$$

Here σ is the mean normal stress

$$\sigma = \frac{1}{3}\mathbf{E}:\boldsymbol{\tau}, \tag{1.59}$$

and $\mathbf{s}$ is a stress deviator. The characteristic property of a deviator is that its first invariant vanishes, i.e.

$$\mathbf{E}:\mathbf{s} = \mathbf{0}. \tag{1.60}$$

1.6 The laws of thermodynamics

The content of the previous sections is standard. The same scheme is used in the majority of the works on mechanics of continuous media, elasticity theory, plasticity theory, theory of fluids and gases. Proceeding now to a brief introduction of the principles and ideas of thermodynamics, it should be pointed out that there exists no conformity of opinions and the authors tried to give their own interpretations of the thermodynamic principles. In this book we will follow the book by Truesdell [180], where, to the author's knowledge, the most modern and compact explanation of the thermodynamic principles is given. A similar, however not identical, interpretation can be found in the book by Sedov [162].

Our consideration is restricted to thermomechanical phenomena only. Then the *first law of thermodynamics* or the law of energy conservation for an arbitrary material volume may be written in the following form

$$\dot{T} + \dot{u} = p + q. \tag{1.61}$$

Here T is the kinetic energy of the material volume

$$T = \frac{1}{2} \int_V \mathbf{v} \cdot \mathbf{v} \rho dV \tag{1.62}$$

and u its internal energy. The latter is assumed to be an additive function of mass of the material volume

$$u = \int_V U \rho dV \tag{1.63}$$

with U being referred to as the *specific internal energy* (per unit mass). The quantities on the right hand side of (1.61) have the following meaning: p is the power of the prescribed body forces and surface tractions of the material volume

$$p = \int_V \mathbf{K} \cdot \mathbf{v} \rho dV + \int_O \boldsymbol{\tau}_n \cdot \mathbf{v} dO \tag{1.64}$$

and q is the rate of heat supply

$$q = \int_V b dV - \int_O \mathbf{n} \cdot \mathbf{h} dO. \tag{1.65}$$

Here b denotes the heat received from external sources and is called the heating supply. The surface integral describes the heat supplied to the volume through its surface, $\mathbf{h}$ being the heat flux.

The law of energy conservation (1.61) is written in an integral form. One can obtain a differential form provided that the fields are differentiable. First, by means of the rule of differentiation of integrals over the material volume we obtain

$$\dot{T} = \int_V \dot{\mathbf{v}} \cdot \mathbf{v} \rho dV, \quad \dot{u} = \int_V \dot{U} \rho dV. \tag{1.66}$$

Then, with help of the Cauchy formula and the Ostrogradsky-Gauss integral theorem we find

$$\begin{aligned} \int_O \mathbf{n} \cdot \mathbf{h} dO &= \int_V \nabla \cdot \mathbf{h} dV, \\ \int_O \boldsymbol{\tau}_n \cdot \mathbf{v} dO &= \int_V \nabla \cdot (\boldsymbol{\tau} \cdot \mathbf{v}) \, dV. \end{aligned} \tag{1.67}$$

The latter integral can be written as follows

$$\int_V \nabla \cdot (\boldsymbol{\tau} \cdot \mathbf{v})\, dV = \int_V [(\nabla \cdot \boldsymbol{\tau}) \cdot \mathbf{v} + \boldsymbol{\tau} : (\mathbf{v}\nabla)]\, dV. \tag{1.68}$$

Substituting (1.66), (1.67) and (1.68) into (1.61) and simplifying the result by means of eqs. (1.56) and (1.57), yields

$$\int_V \left[\rho\dot{U} - \boldsymbol{\tau} : (\nabla\mathbf{v}) - b + \nabla \cdot \mathbf{h}\right] dV = 0. \tag{1.69}$$

Since the latter equation is valid for an arbitrary material volume, the integrand must vanish, yielding the differential form of the first law of thermodynamics

$$\rho\dot{U} = \boldsymbol{\tau} : (\nabla\mathbf{v}) + b - \nabla \cdot \mathbf{h}. \tag{1.70}$$

The *second law of thermodynamics* is written in the form of the Clausius-Duhem inequality which says that the rate of change of the internal entropy of the volume is not less than the rate of the entropy supplied to the volume by external heat sources.

Mathematically, the second law is given by

$$\dot{s} \geq \int_V \frac{b}{\theta} dV - \int_O \frac{\mathbf{n} \cdot \mathbf{h}}{\theta} dO, \tag{1.71}$$

with s being the internal entropy of the material volume and θ being the field of absolute temperature of the continuum ($\theta > 0$). The first term on the right hand side of inequality (1.71) is the entropy supplied to the volume directly while the second term describes the entropy supplied to the volume through its surface.

Internal entropy is supposed to be an additive function of the mass of the material volume

$$s = \int_V S\rho dV, \tag{1.72}$$

where S is referred to as the *specific internal entropy* (per mass unit). By means of the differentiation rule (1.21)

$$\dot{s} = \int_V \dot{S}\rho dV,$$

the Ostrogradsky-Gauss integral theorem

$$\int_O \mathbf{n} \cdot \frac{\mathbf{h}}{\theta} dO = \int_V \nabla \cdot \left(\frac{\mathbf{h}}{\theta}\right) dV = \int_V \left[\frac{\nabla \cdot \mathbf{h}}{\theta} - \frac{\mathbf{h} \cdot \nabla\theta}{\theta^2}\right] dV$$

and due to the fact that the material volume is arbitrary we easily obtain a differential form of the second law

$$\rho \dot{S} \geq \frac{1}{\theta}(b - \nabla \cdot \mathbf{h}) + \frac{1}{\theta^2}\mathbf{h} \cdot (\nabla \theta) \, . \tag{1.73}$$

Another form of the laws of thermodynamics seems more convenient. First, we eliminate the heat flux $\mathbf{h}$ from (1.73) by means of the first law (1.70) and then multiply both sides of the inequality by $\theta > 0$. The result is

$$\rho \theta \dot{S} \geq \rho \dot{U} - \boldsymbol{\tau} : (\nabla \mathbf{v}) + \frac{1}{\theta}\mathbf{h} \cdot (\nabla \theta) \, . \tag{1.74}$$

Next, we introduce the Helmholtz thermodynamic potential F, which is the *specific free energy*, by means of the following equation

$$U = \theta S + F, \quad (F = U - \theta S) \, . \tag{1.75}$$

Substituting (1.75) into (1.74) and (1.70) yields

$$\boldsymbol{\tau} : (\nabla \mathbf{v}) - \rho \dot{F} - \rho S \dot{\theta} - \frac{1}{\theta}\mathbf{h} \cdot (\nabla \theta) \geq 0, \tag{1.76}$$

$$\rho \theta \dot{S} = \boldsymbol{\tau} : (\nabla \mathbf{v}) - \rho \dot{F} - \rho S \dot{\theta} + b - \nabla \cdot \mathbf{h}, \tag{1.77}$$

respectively. Note that equations (1.77) and (1.76) for the first and second laws of thermodynamics respectively contain the same combination of terms

$$\Phi = \boldsymbol{\tau} : (\nabla \mathbf{v}) - \rho \dot{F} - \rho S \dot{\theta}, \tag{1.78}$$

which is referred to here as the *dissipative function.*

The modified second law of thermodynamics, eq. (1.76), is called the *universal dissipative inequality.*

The modified first law of thermodynamics, eq. (1.77), represents a heat conduction equation with additional terms of mechanical origin.

The third law of thermodynamics, also called the Nernst law, cf. [178] and [179], is given in the form of a requirement that

$$S \to 0 \tag{1.79}$$

when the absolute temperature tends to zero, i.e. $\theta \to 0$.

1.7 Thermodynamic processes and constitutive equations

We suppose here that the thermodynamic state of a continuous medium in a small vicinity of a point is determined by its temperature θ, its gradient $\nabla\theta$ and the tensor $\nabla \mathbf{R}$. These quantities are referred to as *defining variables.*

In the framework of the Eulerian approach the above variables are expressed in terms of the actual coordinate $\mathbf{r}$ and the time t. In thermodynamics it turns out to be necessary to observe the *thermodynamic process* at some material point, which is the time-history of the defining variables for this material point with a Lagrangian coordinate $\mathbf{R}$. This history is prescribed by the following functions of time

$$
\begin{aligned}
\theta^{\tau} &= \theta\left[\mathbf{r}\left(\mathbf{R},\tau\right),\tau\right], \\
\left(\nabla\theta\right)^{\tau} &= \nabla\theta\left[\mathbf{r}\left(\mathbf{R},\tau\right),\tau\right], \\
\left(\nabla\mathbf{R}\right)^{\tau} &= \nabla\mathbf{R}\left[\mathbf{r}\left(\mathbf{R},\tau\right),\tau\right], \quad t_0<\tau<t,
\end{aligned} \tag{1.80}
$$

where the Eulerian coordinate $\mathbf{r}$ is replaced by its expression (1.3).

Functions (1.80) prescribe a thermodynamic process at the point $\mathbf{R}$.

Let us now proceed to the definition of a material.

The thermodynamic behaviour of a continuum at a point $\mathbf{r}$ is said to be completely determined provided that the stress tensor $\boldsymbol{\tau}$, the heat flux $\mathbf{h}$, the specific free energy F and the specific internal entropy S are prescribed as operators over the thermodynamic process at this point, cf. [179], i.e.

$$
\begin{aligned}
\boldsymbol{\tau}\left(\mathbf{R},t\right) &= \boldsymbol{\tau}\left\{\mathbf{R},\theta^{\tau},\left(\nabla\theta\right)^{\tau},\left(\nabla\mathbf{R}\right)^{\tau}\right\}, \\
\mathbf{h}\left(\mathbf{R},t\right) &= \mathbf{h}\left\{\mathbf{R},\theta^{\tau},\left(\nabla\theta\right)^{\tau},\left(\nabla\mathbf{R}\right)^{\tau}\right\}, \\
F\left(\mathbf{R},t\right) &= F\left\{\mathbf{R},\theta^{\tau},\left(\nabla\theta\right)^{\tau},\left(\nabla\mathbf{R}\right)^{\tau}\right\}, \\
S\left(\mathbf{R},t\right) &= S\left\{\mathbf{R},\theta^{\tau},\left(\nabla\theta\right)^{\tau},\left(\nabla\mathbf{R}\right)^{\tau}\right\}, \quad t_0<\tau<t.
\end{aligned} \tag{1.81}
$$

It is essential to mark that all introduced operators are operators which carry out transformations in time only. The argument $\mathbf{R}$ indicates an affiliation of the operator to a particular material point. The operators may differ for distinct material points.

Prescribing equations (1.81) is equivalent to the definition of a material. Equations (1.81) are termed *constitutive equations for the material.* Various materials differ from each another in their constitutive equations, or more exactly, in the operators in (1.81).

Let us now discuss the question of how to write down the constitutive equations for a particular real material. The constitutive equations have been said to reflect the material behaviour under deformation and heating. Therefore, it is necessary to possess some experimental information on this material behaviour. The following question arises: is experimental information alone sufficient to formulate the constitutive equations? The answer is negative. The above problem comprises a general problem of system identification whose solution is presently unknown.

Only approximate solutions to this problem exist at present. The general strategy of the solution consists of three steps which are clearly formulated in [134], [135] and [137].

1. Some constitutive equations are written down. This should be done by means of a qualitative information on the material behaviour only.

2. Some experimental situations are theoretically studied. For example, the reaction of a thin-walled tube subject to tension, torsion and heating may be analysed.

3. A comparison of the theoretical predictions and the test results is performed. This allows the parameters' value and the functions in the chosen constitutive equations to be calculated. The problem of identifications may be solved provided that the agreement of the theoretical predictions and the test results is good for the calculated parameters' value and the functions. If this is not the case, then other constitutive equations must be taken and the above procedure repeated until good agreement between the theory and the test is achieved.

Let us illustrate the strategy by considering the simple example of small strains and distortions of some sort of steel.

At first the steel should be assumed to be an isotropic elastic material. For this reason, the so-called Hooke's law should be accepted. In the case when some temperature terms are present, the Duhamel-Neumann law should be taken. In accordance with these laws the stress tensor is an isotropic linear function of the tensor of linear strain. Young's modulus, Poisson's ratio and a thermal expansion coefficient are the factors of this function. It should be noted however that they are no more than some specified variables at this stage and their values are still unknown.

Next, an axial tension of a cylindrical specimen should be studied and expressions for axial stresses and lateral contractions by means of the introduced parameters should be obtained.

The third step implies a comparison of the computed values and the test results. This comparison allows us the possibility to calculate values for Young's modulus, Poisson's ratio and a thermal expansion coefficient. Being substituted into the chosen constitutive equations, they render the constitutive equations for this particular sort of steel in the case of small strains and distortions. The equations derived describe the behaviour of this steel until the yield stress is not achieved. As this limit is achieved, a discrepancy between the theory and test results begins to grow. Another constitutive equation, namely, an equation for an elastoplastic material will be needed and the above three steps are repeated.

It is easy to conclude that the first step is the most important. It turns out that this step is not absolutely arbitrary, i.e. the hypothetical constitutive equations are to satisfy to a series of indisputable restrictions. In what follows, we enumerate and formulate these restrictions.

Let us clarify how arbitrary the operators in eq. (1.81) can be. With this in view we pose the following question: What restrictions do the laws of mechanics and thermodynamics impose on these operators? As the above laws are the general laws of nature, they must hold for any arbitrary differentiable thermodynamic process. The equation of dynamics (1.56) and the equation of the first law of thermodynamics (1.77) do not impose any restriction on the operators (1.81) since for any thermodynamic process these equations can always be satisfied by a proper choice of external body force $\mathbf{K}$ and external heating supply b. The second law of thermodynamics in the form of the dissipative inequality (1.76) has a special status. In its left hand side there appear only operators (1.81) and no external parameter which can be set. Hence, it is the dissipative inequality only which restricts the choice of operators (1.81). These operators must ensure that the dissipative inequality holds for any absolutely arbitrary thermodynamic process in the material.

By analogy with the second law, the third law of thermodynamics imposes a restriction on the specific entropy and on this account the latter operator in eq. (1.81) must unconditionally satisfy the following requirement

$$S \to 0,$$

when the absolute temperature tends to zero (i.e. $\theta \to 0$) regardless of the values of the other defining parameters.

Let us now introduce the following definitions.

A thermodynamic process in some material is termed *reversible* if condition (1.76) holds with the equality sign, and is called *irreversible* if (1.76) holds with the inequality sign.

This definition says that reversibility or irreversibility of a thermodynamic process depends not only upon its character, but also upon the properties of the material in which this thermodynamic process develops. Thus the same thermodynamic process in one material may be reversible while in another material it may be irreversible.

A material is called *homogeneous* if the operators (1.81) do not contain the Lagrangian coordinate $\mathbf{R}$ explicitly, i.e. the operators (1.81) take the same form for any material point of the medium.

Before we proceed to a definition of an isotropic material, it is worthwhile noting that prescribing a history of the tensor $\nabla\mathbf{R}$ is equivalent to prescribing the history of two tensors, namely the strain measure $\mathbf{g}$ and the rotation tensor $\mathbf{B}$. Indeed, if a history of $\nabla\mathbf{R}$ is known, the strain measure is then obtained by means of (1.27). The rotation tensor is obtained according to (1.45), the spin tensor being given by

$$\boldsymbol{\Omega} = -\left[(\nabla\mathbf{R})^{\bullet} \cdot (\nabla\mathbf{R})^{-1}\right]^{a}, \tag{1.82}$$

as seen from (1.24) and (1.30).

Let us prove the converse statement. Let $\mathbf{g}(\tau)$ and $\mathbf{B}(\tau)$ be known for $t_0 < \tau < t$. Then, by virtue of (1.45) we get the spin tensor

$$\boldsymbol{\Omega}(\tau) = -\dot{\mathbf{B}} \cdot \mathbf{B}^{T}. \tag{1.83}$$

That is, by means of (1.24) the skew symmetric part of tensor $\nabla \mathbf{v}$ is determined as follows

$$(\nabla \mathbf{v})^a = \boldsymbol{\Omega}. \tag{1.84}$$

Given $\mathbf{g}$ and $\boldsymbol{\Omega}$ one can easily derive the Jaumann derivative (1.51) of tensor $\mathbf{g}$

$$\mathbf{g}^{\nabla} = \dot{\mathbf{g}} + \boldsymbol{\Omega} \cdot \mathbf{g} - \mathbf{g} \cdot \boldsymbol{\Omega}. \tag{1.85}$$

Substituting for $\dot{\mathbf{g}}$ from (1.32) then yields

$$(\nabla \mathbf{v})^s \cdot \mathbf{g} + \mathbf{g} \cdot (\nabla \mathbf{v})^s = -\mathbf{g}^{\nabla}. \tag{1.86}$$

The only symmetric tensor $(\nabla \mathbf{v})^s$ which is a solution of this equation is given by

$$(\nabla \mathbf{v})^s = -\int_0^{\infty} e^{-\mathbf{g}\lambda} \cdot \mathbf{g}^{\nabla} \cdot e^{-\mathbf{g}\lambda} d\lambda, \tag{1.87}$$

cf. [6]. Finally, in order to determine the tensor $\nabla \mathbf{R}$ one can use eq. (1.30)

$$(\nabla \mathbf{R})^{\bullet} = -(\nabla \mathbf{v}) \cdot (\nabla \mathbf{R}), \tag{1.88}$$

in which $\nabla \mathbf{v}$ is given due to (1.84), (1.87) and the identity

$$(\nabla \mathbf{v}) = (\nabla \mathbf{v})^s + (\nabla \mathbf{v})^a.$$

An initial condition can be taken, for example in the form

$$t = t_0, \quad (\nabla \mathbf{R}) = \mathbf{E}. \tag{1.89}$$

It is known that the solution of the Cauchy problem, eqs. (1.88) and (1.89), is unique, cf. [6].

Therefore, it has been shown that the prescription of the history of the tensors $\mathbf{g}$ and $\mathbf{B}$ is equivalent to prescription of the history of tensor $\nabla \mathbf{R}$.

It is however seen from (1.36) that prescription of the strain tensor $\mathbf{g}$ is equivalent to prescription of mass density ρ and strain of shape change $\boldsymbol{\Lambda}$.

The above-said is considered to be a sufficient substantiation for consideration of the constitutive equations of the following sort

$$\begin{aligned}
\boldsymbol{\tau}(\mathbf{R}, t) &= \boldsymbol{\tau}\{\theta^{\tau}, (\nabla\theta)^{\tau}, \rho^{\tau}, \boldsymbol{\Lambda}^{\tau}, \mathbf{B}^{\tau}\}, \\
\mathbf{h}(\mathbf{R}, t) &= \mathbf{h}\{\theta^{\tau}, (\nabla\theta)^{\tau}, \rho^{\tau}, \boldsymbol{\Lambda}^{\tau}, \mathbf{B}^{\tau}\}, \\
F(\mathbf{R}, t) &= F\{\theta^{\tau}, (\nabla\theta)^{\tau}, \rho^{\tau}, \boldsymbol{\Lambda}^{\tau}, \mathbf{B}^{\tau}\}, \\
S(\mathbf{R}, t) &= S\{\theta^{\tau}, (\nabla\theta)^{\tau}, \rho^{\tau}, \boldsymbol{\Lambda}^{\tau}, \mathbf{B}^{\tau}\}, \quad t_0 < \tau < t.
\end{aligned} \tag{1.90}$$

A material is said to be *isotropic* if its operators $\boldsymbol{\tau}$, $\mathbf{h}$, F and S at point $\mathbf{r}$ at time instant t do not change their values when an arbitrary initial rigid rotation is superimposed onto the body elements. Formally, this is equivalent to replacement of the motion

$$\mathbf{R} = \mathbf{R}\left(\mathbf{r}, t\right) \tag{1.91}$$

by

$$\mathbf{R}_H = \mathbf{R}\left(\mathbf{r}, t\right) \cdot \mathbf{H}, \tag{1.92}$$

where $\mathbf{H}$ is an arbitrary constant orthogonal tensor. In what follows, the motion according to law (1.92) is called *H-motion*.

Let the defining variables in motion (1.91) be θ, $\nabla\theta$, ρ, $\boldsymbol{\Lambda}$ and $\mathbf{B}$ and let the strain rate of shape change and the spin tensor be denoted by $\mathbf{d}$ and $\boldsymbol{\Omega}$, respectively. It is easy to see that

$$\nabla\mathbf{R}_H = \nabla\mathbf{R} \cdot \mathbf{H},$$

and in H-motion the above values are given by

$$\begin{aligned} \theta_H &= \theta,\ (\nabla\theta)_H = \nabla\theta,\ \rho_H = \rho,\ \boldsymbol{\Lambda}_H = \boldsymbol{\Lambda}, \\ \mathbf{B}_H &= \mathbf{B} \cdot \mathbf{H},\ \mathbf{d}_H = \mathbf{d},\ \boldsymbol{\Omega}_H = \boldsymbol{\Omega}. \end{aligned} \tag{1.93}$$

Only the rotation tensor has changed with the incorporation by a constant multiplier $\mathbf{H}$. The requirement of material isotropy implies the following, cf. [178] and [179]

$$\boldsymbol{\tau}_H = \boldsymbol{\tau},\ \mathbf{h}_H = \mathbf{h},\ F_H = F,\ S_H = S, \tag{1.94}$$

where $\boldsymbol{\tau}_H, \mathbf{h}_H, F_H$ and S_H are the operators (1.90) for H-motion (1.92), while $\boldsymbol{\tau}, \mathbf{h}$, F and S are those for motion (1.91).

In what follows, only isotropic materials are considered.

Let us introduce another general requirement for the structure of operators (1.90). It is related to the intuitive concept of a piece of material which can be arbitrarily rotated together with the temperature field and whose reaction (i.e. values of the operators (1.90)) does not depend upon the rotation history in the coordinate system rotating together with this piece.

Mathematically, one considers two motions

$$\mathbf{r} = \mathbf{r}\left(\mathbf{R}, t\right) \tag{1.95}$$

and

$$\mathbf{r}_Q = \mathbf{Q}\left(t\right) \cdot \mathbf{r}\left(\mathbf{R}, t\right), \tag{1.96}$$

and a temperature field $\theta = \theta(\mathbf{R}, t)$. In eq. (1.96) $\mathbf{Q}(t)$ stands for an arbitrary time-varying tensor of rotation. The above requirement states the following, cf. [179]

$$\boldsymbol{\tau}_Q = \mathbf{Q} \cdot \boldsymbol{\tau} \cdot \mathbf{Q}^T, \ \mathbf{h}_Q = \mathbf{Q} \cdot \mathbf{h}, \ F_Q = F, \ S_Q = S, \tag{1.97}$$

where $\boldsymbol{\tau}, \mathbf{h}, F$ and S stand for the values of operators (1.90) for motion (1.95) whereas $\boldsymbol{\tau}_Q, \mathbf{h}_Q, F_Q$ and S_Q stand for motion (1.96) which is referred to as *Q-motion.*

Condition (1.97) expresses the so-called *principle of material frame indifference*, cf. [178] and [179].

Let us link the defining variables for the motions (1.95) and (1.96). As the inverse relations have the form

$$\mathbf{r} = \mathbf{r}_Q \cdot \mathbf{Q}, \ \mathbf{R} = \mathbf{R}(\mathbf{r}, t), \ \mathbf{R} = \mathbf{R}(\mathbf{r}_Q \cdot \mathbf{Q}, t), \tag{1.98}$$

we easily obtain

$$(\nabla \mathbf{R})_Q = \nabla_Q \mathbf{R} = \mathbf{Q} \cdot \nabla \mathbf{R},$$

and

$$\begin{aligned} \theta_Q &= \theta, \ (\nabla\theta)_Q = \mathbf{Q} \cdot \nabla\theta, \ \rho_Q = \rho, \\ \boldsymbol{\Lambda}_Q &= \mathbf{Q} \cdot \boldsymbol{\Lambda} \cdot \mathbf{Q}^T, \ \mathbf{B}_Q = \mathbf{Q} \cdot \mathbf{B}, \\ \mathbf{d}_Q &= \mathbf{Q} \cdot \mathbf{d} \cdot \mathbf{Q}^T, \ \boldsymbol{\Omega}_Q = \mathbf{Q} \cdot \boldsymbol{\Omega} \cdot \mathbf{Q}^T - \dot{\mathbf{Q}} \cdot \mathbf{Q}^T. \end{aligned} \tag{1.99}$$

Recall that the constitutive equations for isotropic materials are insensitive to replacement of tensor $\mathbf{B}$ by $\mathbf{B} \cdot \mathbf{H}$ provided that the values of the other arguments are retained. By means of this replacement in (1.99) and substitution of $\mathbf{Q}(t)$=const and $\mathbf{H} = \mathbf{Q}^T$ we find that equations (1.97) are satisfied provided that the following conditions

$$\begin{aligned} \theta_Q &= \theta, \ (\nabla\theta)_Q = \mathbf{Q} \cdot \nabla\theta, \ \rho_Q = \rho, \\ \boldsymbol{\Lambda}_Q &= \mathbf{Q} \cdot \boldsymbol{\Lambda} \cdot \mathbf{Q}^T, \ \mathbf{B}_Q = \mathbf{Q} \cdot \mathbf{B} \cdot \mathbf{Q}^T \end{aligned} \tag{1.100}$$

hold.

On the other hand, transformations (1.97) and (1.100) correspond to change of basis $\mathbf{i}_k$ on $\mathbf{Q} \cdot \mathbf{i}_k$ only in arguments of the constitutive equations while the tensor and vector components and the scalar values remain unchanged. This means that the constitutive equations do not contain any tensors and vectors except the above tensor and vector arguments, because the components of other tensors and vectors would change under a transformation of the coordinate basis.

In conclusion we say that it is rheology, i.e. a study on the behaviour of loaded materials, cf. [157], [192] and [156], which is concerned with the prescription of the operators in constitutive equations (1.90) for various materials. The method of rheological models developed allows one to derive the constitutive equations with a broad variety of properties. An attractive side of the method of rheological models is that the obtained equations *a priori* satisfy the second law of thermodynamics, the property of material isotropy and the material frame indifference principle.

1.8 Materials with elastic dilatation and the Fourier law of heat conduction

Consider a compressible heat-conducting material.

The mean normal stress is assumed to be a function of relative mass density ρ_0/ρ and temperature θ, i.e.

$$\sigma = \sigma\left(\frac{\rho_0}{\rho}, \theta\right). \tag{1.101}$$

We also assume that the stress deviator vanishes

$$\mathbf{s} = 0 \tag{1.102}$$

and the heat flux is a linear vector function of the temperature gradient

$$\mathbf{h} = -\kappa \nabla\theta, \tag{1.103}$$

where κ denotes the coefficient of thermal conductivity which is a function of temperature, absolute value of its gradient and mass density

$$\kappa = \kappa\left(\frac{\rho_0}{\rho}, \theta, |\nabla\theta|\right). \tag{1.104}$$

Lastly, the specific free energy and specific internal entropy are assumed to be functions (not functionals) of mass density and temperature.

In accordance with (1.101), (1.102) and (1.58) the stress tensor in this material is a spherical tensor and is given by

$$\boldsymbol{\tau} = \mathbf{E}\sigma\left(\frac{\rho_0}{\rho}, \theta\right). \tag{1.105}$$

The rate of change of the free energy is as follows

$$\dot{F}_0 = \frac{\partial F_0}{\partial(\rho_0/\rho)}\left(\frac{\rho_0}{\rho}\right)^{\bullet} + \frac{\partial F_0}{\partial\theta}\dot{\theta}. \tag{1.106}$$

Substituting (1.105) and (1.106) into dissipative inequality (1.76) we arrive at

$$\sigma (\nabla \cdot \mathbf{v}) - \rho \frac{\partial F_0}{\partial (\rho_0/\rho)} \left(\frac{\rho_0}{\rho} \right)^{\bullet} + \rho \left(S_0 + \frac{\partial F_0}{\partial \theta} \right) \dot{\theta} + \frac{\kappa}{\theta} (\nabla \theta)^2 \geq 0.$$

It follows from the mass conservation law that

$$\nabla \cdot \mathbf{v} = \frac{\rho}{\rho_0} \left(\frac{\rho_0}{\rho} \right)^{\bullet}.$$

As a result, the dissipative inequality takes the form

$$\rho \left[\frac{\sigma}{\rho_0} - \frac{\partial F_0}{\partial (\rho_0/\rho)} \right] \left(\frac{\rho_0}{\rho} \right)^{\bullet} - \rho \left(S_0 + \frac{\partial F_0}{\partial \theta} \right) \dot{\theta} + \frac{\kappa}{\theta} (\nabla \theta)^2 \geq 0. \tag{1.107}$$

As already pointed out, the dissipative inequality must hold for any arbitrary thermodynamic process, i.e. in particular for arbitrary velocities $(\rho_0/\rho)^{\bullet}$ and $\dot{\theta}$. The left hand side of (1.107) is however linear in these velocities. If the coefficients of these velocities were not equal to zero, the dissipative inequality could be violated by means of some special choice of the velocities' values and their signs. Therefore, in order to ensure that inequality (1.107) holds, it is necessary and sufficient to require that the coefficients of the velocities vanish and the absolute term is non-negative, i.e.

$$\sigma = \rho_0 \frac{\partial F_0}{\partial (\rho_0/\rho)}, \tag{1.108}$$

$$S_0 = -\frac{\partial F_0}{\partial \theta}, \tag{1.109}$$

$$\kappa (\nabla \theta)^2 \geq 0. \tag{1.110}$$

From these equations one can see that the specific free energy is a potential both for mean normal stress and specific internal entropy. Equation (1.110) requires the thermal conductivity coefficient to be non-negative. If $\kappa > 0$ and $\nabla \theta \neq 0$, condition (1.110) and consequently (1.107) hold with inequality sign. In this case thermodynamic processes in the material considered are irreversible. Thermodynamic processes can be reversible if either $\kappa = 0$ or $\nabla \theta = 0$, i.e. when the heat conduction is absent (the heat flux $\mathbf{h}$, eq. (1.103), vanishes).

The constitutive equations under consideration are widely used to describe the behaviour of gases and compressible fluids. Numerous experiments however show that eqs. (1.101) and (1.103) are valid as a first approximation for the majority of natural and synthetic isotropic structural materials. For this reason, it is generally accepted in rheology, cf. [157], [192] and [156], that eqs. (1.101) and (1.103) may serve as a good approximation for modelling dilatation and heat conduction in nearly all materials.

Many materials differ in the form of dependence of the stress deviator on thermodynamic processes. We restrict ourselves to situations in which the stress deviator is an operator over temperature $\theta(\tau)$, the strain of shape change $\mathbf{\Lambda}(\tau)$, and, possibly, the rotation tensor $\mathbf{B}(\tau)$ and mass density $\rho(\tau)$, i.e.

$$\mathbf{s} = \mathbf{s}\left\{\theta(\tau), \frac{\rho_0}{\rho(\tau)}, \mathbf{\Lambda}(\tau), \mathbf{B}(\tau)\right\}, \quad t_0 < \tau < t, \tag{1.111}$$

while the specific free energy and specific internal entropy are given by

$$\begin{aligned} F &= F_0\left(\frac{\rho_0}{\rho}, \theta\right) + F_*\left\{\theta(\tau), \frac{\rho_0}{\rho(\tau)}, \mathbf{\Lambda}(\tau), \mathbf{B}(\tau)\right\}, \\ S &= S_0\left(\frac{\rho_0}{\rho}, \theta\right) + S_*\left\{\theta(\tau), \frac{\rho_0}{\rho(\tau)}, \mathbf{\Lambda}(\tau), \mathbf{B}(\tau)\right\}. \end{aligned} \tag{1.112}$$

Here F_0 and S_0 are functions of their arguments, and F_* and S_* are operators over temperature, mass density, strain of shape change $\mathbf{\Lambda}(\tau)$ and rotation tensor $\mathbf{B}(\tau)$.

Making use of the arbitrariness of splitting the stress tensor (cf. eq. (1.58)), free energy and entropy (cf. eq. (1.112)) into two terms, let us require that when the shape change and rotation are absent

$$\mathbf{\Lambda}(\tau) = \mathbf{E}, \ \mathbf{B}(\tau) = \mathbf{E}, \quad t_0 < \tau < t, \tag{1.113}$$

the terms $\mathbf{s}$, F_* and S_* vanish, i.e.

$$\mathbf{s}\left\{\theta, \frac{\rho_0}{\rho}, \mathbf{E}, \mathbf{E}\right\} = 0, \ F_*\left\{\theta, \frac{\rho_0}{\rho}, \mathbf{E}, \mathbf{E}\right\} = 0, \ S_*\left\{\theta, \frac{\rho_0}{\rho}, \mathbf{E}, \mathbf{E}\right\} = 0. \tag{1.114}$$

Consider now the constitutive equations (1.101), (1.103), (1.111) and (1.112). Substituting them into the dissipative inequality (1.76) and taking into account eq. (1.58) and the mass conservation law, yields

$$\begin{aligned} &\rho\left[\frac{\sigma}{\rho_0} - \frac{\partial F_0}{\partial(\rho_0/\rho)}\right]\left(\frac{\rho_0}{\rho}\right)^{\bullet} - \rho\left(S_0 + \frac{\partial F_0}{\partial\theta}\right)\dot{\theta} + \frac{\kappa}{\theta}(\nabla\theta)^2 + \\ &\qquad +\mathbf{s} : \mathbf{d} - \rho\dot{F}_* - \rho\dot{\theta}S_* \geq 0, \end{aligned} \tag{1.115}$$

where $\mathbf{d}$ denotes the strain rate of shape change (1.23).

Inequality (1.115) must hold for any process, in particular for process (1.113). In this particular case, all terms in the second line of eq. (1.115) vanish. Because of the arbitrariness of $(\rho_0/\rho)^{\bullet}$, $\dot{\theta}$ and $\nabla\theta$ we arrive at the following group of conditions

$$\sigma = \rho_0 \frac{\partial F_0}{\partial (\rho_0/\rho)} \ , \ S_0 = -\frac{\partial F_0}{\partial \theta} \ , \ \kappa \geq 0. \tag{1.116}$$

These are identical with conditions (1.108), (1.109) and (1.110).

Taking into account (1.116), assuming $\nabla \theta = 0$ and removing restriction (1.113) we obtain the dissipative inequality for the shape change process

$$\mathbf{s} : \mathbf{d} - \rho \dot{F}_* - \rho \dot{\theta} S_* \geq 0. \tag{1.117}$$

It can be easily proved that under conditions (1.116) the first law of thermodynamics takes the following form

$$\rho\theta \left(\dot{S}_0 + \dot{S}_* \right) = \mathbf{s} : \mathbf{d} - \rho \dot{F}_* - \rho \dot{\theta} S_* + b + \nabla \cdot (\kappa \nabla \theta) \, . \tag{1.118}$$

Let us consider a possible particular form of free energy at dilatation. The following expression

$$F_0 = -\frac{k(\theta)}{\rho_0} \left(\ln \frac{\rho_0}{\rho} + 1 - \frac{\rho_0}{\rho} \right) + \frac{m(\theta)}{\rho_0} \left(1 - \frac{\rho_0}{\rho} \right) + \frac{n(\theta)}{\rho_0} \tag{1.119}$$

is an acceptable approximation for modelling volume deformations of solids and fluids. In this case, the mean normal stress and the specific entropy are as follows

$$\sigma = k(\theta) \left(1 - \frac{\rho}{\rho_0} \right) - m(\theta), \tag{1.120}$$

$$S_0 = \frac{1}{\rho_0} \left(\ln \frac{\rho_0}{\rho} + 1 - \frac{\rho_0}{\rho} \right) \frac{dk}{d\theta} - \frac{1}{\rho_0} \left(1 - \frac{\rho_0}{\rho} \right) \frac{dm}{d\theta} - \frac{1}{\rho_0} \frac{dn}{d\theta} \ . \tag{1.121}$$

As seen from (1.120) coefficient $k(\theta)$ links the stress to a change of mass density. Assuming $\sigma = 0$, yields

$$\frac{\rho_0}{\rho} = \frac{k - m}{k}. \tag{1.122}$$

Thus, function $m(\theta)$ determines the change of mass density under temperature variation in a free thermal expansion.

Finally, let us clarify the physical meaning of function $n(\theta)$. To this end, consider a process of heating without deformation ($\rho_0/\rho = 1$, $\mathbf{\Lambda} = \mathbf{E}$, $\mathbf{B} = \mathbf{E}$). In this case

$$S_0 = -\frac{1}{\rho_0} \frac{dn}{d\theta} \ , \tag{1.123}$$

and the heat conduction equation (1.118) takes the following form

$$c(\theta)\dot{\theta} = b + \nabla \cdot (\kappa \nabla \theta), \tag{1.124}$$

where

$$c(\theta) = -\theta \frac{d^2 n}{d\theta^2} \tag{1.125}$$

represents heat capacity without deformation. Equation (1.125) thus clarifies the physical meaning of function $n(\theta)$. Indeed, by means of eq. (1.125) it is easy to find the specific entropy in the case of constrained ($\rho = \rho_0$) heating. By means of eqs. (1.123) and (1.125) we obtain

$$S_0 = -\frac{1}{\rho_0}\frac{dn}{d\theta} = \frac{1}{\rho_0}\int_0^\theta \frac{c(\chi)\,d\chi}{\chi}. \tag{1.126}$$

The lower integration bound is set to zero in order to satisfy the requirement of the third law of thermodynamics. In order to ensure the existence of the integral in the neighbourhood of zero temperature, the thermal capacity of the material must have the following asymptotic form

$$c(\chi) = A\chi^{\alpha}, \quad \alpha > 0, \tag{1.127}$$

where A and α are the material constants.

The question of how to satisfy the requirement of the third law of thermodynamics for a material under deformation must be considered. To this end, eq. (1.121) must be equated to zero and according to (1.126) the last term in eq. (1.121) vanishes as the absolute temperature tends to zero. In order to equate the first two terms in eq. (1.121) to zero it is necessary and sufficient to require that as $\theta \to 0$

$$\frac{dk}{d\theta} \to 0, \quad \frac{dm}{d\theta} \to 0. \tag{1.128}$$

We assume that the real materials possess this property and the property prescribed by eq. (1.127).

1.9 Classical materials

Let us proceed to consideration of some particular dependences (1.111) and (1.112).

1.9.1 *Elastic material*

The stress deviator, free energy and entropy are postulated to be functions (not operators) of the strain of shape change $\mathbf{\Lambda}(t)$, temperature $\theta(t)$ and mass density $\rho(t)$

$$\mathbf{s} = \mathbf{s}(\rho, \mathbf{\Lambda}, \theta), \quad F_* = F_*(\mathbf{\Lambda}, \theta), \quad S_* = S_*(\mathbf{\Lambda}, \theta). \tag{1.129}$$

Tensor $\mathbf{\Lambda}$ is assumed to be prescribed in a fixed basis. In this case the derivative of the free energy is

$$\dot{F}_* = \frac{\partial F_*}{\partial \theta}\dot{\theta} + \frac{\partial F_*}{\partial \mathbf{\Lambda}} : \dot{\mathbf{\Lambda}}.$$

Substituting an explicit expression for $\dot{\mathbf{\Lambda}}$, eq. (1.39), and simplifying the result, we obtain

$$\begin{aligned} \dot{F}_* &= \frac{\partial F_*}{\partial \theta}\dot{\theta} + \frac{\partial F_*}{\partial \mathbf{\Lambda}} : (\mathbf{d}\cdot\mathbf{\Lambda} + \mathbf{\Lambda}\cdot\mathbf{d} + \mathbf{\Omega}\cdot\mathbf{\Lambda} - \mathbf{\Lambda}\cdot\mathbf{\Omega}) = \\ &= \frac{\partial F_*}{\partial \theta}\dot{\theta} - 2\left(\frac{\partial F_*}{\partial \mathbf{\Lambda}}\cdot\mathbf{\Lambda}\right)^s : \mathbf{d} + 2\left(\frac{\partial F_*}{\partial \mathbf{\Lambda}}\cdot\mathbf{\Lambda}\right)^a : \mathbf{\Omega}. \end{aligned} \tag{1.130}$$

Substituting (1.129) and (1.130) into dissipative inequality (1.117) yields

$$\left[\mathbf{s} + 2\rho \mathrm{Dev}\left(\frac{\partial F_*}{\partial \mathbf{\Lambda}}\cdot\mathbf{\Lambda}\right)^s\right] : \mathbf{d} - 2\rho\left(\frac{\partial F_*}{\partial \mathbf{\Lambda}}\cdot\mathbf{\Lambda}\right)^a : \mathbf{\Omega} - \rho\left(S_* + \frac{\partial F_*}{\partial \theta}\right)\dot{\theta} \geq 0. \tag{1.131}$$

This inequality must hold for any arbitrary thermodynamic process and, in particular, for arbitrary $\dot{\theta}$, $\mathbf{\Omega}$ and $\mathbf{d}$. The dissipative inequality (1.131) however is linear in $\dot{\theta}$, $\mathbf{\Omega}$ and $\mathbf{d}$. Thus, the coefficients of these parameters must vanish, otherwise one could violate condition (1.131) by having prescribed some values of $\dot{\theta}$, $\mathbf{\Omega}$ and $\mathbf{d}$. We therefore arrive at the following equations

$$\mathbf{s} = -2\rho \mathrm{Dev}\left(\frac{\partial F_*}{\partial \mathbf{\Lambda}}\cdot\mathbf{\Lambda}\right)^s, \tag{1.132}$$

$$\left(\frac{\partial F_*}{\partial \mathbf{\Lambda}}\cdot\mathbf{\Lambda}\right)^a = 0, \tag{1.133}$$

$$S_* = -\frac{\partial F_*}{\partial \theta}. \tag{1.134}$$

We see from these equations that free energy is a generating function for both the stress deviator and the specific entropy.

The isotropy of the material imposes some restrictions on the form of dependences (1.129). In particular, the free energy must be insensitive to a rotational transformation of the coordinate system. As a scalar, it must be a function of the invariants of tensor $\mathbf{\Lambda}$. According to the definition of the strain of shape change $\mathbf{\Lambda}$, its third main invariant is equal to unity (see Section 1.3). We take the other invariants in the following form

$$a = \mathbf{E} : \mathbf{\Lambda}, \quad b = \frac{1}{2}\mathbf{\Lambda} : \mathbf{\Lambda}. \tag{1.135}$$

Only these two invariants and temperature turn out to be arguments of the free energy of an isotropic material, i.e.

$$F_* = F_* \left(\theta, \, a, \, b\right). \tag{1.136}$$

Its derivative with respect to $\mathbf{\Lambda}$ is as follows

$$\frac{\partial F_*}{\partial \mathbf{\Lambda}} = \frac{\partial F_*}{\partial a}\mathbf{E} + \frac{\partial F_*}{\partial b}\mathbf{\Lambda}.$$

One can see that restriction (1.133) holds, and the stress deviator (1.132) takes the form

$$\mathrm{s} = -2\rho \mathrm{Dev}\left(\frac{\partial F_*}{\partial a}\mathbf{\Lambda} + \frac{\partial F_*}{\partial b}\mathbf{\Lambda}^2\right). \tag{1.137}$$

An acceptable approximation for free energy in the case of finite deformations is the Mooney potential, [99] and [100]

$$F_* = \frac{A}{2\rho_0}(a-3) + \frac{B}{2\rho_0}\left(b - \frac{3}{2}\right), \tag{1.138}$$

where A and B are some functions of temperature. This potential says that conditions (1.114) hold for process (1.113). Besides, for $A>0$ and $B>0$ the free energy (1.138) turns out to be positive for any shape deformation, except the case of $\mathbf{\Lambda} = \mathbf{E}$ in which it vanishes, cf. [99] and [100].

When the free energy is given by (1.138) expression (1.137) takes the especially simple form

$$\mathrm{s} = -\frac{\rho}{\rho_0}\mathrm{Dev}\left(A\mathbf{\Lambda} + B\mathbf{\Lambda}^2\right). \tag{1.139}$$

Also the heat conduction equation essentially simplifies and takes the form

$$\rho\theta(\dot{S}_0 + \dot{S}_*) = b + \nabla \cdot (\kappa \nabla \theta), \tag{1.140}$$

where, due to (1.134) and (1.138), the entropy S_* is given by

$$S_* = -\frac{1}{2\rho_0}\left[(a-3)\frac{\partial A}{\partial \theta} + \left(b - \frac{3}{2}\right)\frac{\partial B}{\partial \theta}\right]. \tag{1.141}$$

Equation (1.139) states that the principle directions of the stress tensor and the strain of shape change coincide. It can be easily shown that this is a result of (1.133), the latter being a direct consequence of the independency of the free energy on the rotation tensor.

1.9.2 Viscous materials

It is postulated that the stress deviator is a tensor function of the strain rate of shape change **d** and may depend upon the following scalar arguments: volume ratio ρ_0/ρ, its time-derivative and temperature.

As mentioned in Section 1.7, for an isotropic material the tensor function **s** of the tensor argument **d** should contain no tensors except its argument tensor. This means that the principle directions of the tensor function and its argument should coincide, otherwise the equation for this tensor function would require a rotation tensor for transfer from the principle directions of the tensor argument to those of the tensorial function.

One of the possible full expressions for a tensor which has the same principle directions as **d** is given by, cf. [99], [157], [162] and [156]

$$\mathbf{H} = 2\left(h\mathbf{E} + \eta\mathbf{d} + \varphi\mathbf{d}^2\right), \tag{1.142}$$

where the scalar coefficients h, η and φ depend upon the invariants of tensor **d** and above scalar arguments.

As **s** is a deviator, its full representation is obtained by evaluating the deviatoric part of eq. (1.142). For an isotropic viscous material the deviator turns out to have the following general form

$$\mathbf{s} = 2\mathrm{Dev}\left(\eta\mathbf{d} + \varphi\mathbf{d}^2\right). \tag{1.143}$$

The parameters η and φ are referred to as the viscosity coefficients.

Constitutive equation (1.143) must be completed by equations for the free energy F_* and the entropy S_*. We assume here that these are identically equal to zero.

Let us next find what restrictions are imposed by the dissipative inequality (1.117) on function (1.143). Substituting (1.143) into (1.117) yields

$$2\left(\eta\mathbf{d} : \mathbf{d} + \varphi\mathbf{d}^2 \colon \mathbf{d}\right) \geq 0.$$

Recalling the equations for the main invariants of deviator **d**

$$\mathbf{d} : \mathbf{d} = 2I_2\left(\mathbf{d}\right), \quad \mathbf{d}^2 \colon \mathbf{d} = 3I_3\left(\mathbf{d}\right),$$

we rewrite the previous inequality in the following form

$$4\eta I_2\left(\mathbf{d}\right) + 6\varphi I_3\left(\mathbf{d}\right) \geq 0. \tag{1.144}$$

It is impossible to derive the restrictions imposed on η and φ which are necessary and sufficient for (1.144) without prescribing the dependence of η and φ on the invariants I_2 and I_3.

Let us consider some particular cases.

Provided that the viscosity coefficients η and φ do not depend on I_2 and I_3 at all, the necessary and sufficient conditions for (1.144) are given by

$$\eta \geq 0, \ \varphi{=}0. \tag{1.145}$$

Indeed, if $\mathbf{d}$ is very large, but $I_3(\mathbf{d}) \neq 0$ we find that the value and the sign of the left hand side of (1.144) is completely determined by the second term. But it is an odd function of $\mathbf{d}$, i.e. it changes its sign when one replaces $\mathbf{d}$ by $-\mathbf{d}$. Thus, one must put $\varphi{=}0$ otherwise (1.144) will not be satisfied. Non-negativeness of the second invariant leads to the first restriction in (1.145).

It can be shown by analogy that if

$$\varphi = I_3 g$$

and if η and g do not depend upon I_2 and I_3, the necessary and sufficient conditions for inequality (1.144) are as follows

$$\eta \geq 0, \ g \geq 0.$$

Finally, if

$$\eta = \frac{\mu}{I_2}, \ \varphi = \frac{\nu}{I_3}$$

and the coefficients μ and ν are independent of I_2 and I_3 we arrive at the following condition

$$2\mu + 3\nu \geq 0.$$

From the latter equation one can see that one of the above coefficients, i.e. μ or ν, may be negative.

Material with the constitutive equation (1.143) is called the *Reiner-Rivlin viscous fluid*, [157] and [156].

The case $\varphi = 0$ is the most common. In this case the governing equation takes the form

$$\mathbf{s} = 2\eta\mathbf{d}, \tag{1.146}$$

and the necessary and sufficient conditions for dissipative inequality (1.144) are given by

$$\eta \geq 0, \tag{1.147}$$

which is insensitive to a particular form of dependence of η on the invariants I_2 and I_3.

Provided that the viscous coefficient η depends only on the temperature, the material with the constitutive equation (1.146) is termed a *Newtonian viscous fluid.*

To conclude, we will write the heat conduction equation for a viscous material. According to (1.143) the heat conduction equation (1.118) takes the following form

$$\rho\theta\dot{S}_0 = 4\eta I_2(\mathbf{d}) + 6\varphi I_3(\mathbf{d}) + b + \nabla\cdot(\kappa\nabla\theta). \tag{1.148}$$

The first two terms on the right hand side characterize the energy dissipation in a viscous material. In agreement with the dissipative inequality (1.144) their sum is non-negative. In the case of a Newtonian viscous fluid the second term disappears, while the first is the second order in **d**.

In some cases it is necessary to consider the inverse tensor function (1.143), i.e. to find the following expression

$$\mathbf{d} = \operatorname{Dev}\left(\beta\mathbf{s} + \gamma\mathbf{s}^2\right), \tag{1.149}$$

with the scalar factors β and γ depending on the invariants of tensor s and the scalars ρ, $\dot{\rho}$ and θ.

In general, an actual inversion (assuming that it is possible) is very difficult because the coefficients η and φ in (1.143) depend on the invariants of tensor **d**. In order to overcome this difficulty we postulate that the constitutive equation is in the form of (1.149) and assume that the free energy and the entropy are equal to zero. The dissipative inequality (1.117) imposes the following restriction on β and γ which are functions of the invariants I_2 and I_3

$$2\beta I_2(\mathbf{s}) + 3\gamma I_3(\mathbf{s}) \geq 0. \tag{1.150}$$

The structure of this inequality is similar to that of eq. (1.144).

1.9.3 *Plastic materials*

In order to formulate the concept of a plastic material it is necessary to introduce a norm in the space of the stress deviator s. The norm is denoted as $N(s)$. It is a non-negative scalar function of tensor s and a function of the tensor's invariants in the case of an isotropic material. In accordance with the norm properties any function $N(\mathrm{s})$ must satisfy the following conditions

$$\begin{aligned} N(\mathbf{s}) &> 0, \;\; \mathbf{s}\neq 0, \\ N(\mathbf{s}) &= 0, \;\; \mathbf{s}=0, \\ N(\alpha\mathbf{s}) &= |\alpha|\, N(\mathbf{s}), \\ N(\mathbf{s}+\mathbf{t}) &\leq N(\mathbf{s}) + N(\mathbf{t}), \end{aligned} \tag{1.151}$$

where α is a real value.

From a physical point of view the norm N characterises, in a generalised sense, the largest absolute value of components of the stress deviator.

Let us proceed to the definition of a plastic material.

We assume that if there is no deformation, i.e. $\mathbf{d} = 0$, then the stress deviator may take any value, however its value should satisfy the following inequality

$$N(\mathbf{s}) \leq \tau_s. \tag{1.152}$$

Next, we assume that under deformation, i.e. $\mathbf{d} \neq 0$, the material behaves like a viscous fluid with a constitutive equation which is similar to (1.143), namely

$$\mathbf{s} = \mu \mathbf{f}(\mathbf{d}). \tag{1.153}$$

Here μ is a positive factor which is not known *a priori*, and $\mathbf{f}$ is a tensor function of the strain rate of shape change $\mathbf{d}$. The dependence of $\mathbf{f}$ on $\mathbf{d}$ is similar to that in (1.143), i.e.

$$\mathbf{f}(\mathbf{d}) = 2\mathrm{Dev}\left(\psi \mathbf{d} + \chi \mathbf{d}^2\right). \tag{1.154}$$

The scalar coefficients ψ and χ are functions of the invariants of $\mathbf{d}$, mass density, its time-derivative and temperature. Finally, we assume that under any deformation (requirement $\mathbf{d} \neq 0$ must hold) the norm of the stress deviator satisfies the following condition

$$N(\mathbf{s}) = \tau_s, \tag{1.155}$$

which is called the *yield criterion*. Here τ_s is referred to as the yield stress which is allowed to be an operator over some thermodynamic process, i.e. some operator over $\mathbf{d}$, ρ and θ.

Equation (1.155) serves to determine the unknown scalar factor μ. Substituting $\mathbf{s}$ from (1.153) into (1.155) and using the third norm property, eq. (1.151), we obtain

$$\mu N(\mathbf{f}) = \tau_s.$$

Substituting for μ in (1.153) then gives

$$\mathbf{s} = \frac{\tau_s}{N(\mathbf{f})} \mathbf{f}(\mathbf{d}). \tag{1.156}$$

It is easy to see that the yield condition (1.155) is also satisfied.

The last assumption is as follows. The specific free energy and the specific entropy for the plastic material are equated to zero.

Therefore, the constitutive equations for a plastic material are given by

$$\left[\begin{array}{ll} \mathbf{d} = 0, & N(\mathbf{s}) \leq \tau_s, \\ \mathbf{d} \neq 0, & \mathbf{s} = \tau_s \mathbf{f}(\mathbf{d}) / N(\mathbf{f}), \end{array}\right. \tag{1.157}$$

$$F_* = 0, \quad S_* = 0. \tag{1.158}$$

Note that any state of stress at which $N(\mathbf{s}) > \tau_s$ is unattainable since, due to (1.157), it cannot be realised neither when $\mathbf{d} \neq 0$ nor when $\mathbf{d} = 0$.

Let us now ascertain what restrictions are imposed by the dissipative inequality (1.117) on the constitutive equations. In accordance with (1.157) we must study two cases: $\mathbf{d} = 0$ and $\mathbf{d} \neq 0$. In the first case inequality (1.117) holds with the equality sign. In the second case (1.117) reduces to the following inequality

$$4\psi I_2(\mathbf{d}) + 6\chi I_3(\mathbf{d}) \geq 0, \tag{1.159}$$

similar to condition (1.144) for the viscous material.

In some cases an inversion of the constitutive equations (1.157), i.e. a functional dependence $\mathbf{d}(\mathbf{s})$, turns out to be necessary. The result of this inversion has the following general form

$$\mathbf{d} = \lambda \mathrm{Dev}\left(r\mathbf{s} + p\mathbf{s}^2\right), \tag{1.160}$$

where p and r are non-trivial scalar functions of the invariants of the stress deviator s, mass density ρ, its time-derivative $\dot{\rho}$ and temperature θ. An undetermined non-negative factor λ satisfies the following conditions

$$\left[\begin{array}{ll} N(\mathbf{s}) < \tau_s, & \lambda = 0, \\ N(\mathbf{s}) = \tau_s, & \lambda \geq 0. \end{array}\right. \tag{1.161}$$

Indeed, if $N(\mathbf{s}) < \tau_s$, then as follows from (1.157) there is no deformation, i.e. $\mathbf{d} = 0$. It is however seen from (1.160) that this is possible only when $\lambda = 0$ as it is written down in the first line of (1.161). Next, if $N(\mathbf{s}) = \tau_s$, conditions (1.157) permit two possibilities: $\mathbf{d} = 0$ and $\mathbf{d} \neq 0$. Representation (1.160) says that this is also possible in the two cases: $\lambda = 0$ and $\lambda \neq 0$. Assuming for determinancy that λ is non-negative, we arrive at the statement in the second line of (1.161).

We proceed now to the dissipative inequality (1.117). Substituting (1.160) and (1.158) into it yields

$$\lambda\left[2rI_2(\mathbf{s}) + 3pI_3(\mathbf{s})\right] \geq 0. \tag{1.162}$$

As long as no plastic deformation occurs $\lambda = 0$, and the dissipative inequality (1.162) holds with the equality sign. When a plastic deformation occurs, then λ is negative, and inequality (1.162) holds under the following condition

$$2rI_2(\mathbf{s}) + 3pI_3(\mathbf{s}) \geq 0, \quad N(\mathbf{s}) = \tau_s. \tag{1.163}$$

The most popular expressions in the mathematical theory of plasticity, for functions and operators which appear in the equations for plastic materials, are given below.

In the overwhelming majority of papers it is proposed that

$$\mathbf{f}(\mathbf{d}) = \mathbf{d}, \tag{1.164}$$

which is analogous to the tensor function (1.146). In this case, the dissipative inequality both in (1.159) and (1.162) holds for any deformation process.

The most commonly used expressions for the norm $N(\mathbf{s})$ in the yield criterion are described in what follows. It is often assumed that this norm is an *intensity of shear stresses*

$$N(\mathbf{s}) = \tau = \sqrt{I_2(\mathbf{s})} = \sqrt{\frac{1}{2}\mathbf{s} : \mathbf{s}}. \tag{1.165}$$

The yield criterion in the following form

$$\tau = \tau_s \tag{1.166}$$

is called the *Mises yield criterion.*

Sometimes, the maximum shear stress

$$N(\mathbf{s}) = T = \max\left[\left|\frac{s_1 - s_2}{2}\right|, \left|\frac{s_2 - s_3}{2}\right|, \left|\frac{s_3 - s_1}{2}\right|\right], \tag{1.167}$$

is used as norm $N(\mathbf{s})$. Here s_1, s_2 and s_3 are the principle values of the stress deviator.

The yield criterion in the form

$$T = \tau_s \tag{1.168}$$

is called the *St. Venant-Tresca yield criterion.*

Finally, let us briefly touch upon some different ways of prescribing the dependence of the yield stress on the history of deformation. It often turns out that τ_s does not depend on the deformation history at all, but it is a function of the temperature and, probably, the mass density, i.e.

$$\tau_s = \tau_s(\theta, \rho). \tag{1.169}$$

Next, the dependence of the yield stress on the deformation history may be prescribed as a function

$$\tau_s = \tau_s(\theta, \rho, q), \tag{1.170}$$

where q is referred to as the *Odqvist parameter*

$$q = \int_0^t \sqrt{I_2(\mathbf{d})}dt. \tag{1.171}$$

The plasticity theory based on equations (1.164), (1.165) and (1.169) is called the St. Venant-Mises theory. Its constitutive equations are given by

$$\left[\begin{array}{ll} \mathbf{d}=0, & \tau \leq \tau_s, \\ \mathbf{d} \neq 0, & \mathbf{s}=\tau_s \mathbf{d}/v, \end{array}\right. \tag{1.172}$$

where

$$v=\sqrt{I_2(\mathbf{d})}=\sqrt{\frac{1}{2}\mathbf{d}:\mathbf{d}} \tag{1.173}$$

is referred to as an *intensity of shear strain rate.*

An inverse form of the constitutive equation is

$$\mathbf{d}=\lambda \mathbf{s}, \tag{1.174}$$

where

$$\left[\begin{array}{ll} \tau \leq \tau_s, & \lambda=0, \\ \tau=\tau_s, & \lambda \geq 0. \end{array}\right. \tag{1.175}$$

To conclude, we write down the heat conduction equation for a plastic material. In the most general case, this is given by

$$\rho\theta\dot{S}_0=\frac{\tau_s}{N(\mathbf{f})}\left[4\psi I_2(\mathbf{d})+6\chi I_3(\mathbf{d})\right]+b+\nabla\cdot(\kappa\nabla\theta). \tag{1.176}$$

In the St. Venant-Mises theory it takes a more simple form

$$\rho\theta\dot{S}_0=\tau_s v+b+\nabla\cdot(\kappa\nabla\theta). \tag{1.177}$$

The first term on the right hand side characterises heat production under plastic deformations. In contrast to the viscous material this value is of first order in components of the strain rate of shape change $\mathbf{d}$.

1.10 Rheological models of materials. Complex materials

Let us introduce the concept of a *rheological model of a material.*

Ignoring for the time being the temperature dependence, we note that equation (1.139) implies elasticity, i.e. it states a functional relation between "force" $\mathbf{s}$ and "strain " $\mathbf{\Lambda}$. The simplest mechanical system which can be described in this way is an elastic spring. For this reason, in rheology, cf. [157], [192] and [156], an elastic spring is considered to be a visual image for the elasticity equation (1.139). This visual image of the constitutive equations is termed a *rheological model of a material.* A rheological model for an elastic material is an elastic spring which is referred to as *the Hooke element* in rheology. The Hooke element is depicted in Fig. 1.2a.

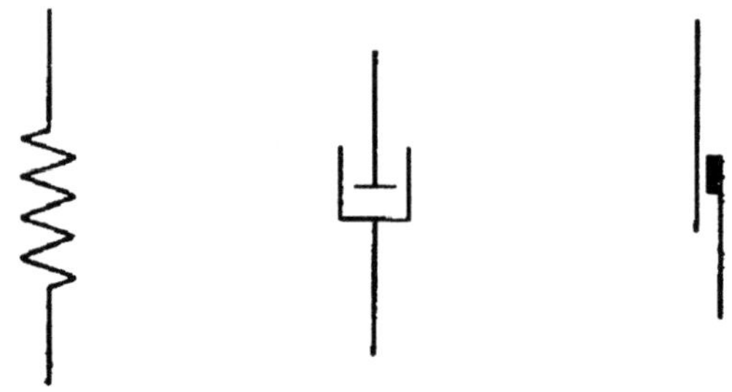

FIGURE 1.2. Rheological models of materials: a) the Hooke element, b) the Newton element and c) the St. Venant element.

Next, the constitutive equation (1.143) is a viscosity relation, since it states a functional dependence between "force" **s** and "velocity" **d**. The simplest mechanical system with the above property is a viscous damper or dashpot. This visual image is chosen as the rheological model for the viscous material. This is shown in Fig. 1.2b and is called *the Newton element.* The simplest realisation of the viscous damper is an oil-filled-cylinder with a piston in it. Due to cylinder clearance, the axial force due to the relative motion of the piston depends upon the velocity and vanishes together with the velocity.

Finally, the constitutive equations (1.157) describe plasticity. In fact, they state that when the norm of the "force" **s** is small, the material experiences no deformations (**d**=0), i.e. it behaves like a rigid body. When the "force" **s** reaches the yield stress, the material may be deformed and the change in the "velocity" **d** does not influence the value (or strictly speaking the norm) of the "force". A dry friction damper is the simplest mechanical system possessing the above properties. This is depicted in Fig. 1.2c and is considered to be the rheological model for a plastic material. This rheological element is called *the St. Venant element.* The simplest realisation of the element is two flat plates pressed together by a constant force. Relative motion of the plates is absent while the tangential forces in the plates do not exceed the forces of dry friction. If relative motion occurs, the force which causes this motion is equal to the force of dry friction and does not depend upon the velocity.

The materials considered possess the following fundamental properties: elasticity, viscosity and plasticity. These properties are visualised by the simplest rheological models, namely, the Hooke element, the Newton element and the St. Venant element.

Let us proceed to considering materials with complex rheological properties. The following way of composing the constitutive equations for materials with complex rheological properties is accepted in rheology. Its basic

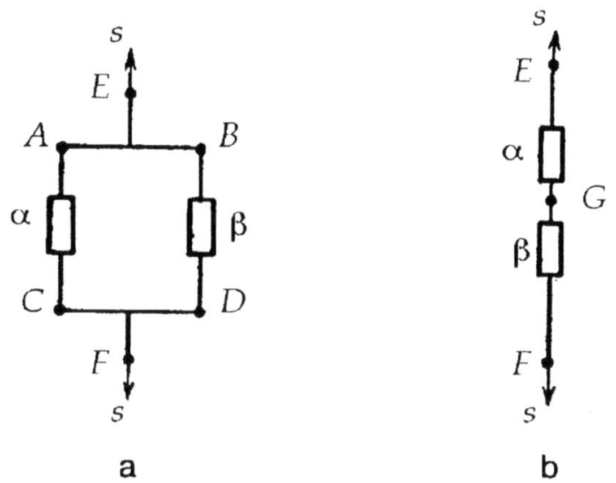

FIGURE 1.3. Connection of the rheological elements in parallel a) and in series b).

concept is a *rheological model.* It is composed of the fundamental rheological elements in series and in parallel. The system of constitutive equations is then derived using some rules. Thus, the rheological model is not only a visualisation of the material behaviour, but also plays an active part, namely it prescribes the structure of the constitutive equation for the material. A large number of materials with broad spectra of mechanical properties is postulated in this way.

Consider now the problem of deriving the constitutive equations for a given rheological model. As pointed out above, a model is composed of the fundamental rheological elements by means of parallel and in series connections. These connections are shown in Fig. 1.3.

We assume that the constitutive equations for each element α and β are given separately in the form of operators such as (1.111) and (1.112), i.e.

$$\begin{array}{ll} \mathbf{s}_\alpha = \mathbf{s}_\alpha \{\theta,\, \rho_0/\rho,\, \mathbf{\Lambda}_\alpha, \mathbf{B}\}\,, & \mathbf{s}_\beta = \mathbf{s}_\beta \{\theta,\, \rho_0/\rho,\, \mathbf{\Lambda}_\beta, \mathbf{B}\}\,, \\ F_{*\alpha} = F_{*\alpha} \{\theta,\, \rho_0/\rho,\, \mathbf{\Lambda}_\alpha, \mathbf{B}\}\,, & F_{*\beta} = F_{*\beta} \{\theta,\, \rho_0/\rho,\, \mathbf{\Lambda}_\beta, \mathbf{B}\}\,, \\ S_{*\alpha} = S_{*\alpha} \{\theta,\, \rho_0/\rho,\, \mathbf{\Lambda}_\alpha, \mathbf{B}\}\,, & S_{*\beta} = S_{*\beta} \{\theta,\, \rho_0/\rho,\, \mathbf{\Lambda}_\beta, \mathbf{B}\}\,. \end{array} \qquad (1.178)$$

Assume that the rotation tensor $\mathbf{B}$, temperature θ, and mass density ρ in both constitutive equations (1.178) coincide regardless of the type of connection, while the strains of shape change are in general different, i.e. $\mathbf{\Lambda}_\alpha(\tau) \neq \mathbf{\Lambda}_\beta(\tau)$. These tensors are assumed to be related to the tensors $\mathbf{d}_\alpha$ and $\mathbf{d}_\beta$ by means of eq. (1.39), i.e.

$$\begin{aligned} \dot{\mathbf{\Lambda}}_\alpha &= -\mathbf{d}_\alpha \cdot \mathbf{\Lambda}_\alpha - \mathbf{\Lambda}_\alpha \cdot \mathbf{d}_\alpha - \mathbf{\Omega} \cdot \mathbf{\Lambda}_\alpha + \mathbf{\Lambda}_\alpha \cdot \mathbf{\Omega}, \\ \dot{\mathbf{\Lambda}}_\beta &= -\mathbf{d}_\beta \cdot \mathbf{\Lambda}_\beta - \mathbf{\Lambda}_\beta \cdot \mathbf{d}_\beta - \mathbf{\Omega} \cdot \mathbf{\Lambda}_\beta + \mathbf{\Lambda}_\beta \cdot \mathbf{\Omega}, \end{aligned} \qquad (1.179)$$

where the spin tensor $\boldsymbol{\Omega}$ needs to be obtained from (1.45) in terms of the rotation tensor $\mathbf{B}$.

We suppose that the initial condition (1.40) holds for any tensor $\boldsymbol{\Lambda}_\alpha$ and $\boldsymbol{\Lambda}_\beta$, i.e.

$$\boldsymbol{\Lambda}_\alpha(t_0) = \mathbf{E}, \quad \boldsymbol{\Lambda}_\beta(t_0) = \mathbf{E}. \tag{1.180}$$

Let us introduce the connection axioms.

1. Axiom of a connection in parallel as shown in Fig. 1.3a

$$\mathbf{d} = \mathbf{d}_\alpha = \mathbf{d}_\beta, \ \mathbf{s} = \mathbf{s}_\alpha + \mathbf{s}_\beta. \tag{1.181}$$

2. Axiom of a connection in series as shown in Fig. 1.3b

$$\mathbf{d} = \mathbf{d}_\alpha + \mathbf{d}_\beta, \ \mathbf{s} = \mathbf{s}_\alpha = \mathbf{s}_\beta. \tag{1.182}$$

For both connections

$$F_* = F_{*\alpha} + F_{*\beta}, \ S_* = S_{*\alpha} + S_{*\beta}. \tag{1.183}$$

Finally, recall that for a given $\mathbf{d}$ tensor $\boldsymbol{\Lambda}$ is a solution of the following kinematic equations, cf. (1.39)

$$\dot{\boldsymbol{\Lambda}} = -\mathbf{d}\cdot\boldsymbol{\Lambda} - \boldsymbol{\Lambda}\cdot\mathbf{d} - \boldsymbol{\Omega}\cdot\boldsymbol{\Lambda} + \boldsymbol{\Lambda}\cdot\boldsymbol{\Omega} \tag{1.184}$$

subject to the following initial condition, cf. (1.40)

$$\boldsymbol{\Lambda}(t_0) = \mathbf{E}. \tag{1.185}$$

Analysis of one-dimensional mechanical systems, i.e. the rheological models, may serve some logical substantiation for just these very axioms. We start with the connection in parallel, see Fig. 1.3a. Provided that the rigid rods AB and CD may move only vertically and translationally, the velocity d of the point E relative to F coincides with the velocity d_α of the point A relative to C and d_β of the point B relative to D. This observation allows us to write down the following kinematic equation

$$d = d_\alpha = d_\beta,$$

whose structure is similar to that of the first tensor equation (1.181). Further, the force s applied to points E and F is the sum of the force in the first element, i.e. s_α, and the second element, i.e. s_β. This yields the equation

$$s = s_\alpha + s_\beta,$$

whose structure is again similar to that of the second tensor equation (1.181).

By analogy, for a connection in series (see Fig. 1.3b) the vertical velocity of the point E relative to F is the sum of velocity d_α of the point E relative to G and d_β of the point G relative to F, i.e.

$$d = d_\alpha + d_\beta.$$

This is analogous to the first tensor equation in (1.182). Further, in the one-dimensional model of Fig. 1.3b the forces s applied to points E and F are equal to the forces in the elements s_α and s_β, i.e.

$$s = s_\alpha = s_\beta,$$

which is analogous to the second tensor equation in (1.182).

Finally, equations (1.183) are a logical consequence of additivity of free energy and entropy for mechanical elements of the models.

The connection axioms allow one to derive constitutive equations for any arbitrary complicated rheological model composed of the fundamental elements such as Hooke's element, Newton's element and St. Venant's element by means of connections in series and in parallel.

Theorem 1 states: Constitutive equations for any rheological model satisfy the dissipative inequality.

The theorem will be proved by the method of mathematical induction.

Consider two arbitrary rheological models, say α and β. Let the constitutive equations for materials with these rheological models be given by eq. (1.178). Assume that inequality (1.117) holds separately for each of these two materials, i.e.

$$\mathbf{s}_\alpha : \mathbf{d}_\alpha - \rho \dot{F}_{*\alpha} - \rho \dot{\theta} S_{*\alpha} \geq 0, \tag{1.186}$$

$$\mathbf{s}_\beta : \mathbf{d}_\beta - \rho \dot{F}_{*\beta} - \rho \dot{\theta} S_{*\beta} \geq 0. \tag{1.187}$$

Prove that inequality (1.117)

$$\mathbf{s} : \mathbf{d} - \rho \dot{F}_{*} - \rho \dot{\theta} S_{*} \geq 0 \tag{1.188}$$

holds for materials with a rheological model composed of the rheological models of materials α and β by means of connections in series and in parallel.

We first address a parallel connection. We substitute expressions for $\mathbf{s}$, $\mathbf{d}$, F_* and S_* due to the axioms of parallel connections, eqs. (1.181) and (1.183), into (1.188). Combining the terms, we obtain

$$\mathbf{s}_\alpha \colon \mathbf{d}_\alpha - \rho \dot{F}_{*\alpha} - \rho \dot{\theta} S_{*\alpha} + \mathbf{s}_\beta \colon \mathbf{d}_\beta - \rho \dot{F}_{*\beta} - \rho \dot{\theta} S_{*\beta} \geq 0. \tag{1.189}$$

The sum of the first three terms on the left hand side is non-negative due to assumption (1.186), while the rest is non-negative due to (1.187). Hence, condition (1.188) holds.

Consider now a connection in series. We substitute expressions for $\mathbf{s}$, $\mathbf{d}$, F_* and S_* due to the axioms of connection in series, eqs. (1.182) and (1.183), into (1.188). Combining the terms, we again arrive at inequality (1.189) which holds by virtue of assumptions (1.186) and (1.187).

It only remains to remember that the dissipative inequality, eqs. (1.186) and (1.187), holds for any thermodynamic process for each fundamental element: elastic, viscous and plastic.

Therefore, the theorem is proved.

We now prove another two theorems.

Theorem 2. Constitutive equations obtained for a rheological model correspond to an isotropic material.

Theorem 3. Constitutive equations obtained for a rheological model are frame-indifferent.

Statements of these theorems are concerned with the system of constitutive equations (1.58), (1.101), (1.103), (1.111) and (1.112). It is easy to understand that the proof should touch upon only the variables and tensors describing shape distortion. Mathematically, the following equations for H- and Q-motions

$$\begin{aligned} &\mathbf{s}_H = \mathbf{s}, && F_{*H} = F_*, && S_{*H} = S_*, \\ &\mathbf{s}_Q = \mathbf{Q}\cdot\mathbf{s}\cdot\mathbf{Q}^T, && F_{*Q} = F_*, && S_{*Q} = S_* \end{aligned} \tag{1.190}$$

must hold.

The theorem will be proved by the method of mathematical induction. Consider two arbitrary rheological models: α and β. Denote the operators governing the stress deviators as follows

$$\begin{aligned} \mathbf{s}_\alpha &= \mathbf{Z}_\alpha\{\theta,\ \rho_0/\rho,\ \mathbf{\Lambda}_\alpha, \mathbf{B}\}, \\ \mathbf{s}_\beta &= \mathbf{Z}_\beta\{\theta,\ \rho_0/\rho,\ \mathbf{\Lambda}_\beta, \mathbf{B}\}. \end{aligned}$$

Let us allow that these operators as well as operators F_* and S_* (1.112) prescribing the free energy and the entropy of shape distortion for these models, describe isotropic materials and satisfy the material frame indifference principle.

Due to (1.93) and (1.190) the isotropy assumption means that the above operators satisfy the following conditions

$$\begin{aligned} &\mathbf{Z}_\gamma\{\theta,\ \rho_0/\rho,\ \mathbf{\Lambda}_\gamma, \mathbf{B}\cdot\mathbf{H}\} = \mathbf{Z}_\gamma\{\theta,\ \rho_0/\rho,\ \mathbf{\Lambda}_\gamma, \mathbf{B}\}, \\ &F_{*\gamma}\{\theta,\ \rho_0/\rho,\ \mathbf{\Lambda}_\gamma, \mathbf{B}\cdot\mathbf{H}\} = F_{*\gamma}\{\theta,\ \rho_0/\rho,\ \mathbf{\Lambda}_\gamma, \mathbf{B}\}, \\ &S_{*\gamma}\{\theta,\ \rho_0/\rho,\ \mathbf{\Lambda}_\gamma, \mathbf{B}\cdot\mathbf{H}\} = S_{*\gamma}\{\theta,\ \rho_0/\rho,\ \mathbf{\Lambda}_\gamma, \mathbf{B}\}, \\ &\gamma = \alpha,\ \beta. \end{aligned} \tag{1.191}$$

By virtue of (1.97), (1.99) and (1.190) the frame-indifference assumption corresponds to the conditions

$$\begin{aligned}
&\mathbf{Z}_\gamma\left\{\theta,\,\rho_0/\rho,\,\mathbf{Q}\cdot\mathbf{\Lambda}_\gamma\cdot\mathbf{Q}^T,\mathbf{Q}\cdot\mathbf{B}\right\}=\mathbf{Q}\cdot\mathbf{Z}_\gamma\left\{\theta,\,\rho_0/\rho,\,\mathbf{\Lambda}_\gamma,\mathbf{B}\right\}\cdot\mathbf{Q}^T,\\
&F_{*\gamma}\left\{\theta,\,\rho_0/\rho,\,\mathbf{Q}\cdot\mathbf{\Lambda}_\gamma\cdot\mathbf{Q}^T,\mathbf{Q}\cdot\mathbf{B}\right\}=F_{*\gamma}\left\{\theta,\,\rho_0/\rho,\,\mathbf{\Lambda}_\gamma,\mathbf{B}\right\},\\
&S_{*\gamma}\left\{\theta,\,\rho_0/\rho,\,\mathbf{Q}\cdot\mathbf{\Lambda}_\gamma\cdot\mathbf{Q}^T,\mathbf{Q}\cdot\mathbf{B}\right\}=S_{*\gamma}\left\{\theta,\,\rho_0/\rho,\,\mathbf{\Lambda}_\gamma,\mathbf{B}\right\},\\
&\gamma=\alpha,\,\beta.
\end{aligned} \tag{1.192}$$

Let us prove that any material with a rheological model composed by means of parallel and in series connections of models α and β is isotropic and its constitutive equations satisfy the material frame indifference principle.

1. Consider a parallel connection of elements α and β. Due to equations (1.179) and (1.184) and initial conditions (1.180) and (1.185), the equation

$$\mathbf{d}_\alpha=\mathbf{d}_\beta=\mathbf{d},$$

renders the following result

$$\mathbf{\Lambda}_\alpha=\mathbf{\Lambda}_\beta=\mathbf{\Lambda}. \tag{1.193}$$

Further, by virtue of (1.190) and (1.193) we obtain the following equations

$$\begin{aligned}
\mathbf{s}&=\mathbf{Z}_\alpha\left\{\theta,\,\rho_0/\rho,\mathbf{\Lambda},\mathbf{B}\right\}+\mathbf{Z}_\beta\left\{\theta,\,\rho_0/\rho,\mathbf{\Lambda},\mathbf{B}\right\},\\
F_*&=F_{*\alpha}\left\{\theta,\,\rho_0/\rho,\mathbf{\Lambda},\mathbf{B}\right\}+F_{*\beta}\left\{\theta,\,\rho_0/\rho,\mathbf{\Lambda},\mathbf{B}\right\},\\
S_*&=S_{*\alpha}\left\{\theta,\,\rho_0/\rho,\mathbf{\Lambda},\mathbf{B}\right\}+S_{*\beta}\left\{\theta,\,\rho_0/\rho,\mathbf{\Lambda},\mathbf{B}\right\}.
\end{aligned} \tag{1.194}$$

It is seen that $\mathbf{s}$, F_* and S_* are operators over θ, ρ, $\mathbf{\Lambda}$ and $\mathbf{B}$.

a. Let us prove that operators (1.194) describe an isotropic material. To this end, we obtain their values in H-motion. Denoting them by subscript H we obtain by means of (1.93)

$$\begin{aligned}
\mathbf{s}_H&=\mathbf{Z}_\alpha\left\{\theta,\rho_0/\rho,\mathbf{\Lambda},\mathbf{B}\cdot\mathbf{H}\right\}+\mathbf{Z}_\beta\left\{\theta,\rho_0/\rho,\mathbf{\Lambda},\mathbf{B}\cdot\mathbf{H}\right\},\\
F_{*H}&=F_{*\alpha}\left\{\theta,\rho_0/\rho,\mathbf{\Lambda},\mathbf{B}\cdot\mathbf{H}\right\}+F_{*\beta}\left\{\theta,\rho_0/\rho,\mathbf{\Lambda},\mathbf{B}\cdot\mathbf{H}\right\},\\
S_{*H}&=S_{*\alpha}\left\{\theta,\rho_0/\rho,\mathbf{\Lambda},\mathbf{B}\cdot\mathbf{H}\right\}+S_{*\beta}\left\{\theta,\rho_0/\rho,\mathbf{\Lambda},\mathbf{B}\cdot\mathbf{H}\right\}.
\end{aligned} \tag{1.195}$$

Finally, taking into account assumptions (1.191) we arrive at the following equations

$$\mathbf{s}_H=\mathbf{s},\;\;F_{*H}=F_*,\;\;S_{*H}=S_*, \tag{1.196}$$

which indicate that the material is isotropic.

b. Let us write down the equations for operators (1.194) in Q-motion. By virtue of (1.99) we have

$$\begin{aligned}
\mathbf{s}_Q&=\mathbf{Z}_\alpha\left\{\theta,\tfrac{\rho_0}{\rho},\mathbf{Q}\cdot\mathbf{\Lambda}\cdot\mathbf{Q}^T,\mathbf{Q}\cdot\mathbf{B}\right\}+\mathbf{Z}_\beta\left\{\theta,\tfrac{\rho_0}{\rho},\mathbf{Q}\cdot\mathbf{\Lambda}\cdot\mathbf{Q}^T,\mathbf{Q}\cdot\mathbf{B}\right\},\\
F_{*Q}&=F_{*\alpha}\left\{\theta,\tfrac{\rho_0}{\rho},\mathbf{Q}\cdot\mathbf{\Lambda}\cdot\mathbf{Q}^T,\mathbf{Q}\cdot\mathbf{B}\right\}+F_{*\beta}\left\{\theta,\tfrac{\rho_0}{\rho},\mathbf{Q}\cdot\mathbf{\Lambda}\cdot\mathbf{Q}^T,\mathbf{Q}\cdot\mathbf{B}\right\},\\
S_{*Q}&=S_{*\alpha}\left\{\theta,\tfrac{\rho_0}{\rho},\mathbf{Q}\cdot\mathbf{\Lambda}\cdot\mathbf{Q}^T,\mathbf{Q}\cdot\mathbf{B}\right\}+S_{*\beta}\left\{\theta,\tfrac{\rho_0}{\rho},\mathbf{Q}\cdot\mathbf{\Lambda}\cdot\mathbf{Q}^T,\mathbf{Q}\cdot\mathbf{B}\right\}.
\end{aligned}$$

Taking into account assumptions (1.192) we obtain the following equations

$$\mathbf{s}_Q = \mathbf{Q} \cdot \mathbf{s} \cdot \mathbf{Q}^T, \quad F_{*Q} = F_*, \quad S_{*Q} = S_*. \tag{1.197}$$

Comparison with (1.190) shows that the constitutive equations satisfy the principle of material frame indifference.

2. Consider now a connection of two elements α and β in series. The condition for a connection in series are as follows

$$\begin{aligned}
&\mathbf{d}_\alpha + \mathbf{d}_\beta = \mathbf{d}, \\
&\dot{\mathbf{\Lambda}}_\alpha = -\left(\mathbf{d}_\alpha + \mathbf{\Omega}\right) \cdot \mathbf{\Lambda}_\alpha - \mathbf{\Lambda}_\alpha \cdot \left(\mathbf{d}_\alpha - \mathbf{\Omega}\right), \\
&\dot{\mathbf{\Lambda}}_\beta = -\left(\mathbf{d}_\beta + \mathbf{\Omega}\right) \cdot \mathbf{\Lambda}_\beta - \mathbf{\Lambda}_\beta \cdot \left(\mathbf{d}_\beta - \mathbf{\Omega}\right), \\
&\mathbf{Z}_\alpha \left\{\theta, \rho_0/\rho, \mathbf{\Lambda}_\alpha, \mathbf{B}\right\} = \mathbf{Z}_\beta \left\{\theta, \rho_0/\rho, \mathbf{\Lambda}_\beta, \mathbf{B}\right\}.
\end{aligned} \tag{1.198}$$

This is a system of equations for unknown $\mathbf{\Lambda}_\alpha$, $\mathbf{\Lambda}_\beta$, $\mathbf{d}_\alpha$ and $\mathbf{d}_\beta$ for given $\mathbf{\Lambda}$, $\mathbf{B}$, $\mathbf{\Omega}$, $\mathbf{d}$, ρ and θ. The initial condition is given by eq. (1.180). The stress deviator for this connection is obtained from condition (1.182). For example, we have

$$\mathbf{s} = \mathbf{Z}_\alpha \left\{\theta, \rho_0/\rho, \mathbf{\Lambda}_\alpha, \mathbf{B}\right\}. \tag{1.199}$$

Further, by means of (1.183) we obtain

$$\begin{aligned}
F_* &= F_{*\alpha} \left\{\theta, \rho_0/\rho, \mathbf{\Lambda}_\alpha, \mathbf{B}\right\} + F_{*\beta} \left\{\theta, \rho_0/\rho, \mathbf{\Lambda}_\beta, \mathbf{B}\right\}, \\
S_* &= S_{*\alpha} \left\{\theta, \rho_0/\rho, \mathbf{\Lambda}_\alpha, \mathbf{B}\right\} + S_{*\beta} \left\{\theta, \rho_0/\rho, \mathbf{\Lambda}_\beta, \mathbf{B}\right\}.
\end{aligned} \tag{1.200}$$

We assume that this Cauchy's problem, (1.198) and (1.180), has only one solution. Then $\mathbf{\Lambda}_\alpha$ and $\mathbf{\Lambda}_\beta$ are some functionals over $\mathbf{\Lambda}$, $\mathbf{B}$, ρ and θ. By virtue of (1.199) and (1.200) $\mathbf{s}$, $F_{*\alpha}$ and $S_{*\beta}$ are also functionals over the same arguments.

a. Let us write down the system of equations (1.198) for obtaining $\mathbf{\Lambda}_\alpha$ and $\mathbf{\Lambda}_\beta$ in H -motion.

By virtue of (1.93) and (1.94) we have

$$\begin{aligned}
&\mathbf{d}_{\alpha H} + \mathbf{d}_{\beta H} = \mathbf{d}, \\
&\dot{\mathbf{\Lambda}}_{\alpha H} = -\left(\mathbf{d}_{\alpha H} + \mathbf{\Omega}\right) \cdot \mathbf{\Lambda}_{\alpha H} - \mathbf{\Lambda}_{\alpha H} \cdot \left(\mathbf{d}_{\alpha H} - \mathbf{\Omega}\right), \\
&\dot{\mathbf{\Lambda}}_{\beta H} = -\left(\mathbf{d}_{\beta H} + \mathbf{\Omega}\right) \cdot \mathbf{\Lambda}_{\beta H} - \mathbf{\Lambda}_{\beta H} \cdot \left(\mathbf{d}_{\beta H} - \mathbf{\Omega}\right)
\end{aligned} \tag{1.201}$$

and

$$\mathbf{Z}_\alpha \left\{\theta,\ \rho_0/\rho,\ \mathbf{\Lambda}_{\alpha H}, \mathbf{B} \cdot \mathbf{H}\right\} = \mathbf{Z}_\beta \left\{\theta,\ \rho_0/\rho,\ \mathbf{\Lambda}_{\beta H}, \mathbf{B} \cdot \mathbf{H}\right\}.$$

Due to the isotropy of the elements, eq. (1.191), the latter equation is replaced by the following one

$$\mathbf{Z}_\alpha \left\{\theta, \rho_0/\rho, \mathbf{\Lambda}_{\alpha H}, \mathbf{B}\right\} = \mathbf{Z}_\beta \left\{\theta, \rho_0/\rho, \mathbf{\Lambda}_{\beta H}, \mathbf{B}\right\}. \tag{1.202}$$

One sees now that the system of equations (1.201) and (1.202) coincides with the system (1.198). Also their initial conditions coincide. Further, because of the uniqueness theorem, their solutions coincide, i.e.

$$\mathbf{\Lambda}_{\alpha H} = \mathbf{\Lambda}_{\alpha}, \quad \mathbf{\Lambda}_{\beta H} = \mathbf{\Lambda}_{\beta}. \tag{1.203}$$

Let us find now the stress deviator, the free energy and the entropy for H-motion. To this aim we substitute (1.203) and (1.93) into (1.199) and (1.200), to get

$$\begin{aligned}
\mathbf{s}_H &= \mathbf{Z}_{\alpha}\left\{\theta, \rho_0/\rho, \mathbf{\Lambda}_{\alpha}, \mathbf{B}\cdot\mathbf{H}\right\}, \\
F_{*H} &= F_{*\alpha}\left\{\theta, \rho_0/\rho, \mathbf{\Lambda}_{\alpha}, \mathbf{B}\cdot\mathbf{H}\right\} + F_{*\beta}\left\{\theta, \rho_0/\rho, \mathbf{\Lambda}_{\beta}, \mathbf{B}\cdot\mathbf{H}\right\}, \\
S_{*H} &= S_{*\alpha}\left\{\theta, \rho_0/\rho, \mathbf{\Lambda}_{\alpha}, \mathbf{B}\cdot\mathbf{H}\right\} + S_{*\beta}\left\{\theta, \rho_0/\rho, \mathbf{\Lambda}_{\beta}, \mathbf{B}\cdot\mathbf{H}\right\}.
\end{aligned}$$

Finally, allowing for assumptions (1.191) the following equations are obtained

$$\mathbf{s}_H = \mathbf{s}, \;\; F_{*H} = F_{*}, \;\; S_{*H} = S_{*},$$

which indicate that the operators (1.199) and (1.200) together with eqs. (1.198) and initial conditions (1.180) correspond to an isotropic material.

b. Let us prove that operators (1.199) and (1.200) together with equations (1.198) and initial conditions (1.180) satisfy the principle of material frame indifference. With this in view, we write down equations (1.198) for Q-motion. Accounting for conditions (1.99), we obtain

$$\begin{aligned}
&\mathbf{d}_{\alpha Q} + \mathbf{d}_{\beta Q} = \mathbf{Q}\cdot\mathbf{d}\cdot\mathbf{Q}^T, \\
&\dot{\mathbf{\Lambda}}_{\alpha Q} = -\left(\mathbf{d}_{\alpha Q} + \mathbf{Q}\cdot\boldsymbol{\Omega}\cdot\mathbf{Q}^T - \dot{\mathbf{Q}}\cdot\mathbf{Q}^T\right)\cdot\mathbf{\Lambda}_{\alpha Q} - \\
&-\mathbf{\Lambda}_{\alpha Q}\cdot\left(\mathbf{d}_{\alpha Q} - \mathbf{Q}\cdot\boldsymbol{\Omega}\cdot\mathbf{Q}^T + \dot{\mathbf{Q}}\cdot\mathbf{Q}^T\right), \\
&\dot{\mathbf{\Lambda}}_{\beta Q} = -\left(\mathbf{d}_{\beta Q} + \mathbf{Q}\cdot\boldsymbol{\Omega}\cdot\mathbf{Q}^T - \dot{\mathbf{Q}}\cdot\mathbf{Q}^T\right)\cdot\mathbf{\Lambda}_{\beta Q} - \\
&-\mathbf{\Lambda}_{\beta Q}\cdot\left(\mathbf{d}_{\beta Q} - \mathbf{Q}\cdot\boldsymbol{\Omega}\cdot\mathbf{Q}^T + \dot{\mathbf{Q}}\cdot\mathbf{Q}^T\right), \\
&\mathbf{Z}_{\alpha}\left\{\theta,\, \rho_0/\rho,\, \mathbf{\Lambda}_{\alpha Q}, \mathbf{Q}\cdot\mathbf{B}\right\} = \mathbf{Z}_{\beta}\left\{\theta,\, \rho_0/\rho,\, \mathbf{\Lambda}_{\beta Q}, \mathbf{Q}\cdot\mathbf{B}\right\}.
\end{aligned} \tag{1.204}$$

The initial conditions, cf. (1.180), are given by

$$\mathbf{\Lambda}_{\alpha Q}\left(t_0\right) = \mathbf{E}, \;\; \mathbf{\Lambda}_{\beta Q}\left(t_0\right) = \mathbf{E}. \tag{1.205}$$

It has been assumed that the solution of the system (1.198) and (1.180) is unique. The solution of the system (1.204) and (1.205) is unique also, since this system is a particular case of the first one. One easily finds by direct substitution that the solution of the system (1.204) and (1.205) is as follows

$$\begin{aligned}
\mathbf{\Lambda}_{\alpha Q} &= \mathbf{Q}\cdot\mathbf{\Lambda}_{\alpha}\cdot\mathbf{Q}^T, \;\; \mathbf{\Lambda}_{\beta Q} = \mathbf{Q}\cdot\mathbf{\Lambda}_{\beta}\cdot\mathbf{Q}^T, \\
\mathbf{d}_{\alpha Q} &= \mathbf{Q}\cdot\mathbf{d}_{\alpha}\cdot\mathbf{Q}^T, \;\; \mathbf{d}_{\beta Q} = \mathbf{Q}\cdot\mathbf{d}_{\beta}\cdot\mathbf{Q}^T,
\end{aligned} \tag{1.206}$$

where $\mathbf{\Lambda}_\alpha$, $\mathbf{\Lambda}_\beta$, $\mathbf{d}_\alpha$ and $\mathbf{d}_\beta$ are governed by system of equations (1.198) and the initial conditions (1.180).

Indeed, the first three equations of system (1.204) and initial conditions (1.205) are satisfied, while the latter equation takes the form

$$\mathbf{Z}_\alpha\left\{\theta, \rho_0/\rho, \mathbf{Q}\cdot\mathbf{\Lambda}_\alpha\cdot\mathbf{Q}^T, \mathbf{Q}\cdot\mathbf{B}\right\} = \mathbf{Z}_\beta\left\{\theta, \rho_0/\rho, \mathbf{Q}\cdot\mathbf{\Lambda}_\beta\cdot\mathbf{Q}^T, \mathbf{Q}\cdot\mathbf{B}\right\}.$$

By virtue of assumptions (1.192) this is equivalent to the following equation

$$\mathbf{Q}\cdot\mathbf{Z}_\alpha\left\{\theta, \rho_0/\rho, \mathbf{\Lambda}_\alpha, \mathbf{B}\right\}\cdot\mathbf{Q}^T = \mathbf{Q}\cdot\mathbf{Z}_\beta\left\{\theta, \rho_0/\rho, \mathbf{\Lambda}_\beta, \mathbf{B}\right\}\cdot\mathbf{Q}^T$$

or

$$\mathbf{Z}_\alpha\left\{\theta, \rho_0/\rho, \mathbf{\Lambda}_\alpha, \mathbf{B}\right\} = \mathbf{Z}_\beta\left\{\theta, \rho_0/\rho, \mathbf{\Lambda}_\beta, \mathbf{B}\right\}.$$

This equation is the forth equation of the system (1.198) for determining $\mathbf{\Lambda}_\alpha$ and $\mathbf{\Lambda}_\beta$, and hence it is satisfied.

Let us obtain now equations for the stress deviator, free energy and entropy in Q-motion.

By virtue of (1.99), (1.206), (1.199) and (1.200) we have

$$\begin{aligned}
&\mathbf{s}_Q = \mathbf{Z}_\alpha\left\{\theta, \rho_0/\rho, \mathbf{Q}\cdot\mathbf{\Lambda}_\alpha\cdot\mathbf{Q}^T, \mathbf{Q}\cdot\mathbf{B}\right\},\\
&F_{*Q} = F_{*\alpha}\left\{\theta, \rho_0/\rho, \mathbf{Q}\cdot\mathbf{\Lambda}_\alpha\cdot\mathbf{Q}^T, \mathbf{Q}\cdot\mathbf{B}\right\} +\\
&+F_{*\beta}\left\{\theta, \rho_0/\rho, \mathbf{Q}\cdot\mathbf{\Lambda}_\beta\cdot\mathbf{Q}^T, \mathbf{Q}\cdot\mathbf{B}\right\},\\
&S_{*Q} = S_{*\alpha}\left\{\theta, \rho_0/\rho, \mathbf{Q}\cdot\mathbf{\Lambda}_\alpha\cdot\mathbf{Q}^T, \mathbf{Q}\cdot\mathbf{B}\right\} +\\
&+S_{*\beta}\left\{\theta, \rho_0/\rho, \mathbf{Q}\cdot\mathbf{\Lambda}_\beta\cdot\mathbf{Q}^T, \mathbf{Q}\cdot\mathbf{B}\right\}.
\end{aligned}$$

Taking into account assumptions (1.192) we easily obtain

$$\mathbf{s}_Q = \mathbf{Q}\cdot\mathbf{s}\cdot\mathbf{Q}^T, \quad F_{*Q} = F_*, \quad S_{*Q} = S_*,$$

which was the objective of the proof.

We thus proved that, if the constitutive equations for the models α and β correspond to an isotropic material and satisfy the material frame indifference principle, then the constitutive equations for parallel and in series connections of the models α and β possess the same properties.

The proof of the above theorems is completed by the statement that for the fundamental rheological models, such as elastic, viscous and plastic elements, conditions (1.191) and (1.192) are satisfied. This statement can be easily proved by direct substitution.

Depending on the structure and the set of fundamental rheological elements used to compose a rheological model, materials can be divided into a few large groups.

1. *Materials with pure properties.* Rheological models for these materials consist of the same fundamental elements of one sort, e.g. either Hooke's elements, or Newton's elements or St. Venant's elements.

2. *Viscoelastic materials.* For these materials, rheological models are composed of a set of Hooke elements and Newton elements.

3. *Elastoplastic materials.* For these materials, rheological models consist of a set of Hooke elements and St. Venant elements.

4. *Viscoplastic materials.* Sets of the Newton elements and the St. Venant elements are used to compose rheological models for these materials.

5. *Viscoelastoplastic materials.* Rheological models for these materials contain all fundamental rheological elements.

To conclude, we draw attention to the fact that there exists no one-to-one correspondence between the rheological models and the constitutive equations. For each rheological model one can obtain its constitutive equations. However, a rheological model exists not for any arbitrary constitutive equation, even if the latter satisfies the dissipative inequality. In other words, the set of rheological models is a subset of the set of constitutive equations. Despite this fact, the method of rheological models is extraordinarily useful since it allows one to construct particular constitutive equations with various mechanical properties in a simple way.

The procedure of prescribing the constitutive equations described in Sections 1.8, 1.9 and 1.10 enables various generalisations. For example, the assumption that the dilatation is governed by an elasticity equation and the thermal conductivity is governed by the Fourier law is not used in [134], where the most general governing equations like eqs. (1.81) and (1.90) are utilised. A generalisation of the method of rheological models for the general case has been proposed there. The constitutive equations of thermoelasticity, thermoviscosity and thermoplasticity are taken as the basic equations which are combined according to the generalised rheological model. The difference of the latter model from the rheological models of this book, as well as the models of the books [192] and [156], is that not only the stress tensor and the strain rate tensor, but also the heat flux vector and the temperature gradient are included in the modelling. We will not consider very rigorous details of this description as these generalisations will not be applied to the dynamics of materials which is the main topic of this book.

In concluding this Section it is worthwhile noting that other approaches are often applied in the literature to describe plasticity theory and viscoelasticity theory under finite deformations. However the method of rheological models has never been mentioned. The recent papers [51] and [79] are evidence of this fact. The authors use a decomposition of the total strain on elastic and inelastic, i.e. viscous or plastic, components. This way may be interpreted by means of the method of rheological models. An example of such analysis will be shown at Section 1.12.

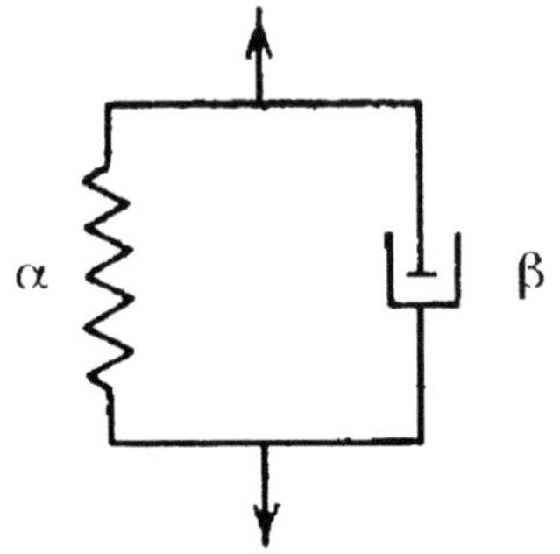

FIGURE 1.4. Rheological model of the Kelvin-Voigt material.

1.11 Examples of complex materials

1.11.1 *Example 1. The Kelvin-Voigt material*

This is a viscoelastic material with the rheological model depicted in Fig. 1.4. Its rheological model represents a parallel connection of a Hooke element and a Newton element. According to (1.137) and (1.143) we have

$$\begin{aligned} \mathbf{s}_\alpha &= -2\rho \operatorname{Dev}\left(\frac{\partial F_{*\alpha}}{\partial a_\alpha}\mathbf{\Lambda}_\alpha + \frac{\partial F_{*\alpha}}{\partial b_\alpha}\mathbf{\Lambda}_\alpha^2\right), \\ \mathbf{s}_\beta &= 2\operatorname{Dev}\left(\eta \mathbf{d}_\beta + \varphi \mathbf{d}_\beta^2\right), \end{aligned} \tag{1.207}$$

where $F_{*\alpha}$ denotes free energy of the Hooke element, and η and φ satisfy condition (1.144).

Taking account of the parallel connection axioms (1.181), equations and initial conditions for $\mathbf{\Lambda}_\alpha$, $\mathbf{\Lambda}_\beta$ and $\mathbf{\Lambda}$ we find that

$$\begin{aligned} \mathbf{d}_\alpha &= \mathbf{d}_\beta = \mathbf{d},\ \mathbf{\Lambda}_\alpha = \mathbf{\Lambda}_\beta = \mathbf{\Lambda},\ \ \mathbf{s} = \mathbf{s}_\alpha + \mathbf{s}_\beta, \\ F_* &= F_{*\alpha},\ S_* = S_{*\alpha}. \end{aligned}$$

Therefore, the equations for the material are as follows

$$\mathbf{s} = \operatorname{Dev}\left[-2\rho\left(\frac{\partial F_*}{\partial a}\mathbf{\Lambda} + \frac{\partial F_*}{\partial b}\mathbf{\Lambda}^2\right) + 2\left(\eta\mathbf{d} + \varphi\mathbf{d}^2\right)\right], \tag{1.208}$$

$$F_* = F_{*\alpha}(a, b, \theta),\ \ S_* = S_{*\alpha}(a, b, \theta), \tag{1.209}$$

where a and b are invariants of tensor $\mathbf{\Lambda}$, cf. (1.135). From these equations one sees that the stress deviator depends upon the strain of shape change $\mathbf{\Lambda}$ and the strain rate of shape change $\mathbf{d}$. Equation (1.208) however is not the most general expression for an isotropic tensor function of two tensor arguments. It is known [172], that the most general expression (under the assumption that $\mathbf{s}$ is a deviator) is given by

$$\mathbf{s} = \mathrm{Dev}\left[\left(\varphi_1 \mathbf{\Lambda} + \varphi_2 \mathbf{\Lambda}^2\right) + \varphi_3 \mathbf{d} + \varphi_4 \mathbf{d}^2 + \varphi_5 \left(\mathbf{\Lambda}\cdot\mathbf{d} + \mathbf{d}\cdot\mathbf{\Lambda}\right) + \right.$$

$$\left. + \varphi_6 \left(\mathbf{\Lambda}\cdot\mathbf{d}^2 + \mathbf{d}^2\cdot\mathbf{\Lambda}\right) + \varphi_7 \left(\mathbf{\Lambda}^2\cdot\mathbf{d} + \mathbf{d}\cdot\mathbf{\Lambda}^2\right) + \varphi_8 \left(\mathbf{\Lambda}^2\cdot\mathbf{d}^2 + \mathbf{d}^2\cdot\mathbf{\Lambda}^2\right)\right], \tag{1.210}$$

where the scalar coefficients φ_i depend on the invariants of tensors $\mathbf{\Lambda}$ and $\mathbf{d}$, their cross-invariants and other scalar parameters such as ρ, $\dot{\rho}$ and θ.

Substituting (1.210) and (1.209) into dissipative inequality (1.107) yields

$$\left[\left(\varphi_1 - 2\rho\frac{\partial F_*}{\partial a}\right)\mathbf{\Lambda} + \left(\varphi_2 - 2\rho\frac{\partial F_*}{\partial b}\right)\mathbf{\Lambda}^2\right] : \mathbf{d} + 2\left(\varphi_3 \mathbf{E} + \varphi_5 \mathbf{\Lambda} + \varphi_7 \mathbf{\Lambda}^2\right) : \mathbf{d}^2 +$$

$$+ 2\left(\varphi_4 \mathbf{E} + \varphi_6 \mathbf{\Lambda} + \varphi_8 \mathbf{\Lambda}^2\right) : \mathbf{d}^3 - \rho\dot{\theta}\left(\frac{\partial F_*}{\partial \theta} + S_*\right) \geq 0. \tag{1.211}$$

It is difficult to point out the necessary and sufficient conditions for this inequality, but it is easy to see that if $F_* = 0$, $S_* = 0$ and $\varphi_i = 0$ $(i \neq 7)$ the latter equation takes the form

$$2\varphi_7 \left(\mathbf{\Lambda}\cdot\mathbf{d}\right) : \left(\mathbf{\Lambda}\cdot\mathbf{d}\right)^T \geq 0.$$

Thus, provided that

$$\varphi_7 \geq 0$$

the dissipative inequality holds for any thermodynamical process. In this particular case the constitutive equations are as follows

$$\mathbf{s} = \varphi_7 \mathrm{Dev}\left(\mathbf{\Lambda}^2\cdot\mathbf{d} + \mathbf{d}\cdot\mathbf{\Lambda}^2\right), \quad F_* = 0, \quad S_* = 0. \tag{1.212}$$

Analogously, the constitutive equations

$$\mathbf{s} = \gamma\left(\mathbf{\Lambda} : \mathbf{d}\right)\mathrm{Dev}\mathbf{\Lambda}, \quad F_* = 0, \quad S_* = 0 \tag{1.213}$$

satisfy the dissipative inequality under the following condition

$$\gamma \geq 0.$$

However, neither eq. (1.212) nor eq. (1.213) is a particular case of eq. (1.208) derived by means of the rheological model. This fact demonstrates the above-said relation between the set of constitutive equations and the set of rheological models.

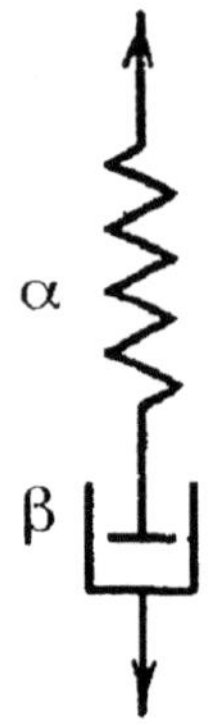

FIGURE 1.5. Rheological model of the Maxwell material.

1.11.2 Example 2. The Maxwell material

This viscoelastic material whose rheological model is depicted in Fig. 1.5 consists of a Hooke element and a Newton element in series.

Let us show how to obtain the dependence of the stress deviator $\mathbf{s}$ on the deformation history for this material. In accordance with the axioms of an in series connection we have

$$\mathbf{s} = \mathbf{s}_\alpha = \mathbf{s}_\beta, \quad \mathbf{d} = \mathbf{d}_\alpha + \mathbf{d}_\beta, \quad F_* = F_{*\alpha}, \quad S_* = S_{*\alpha}.$$

Use is made of the general form for the elements' equations (1.137) and (1.149). The elements' connection yields

$$\mathbf{s} = -2\rho \mathrm{Dev}\left(\frac{\partial F_{*\alpha}}{\partial a_\alpha}\mathbf{\Lambda}_\alpha + \frac{\partial F_{*\alpha}}{\partial b_\alpha}\mathbf{\Lambda}_\alpha^2\right), \tag{1.214}$$

$$\mathbf{d}_\beta = \mathrm{Dev}\left(\beta\mathbf{s} + \gamma\mathbf{s}^2\right), \tag{1.215}$$

where β and γ satisfy condition (1.150).

Further, the kinematic condition of in series connection of the elements requires that

$$\mathbf{d}_\alpha = \mathbf{d} - \mathbf{d}_\beta,$$

which, due to (1.215), takes the form

$$\mathbf{d}_\alpha = \mathbf{d} - \mathrm{Dev}\left(\beta\mathbf{s} + \gamma\mathbf{s}^2\right). \tag{1.216}$$

Finally, substituting this expression into differential equation (1.179) for tensor $\mathbf{\Lambda}$, we obtain

$$\dot{\mathbf{\Lambda}}_\alpha = -\left[\mathbf{d} + \mathbf{\Omega} - \mathrm{Dev}\left(\beta\mathbf{s} + \gamma\mathbf{s}^2\right)\right]\cdot\mathbf{\Lambda}_\alpha - \mathbf{\Lambda}_\alpha\cdot\left[\mathbf{d} - \mathbf{\Omega} - \mathrm{Dev}\left(\beta\mathbf{s} + \gamma\mathbf{s}^2\right)\right]. \tag{1.217}$$

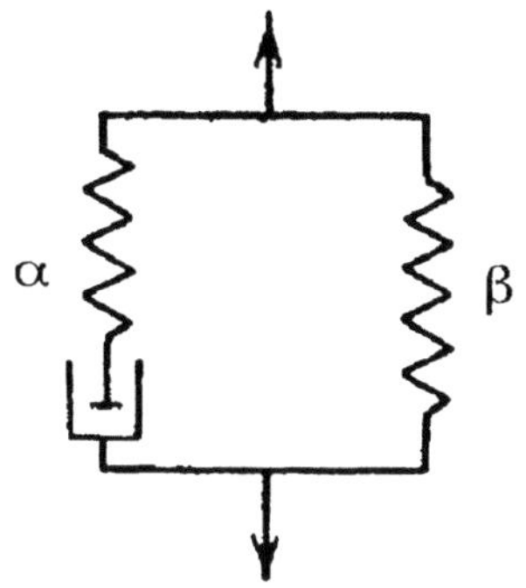

FIGURE 1.6. The Poynting-Thomson rheological model.

Here tensors $\mathbf{d}$ and $\mathbf{\Omega}$ are considered to be given functions of time. Given the process history $\mathbf{\Lambda} = \mathbf{\Lambda}(t)$ and $\mathbf{B} = \mathbf{B}(t)$, the tensors $\mathbf{d}$ and $\mathbf{\Omega}$ can be obtained either by means of (1.83) and the following equation

$$\mathbf{d} = -\int_0^{\infty} e^{-\mathbf{\Lambda}\lambda} \cdot \mathbf{\Lambda}^{\nabla} \cdot e^{-\mathbf{\Lambda}\lambda} d\lambda, \tag{1.218}$$

cf. eq. (1.87), or directly from eq. (1.184). The stress deviator $\mathbf{s}$ should be expressed in terms of tensor $\mathbf{\Lambda}_\alpha$ by means of (1.214) and substituted into (1.217).

Thus, equation (1.217), together with the initial condition (1.180) determines $\mathbf{\Lambda}_\alpha$ as a function of time, i.e. $\mathbf{\Lambda}_\alpha = \mathbf{\Lambda}_\alpha(t)$. Substitution of the latter function into (1.214) yields

$$\mathbf{s} = \mathbf{s}(t). \tag{1.219}$$

Therefore, determination of the stress deviator for the Maxwell material is concerned with solving a nonlinear ordinary differential equation (1.217). Similar equations have been obtained in [40] and [96] by somewhat different arguments.

1.11.3 *Example 3. The Poynting-Thomson viscoelastic material*

The name of this material is taken from [157] and [156]. Its rheological model depicted in Fig. 1.6 consists of a Maxwell model, Fig. 1.5, and a Hooke's element, Fig. 1.2a, in parallel.

In accordance with the rules of parallel connections we have

$$\mathbf{s} = -2\rho \mathrm{Dev}\left(\frac{\partial F_{*\alpha}}{\partial a_\alpha}\mathbf{\Lambda}_\alpha + \frac{\partial F_{*\alpha}}{\partial b_\alpha}\mathbf{\Lambda}_\alpha^2 + \frac{\partial F_{*\beta}}{\partial a}\mathbf{\Lambda} + \frac{\partial F_{*\beta}}{\partial b}\mathbf{\Lambda}^2\right), \tag{1.220}$$

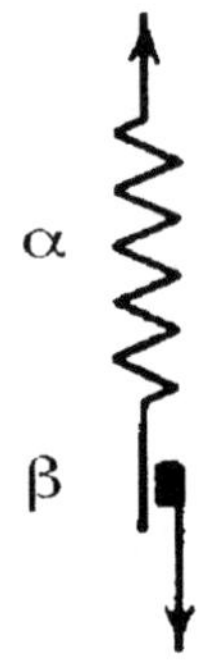

FIGURE 1.7. The Prandtl rheological model.

$$F_* = F_{*1\alpha}(\theta, a_\alpha, b_\alpha) + F_{*\beta}(\theta, a, b), \tag{1.221}$$

$$S_* = -\frac{\partial F_*}{\partial \theta}, \tag{1.222}$$

where a and b are the invariants (1.135) of tensor $\mathbf{\Lambda}$ and a_α and b_α are the corresponding invariants of tensor $\mathbf{\Lambda}_\alpha$.

The governing differential equation for $\mathbf{\Lambda}_\alpha$, eq. (1.217), is given by

$$\begin{aligned} \dot{\mathbf{\Lambda}}_\alpha = &- \left[\mathbf{d} + \mathbf{\Omega} - \mathrm{Dev}\left(\beta_\alpha \mathbf{s}_\alpha + \gamma_\alpha \mathbf{s}_\alpha^2\right)\right] \cdot \mathbf{\Lambda}_\alpha - \\ &- \mathbf{\Lambda}_\alpha \cdot \left[\mathbf{d} - \mathbf{\Omega} - \mathrm{Dev}\left(\beta_\alpha \mathbf{s}_\alpha + \gamma_\alpha \mathbf{s}_\alpha^2\right)\right]. \end{aligned} \tag{1.223}$$

Here $\mathbf{s}_\alpha$ is given by an equation similar to eq. (1.214)

$$\mathbf{s}_\alpha = -2\rho \mathrm{Dev}\left(\frac{\partial F_{*\alpha}}{\partial a_\alpha} \mathbf{\Lambda}_\alpha + \frac{\partial F_{*\alpha}}{\partial b_\alpha} \mathbf{\Lambda}_\alpha^2\right),$$

while β_α and γ_α are functions of the invariants of $\mathbf{s}_\alpha$ and temperature.

1.11.4 *Example 4. Elastoplastic materials with the Prandtl rheological model*

Its rheological model is a Hooke's element and a St. Venant's element connected in series, cf. Fig. 1.7. The connection conditions are given by

$$\begin{aligned} &\mathbf{d}_\alpha = \mathbf{d} - \mathbf{d}_\beta, \ \mathbf{s}_\alpha = \mathbf{s}_\beta = \mathbf{s}, \\ &S_* = S_{*\alpha}, \ \ F_* = F_{*\alpha}. \end{aligned} \tag{1.224}$$

The element's equations are taken in a general form, i.e. eqs. (1.137), (1.160) and (1.161) such that

$$\mathbf{s} = -2\rho \mathrm{Dev}\left(\frac{\partial F_{*\alpha}}{\partial a_\alpha} \mathbf{\Lambda}_\alpha + \frac{\partial F_{*\alpha}}{\partial b_\alpha} \mathbf{\Lambda}_\alpha^2\right), \tag{1.225}$$

$$\mathbf{d}_\beta = \lambda \mathrm{Dev}\left(r\mathbf{s} + p\mathbf{s}^2\right), \tag{1.226}$$

$$\left[\begin{array}{ll} N(\mathbf{s}) < \tau_s, & \lambda = 0, \\ N(\mathbf{s}) = \tau_s, & \lambda \geq 0. \end{array}\right. \tag{1.227}$$

We substitute $\mathbf{d}_\beta$ (1.226) into (1.224) and then into equation (1.179) for tensor $\mathbf{\Lambda}_\alpha$, to obtain

$$\dot{\mathbf{\Lambda}}_\alpha = -\left[\mathbf{d} + \mathbf{\Omega} - \lambda \mathrm{Dev}\left(r\mathbf{s} + p\mathbf{s}^2\right)\right] \cdot \mathbf{\Lambda}_\alpha - \mathbf{\Lambda}_\alpha \cdot \left[\mathbf{d} - \mathbf{\Omega} - \lambda \mathrm{Dev}\left(r\mathbf{s} + p\mathbf{s}^2\right)\right]. \tag{1.228}$$

In this equation $\mathbf{d}$ and $\mathbf{\Omega}$ are considered to be given functions of time and s is expressed in terms of $\mathbf{\Lambda}_\alpha$ by means of (1.225). The undetermined factor λ should be found by means of conditions (1.227) which, after substitution of $\mathbf{s}$ due to (1.225), take the form

$$\left[\begin{array}{ll} N_\Lambda < \tau_s/\rho, & \lambda = 0, \\ N_\Lambda = \tau_s/\rho, & \lambda \geq 0. \end{array}\right. \tag{1.229}$$

Here we introduced the following denotation

$$N_\Lambda = N\left[2\mathrm{Dev}\left(\frac{\partial F_*}{\partial a_\alpha}\mathbf{\Lambda}_\alpha + \frac{\partial F_*}{\partial b_\alpha}\mathbf{\Lambda}_\alpha^2\right)\right]. \tag{1.230}$$

Because of the material isotropy function N_Λ depends apparently upon the invariants of tensor $\mathbf{\Lambda}_\alpha$ and temperature, i.e.

$$N_\Lambda = N_\Lambda\left(a_\alpha, b_\alpha, \theta\right).$$

Only if no plastic deformation occurs, condition (1.229) determines the factor λ uniquely. In the case of plastic deformations the factor λ must be so chosen that the solution of (1.228) satisfies the yield condition

$$N_\Lambda = \frac{\tau_s}{\rho}. \tag{1.231}$$

Let us obtain an expression for λ. To this end, we take derivative with respect to time of both sides of (1.231). The result is

$$\frac{\partial N_\Lambda}{\partial \mathbf{\Lambda}_\alpha} : \dot{\mathbf{\Lambda}}_\alpha + \frac{\partial N_\Lambda}{\partial \theta}\dot{\theta} = \left(\frac{\tau_s}{\rho}\right)^{\bullet},$$

where

$$\frac{\partial N_\Lambda}{\partial \mathbf{\Lambda}_\alpha} = \frac{\partial N_\Lambda}{\partial a_\alpha}\mathbf{E} + \frac{\partial N_\Lambda}{\partial b_\alpha}\mathbf{\Lambda}_\alpha.$$

Substituting for $\dot{\mathbf{\Lambda}}_\alpha$ from (1.228) and τ_s from (1.170) we arrive at the following equation for λ

$$
\begin{aligned}
&-2\left(\frac{\partial N_\Lambda}{\partial \mathbf{\Lambda}_\alpha}\cdot\mathbf{\Lambda}_\alpha\right):\left[\mathbf{d}-\lambda \mathrm{Dev}\left(r\mathbf{s}+p\mathbf{s}^2\right)\right]+\frac{\partial N_\Lambda}{\partial\theta}\dot{\theta}= \\
&=\frac{1}{\rho}\frac{\partial \tau_s}{\partial\theta}\dot{\theta}+\left[\frac{\partial}{\partial\rho}\left(\frac{\tau_s}{\rho}\right)\right]\dot{\rho}+\frac{\lambda}{\rho}\frac{\partial\tau_s}{\partial q}I\left(\mathbf{s}\right),
\end{aligned}
\qquad (1.232)
$$

where

$$
I\left(\mathbf{s}\right)=\sqrt{I_2\left[\mathrm{Dev}\left(r\mathbf{s}+p\mathbf{s}^2\right)\right]}. \qquad (1.233)
$$

Equation (1.232) is linear in λ and has the following solution

$$
\lambda=\frac{2\left(\frac{\partial N_\Lambda}{\partial \mathbf{\Lambda}_\alpha}\cdot\mathbf{\Lambda}_\alpha\right):\mathbf{d}+\left(\frac{1}{\rho}\frac{\partial\tau_s}{\partial\theta}-\frac{\partial N_\Lambda}{\partial\theta}\right)\dot{\theta}+\left[\frac{\partial}{\partial\rho}\left(\frac{\tau_s}{\rho}\right)\right]\dot{\rho}}{2\left(\frac{\partial N_\Lambda}{\partial \mathbf{\Lambda}_\alpha}\cdot\mathbf{\Lambda}_\alpha\right):\mathrm{Dev}\left(r\mathbf{s}+p\mathbf{s}^2\right)-\frac{1}{\rho}\frac{\partial\tau_s}{\partial q}I\left(\mathbf{s}\right)}. \qquad (1.234)
$$

The solution is acceptable only if it is nonnegative and satisfies an additional condition (1.231). If (1.234) renders a negative value of λ this means that the condition in the second line in (1.229) is violated. In this case the only alternative is the first line in (1.229), i.e. one must take $\lambda = 0$.

Hence, in order to integrate equations (1.228) one should take into account conditions (1.229) in which λ is due to (1.234) when plastic deformation occurs. After the integral has been found, the stress deviator is obtained from (1.225). Thus, the stress deviator is determined at any instant of time for a given history of deformation $\mathbf{\Lambda} = \mathbf{\Lambda}(t)$.

Expression (1.234) for λ is rather complicated. In an important particular case, in which (i) the plasticity element is governed by the equations of the St. Venant-Mises theory, (ii) the process is isothermic, (iii) the yield stress is a linear function of mass density ($\tau_s = \rho\tau_s^*$) and (iv) the elasticity element is governed by the Mooney equation (1.139) with B=0, equation (1.234) simplifies and reduces to the following form

$$
\lambda=\frac{\rho_0}{\rho A}\frac{\left[\left(\mathrm{Dev}\mathbf{\Lambda}_\alpha\right)\cdot\mathbf{\Lambda}_\alpha\right]:\mathbf{d}}{\left[\left(\mathrm{Dev}\mathbf{\Lambda}_\alpha\right)\cdot\mathbf{\Lambda}_\alpha\right]:\left(\mathrm{Dev}\mathbf{\Lambda}_\alpha\right)}. \qquad (1.235)
$$

1.11.5 *Example 5. A viscoelastoplastic material*

The model for this material is composed of a Hooke, a Newton and a St. Venant element all in series. Its rheological model is depicted in Fig. 1.8.

According to the axioms of parallel connection we have

$$
\mathbf{d}_\alpha=\mathbf{d}-\mathbf{d}_\beta-\mathbf{d}_\gamma,\ \ \mathbf{s}=\mathbf{s}_\alpha=\mathbf{s}_\beta=\mathbf{s}_\gamma,\ \ F_*=F_{*\alpha},\ S_*=S_{*\alpha}. \qquad (1.236)
$$

The elements' equations are as follows

FIGURE 1.8. Rheological model of a viscoelastoplastic material.

$$\mathrm{s} = -2\rho \mathrm{Dev}\left(\frac{\partial F_{*\alpha}}{\partial a_\alpha}\mathbf{\Lambda}_\alpha + \frac{\partial F_{*\alpha}}{\partial b_\alpha}\mathbf{\Lambda}_\alpha^2\right), \quad (1.237)$$

$$\mathbf{d}_\beta = \mathrm{Dev}\left(\beta \mathbf{s} + \gamma \mathbf{s}^2\right), \quad (1.238)$$

$$\mathbf{d}_\gamma = \lambda \mathrm{Dev}\left(r\mathbf{s} + p\mathbf{s}^2\right), \quad (1.239)$$

$$\left[\begin{array}{l} N(\mathbf{s}) < \tau_s, \ \ \lambda = 0, \\ N(\mathbf{s}) = \tau_s, \ \ \lambda \geq 0. \end{array}\right. \quad (1.240)$$

By means of (1.238), (1.239) and kinematic condition (1.236) we arrive at the following equation for $\mathbf{\Lambda}_\alpha$

$$\begin{aligned} \dot{\mathbf{\Lambda}}_\alpha &= -\left\{\mathbf{d} + \mathbf{\Omega} - \mathrm{Dev}\left[\beta \mathbf{s} + \gamma \mathbf{s}^2 + \lambda\left(r\mathbf{s} + p\mathbf{s}^2\right)\right]\right\} \cdot \mathbf{\Lambda}_\alpha - \\ &\quad -\mathbf{\Lambda}_\alpha \cdot \left\{\mathbf{d} - \mathbf{\Omega} - \mathrm{Dev}\left[\beta \mathbf{s} + \gamma \mathbf{s}^2 + \lambda\left(r\mathbf{s} + p\mathbf{s}^2\right)\right]\right\}, \quad (1.241) \end{aligned}$$

where the stress deviator $\mathbf{s}$ is obtained from (1.237). Next, an equation for λ is as follows

$$\left[\begin{array}{l} N_\Lambda < \tau_s/\rho, \ \ \lambda = 0, \\ N_\Lambda = \tau_s/\rho, \ \ \lambda \geq 0, \end{array}\right. \quad (1.242)$$

where N_Λ is given by (1.230). Finally, under plastic deformations one can use the following equation for λ

$$\lambda = \left[2\left(\frac{\partial N_\Lambda}{\partial \mathbf{\Lambda}_\alpha} \cdot \mathbf{\Lambda}_\alpha\right) : \mathrm{Dev}\left(r\mathbf{s} + p\mathbf{s}^2\right) - \frac{1}{\rho}\frac{\partial \tau_s}{\partial q} I(\mathbf{s})\right]^{-1} \times$$

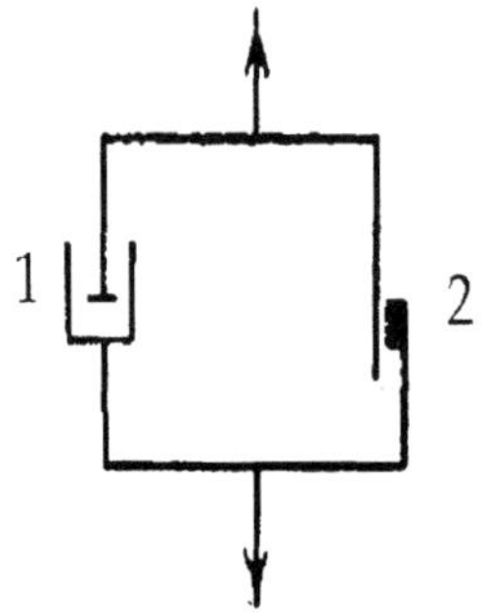

FIGURE 1.9. The rheological model of the Bingham viscoplastic material.

$$\times\left\{2\left(\frac{\partial N_\Lambda}{\partial \mathbf{\Lambda}_\alpha}\cdot\mathbf{\Lambda}_\alpha\right):\left[\mathbf{d}-\mathrm{Dev}\left(\beta\mathbf{s}+\gamma\mathbf{s}^2\right)\right]+\right.$$
$$\left.+\left(\frac{1}{\rho}\frac{\partial\tau_s}{\partial\theta}-\frac{\partial N_\Lambda}{\partial\theta}\right)\dot{\theta}+\left[\frac{\partial}{\partial\rho}\left(\frac{\tau_s}{\rho}\right)\right]\dot{\rho}\right\}.$$

which is applicable only if $\lambda \geq 0$.

1.11.6 Example 6. The Bingham viscoplastic material

The rheological model of this material consists of a Newton's elements and a St. Venant's element in parallel and is shown in Fig. 1.9. The conditions for the elements' connections and the governing equations due to eqs. (1.181), (1.183), (1.143) and (1.157) are as follows

$$\mathbf{d}_1=\mathbf{d}_2=\mathbf{d},\ \ \mathbf{s}=\mathbf{s}_1+\mathbf{s}_2\ ,S_*=F_*=0,$$

$$\mathbf{s}_1=2\mathrm{Dev}\left(\eta\mathbf{d}+\varphi\mathbf{d}^2\right),$$

$$\left[\begin{array}{ll}\mathbf{d}=0 & N(\mathbf{s}_2)<K,\\ \mathbf{d}\neq 0, & N(\mathbf{s}_2)=K,\ \ \mathbf{s}_2=\mu\mathbf{f}\left(\mathbf{d}\right),\end{array}\right. \tag{1.243}$$

$$\mathbf{f}\left(\mathbf{d}\right)=2\mathrm{Dev}\left(\psi\mathbf{d}+\chi\mathbf{d}^2\right),$$

The yield condition renders

$$\mu=\frac{K}{N\left[\mathbf{f}\left(\mathbf{d}\right)\right]}.$$

The following equality is assumed to hold for the viscous material

$$\lim_{\mathbf{d}\to 0}\mathbf{s}_1=0\,.$$

It allows us to transform the governing equation (1.243) to the following form

$$\left[\begin{array}{ll}\mathbf{d}=0 & N(\mathbf{s})\leq K,\\ \mathbf{d}\neq 0, & \mathbf{s}=2\mathrm{Dev}\left[\eta\mathbf{d}+\varphi\mathbf{d}^2+\mu\left(\psi\mathbf{d}+\chi\mathbf{d}^2\right)\right].\end{array}\right. \tag{1.244}$$

In contrast to the equation for the viscous material, $\mathbf{s}$ has a discontinuity at $\mathbf{d}\to 0$ while, contrary to the plastic material, the yield condition $N(\mathbf{s})=K$ at $\mathbf{d}\neq 0$ is absent.

An inverse form of the constitutive equation is of considerable interest. Unfortunately, a general inversion is difficult. Let us treat an important particular case in which the equations of viscosity and plasticity are linear in strain rate of shape change $\mathbf{d}$ and the Mises yield condition is taken

$$\left[\begin{array}{ll}\mathbf{d}=0 & \tau(\mathbf{s})\leq K,\\ \mathbf{d}\neq 0, & \mathbf{s}=\left(2\eta+K/\tau\left(\mathbf{d}\right)\right)\mathbf{d}.\end{array}\right. \tag{1.245}$$

When $\mathbf{d}\neq 0$ we have

$$\tau(\mathbf{s})=2\eta\tau(\mathbf{d})+K$$

or

$$\tau(\mathbf{d})=\frac{\tau(\mathbf{s})-K}{2\eta}$$

and therefore

$$\mathbf{d}=\frac{\tau(\mathbf{s})-K}{2\eta\tau(\mathbf{s})}\mathbf{s},\ \ \tau(\mathbf{s})>K.$$

It is easy to see now that the inverted form of eq. (1.245) has the form

$$\left[\begin{array}{ll}\tau(\mathbf{s})\leq K, & \mathbf{d}=0,\\ \tau(\mathbf{s})\geq K, & \mathbf{d}=\mathbf{s}\left[\tau(\mathbf{s})-K\right]/2\eta\tau(\mathbf{s})\end{array}\right.$$

or

$$\mathbf{d}=\frac{L\left[\tau(\mathbf{s})-K\right]}{2\eta\tau(\mathbf{s})}\mathbf{s}\ , \tag{1.246}$$

where

$$L\left(x\right)=\left[\begin{array}{ll}0, & x\leq 0,\\ x, & x\geq 0.\end{array}\right.$$

In contrast to the equation for a viscous material, the scalar coefficient in (1.246) vanishes at $\tau(\mathbf{s})=K$. Contrary to the equation for an ideal plastic material, a state $\tau(\mathbf{s})>K$ is allowed.

1.11.7 Example 7. The Bingham elastoviscoplastic material

The rheological model for this material consists of a Bingham's viscoplastic material and an elastic Hooke's element in series, see Fig. 1.10. By analogy

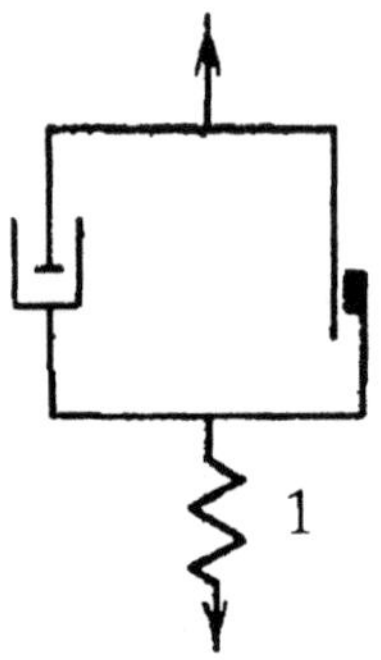

FIGURE 1.10. The rheological model of the Bingham elastoviscoplastic material.

with the Maxwell model we find that the constitutive equations are given by

$$\mathrm{s} = -2\rho \mathrm{Dev}\left(\frac{\partial F_*}{\partial a_1}\mathbf{\Lambda}_1 + \frac{\partial F_*}{\partial b_1}\mathbf{\Lambda}_1^2\right),$$

$$S_* = -\frac{\partial F_*}{\partial \theta}, \; F_* = F_*(\theta, a_1, b_1), \tag{1.247}$$

with F_* being free energy of the elastic element. The shape change strain $\mathbf{\Lambda}_1$ of the elastic element is governed by the following equation

$$\dot{\mathbf{\Lambda}}_1 = -\left[\mathbf{d} + \mathbf{\Omega} - \frac{L[\tau(\mathrm{s}) - K]}{2\eta\tau(\mathrm{s})}\mathrm{s}\right] \cdot \mathbf{\Lambda}_1 - \mathbf{\Lambda}_1 \cdot \left[\mathbf{d} - \mathbf{\Omega} - \frac{L[\tau(\mathrm{s}) - K]}{2\eta\tau(\mathrm{s})}\mathrm{s}\right]. \tag{1.248}$$

It goes without saying that this equation corresponds to a simplified version of the Bingham model in the form of eq. (1.246). The stress deviator appearing in eq. (1.248) is to be expressed in terms of $\mathbf{\Lambda}_1$ due to eq. (1.247). Initial conditions are to be added. Let us assume that the material deformation starts from the natural state, then the following initial condition

$$t = t_0, \; \mathbf{\Lambda}_1 = 0$$

is to be prescribed. Other examples of the constitutive equations can be found in [133].

1.12 Discussion about the decomposition methods

These examples are remarkable as they utilize an additive decomposition of the tensor of rate of finite viscoelastic, elastoplastic and viscoelastoplastic strain on elastic, viscous, plastic and viscoplastic parts. However, other methods of decompositions are used in the literature, cf. [166], [167], [33] and [13].

In what follows, we study an example of isothermal deformation of an incompressible viscoelastic Maxwell material and compare different decompositions.

The following three methods of decomposition will be considered and compared:

1. An additive decomposition of the strain rate of shape change on elastic and viscous components, cf. (1.182)

$$\mathbf{d} = \mathbf{d}_\alpha + \mathbf{d}_\beta, \tag{1.249}$$

where, by analogy with Example 2, $\mathbf{d}_\alpha$ and $\mathbf{d}_\beta$ are assumed to be the rates of elastic and viscous strains, respectively.

2. An additive decomposition of a strain tensor , e.g. the Almansi strain tensor, in the following form

$$\boldsymbol{\varepsilon} = \boldsymbol{\varepsilon}_\alpha + \boldsymbol{\varepsilon}_\beta, \tag{1.250}$$

where $\boldsymbol{\varepsilon}_\alpha$ and $\boldsymbol{\varepsilon}_\beta$ are tensors of elastic and viscous strains, respectively. This decomposition is especially popular in infinitesimal continuum mechanics and can be generalised to viscoplasticity. For example, $\boldsymbol{\varepsilon}_\alpha$ describes deformation of the elastic background material, while $\boldsymbol{\varepsilon}_\beta$ describes the imposed viscoplastic deformation, cf. [53].

3. A multiplicative decomposition of the deformation gradient, cf. [166], [167] and [33]

$$\mathbf{F} = \nabla\mathbf{R} = \mathbf{F}_\alpha \cdot \mathbf{F}_\beta. \tag{1.251}$$

Let us recall the relations between the introduced parameters. Equation (1.30) renders a fundamental formula

$$\mathbf{d} = -\left(\dot{\mathbf{F}} \cdot \mathbf{F}^{-1}\right)^s. \tag{1.252}$$

The strain tensor is defined as follows

$$\boldsymbol{\varepsilon} = \frac{1}{2}\left(\mathbf{E} - \mathbf{g}\right), \tag{1.253}$$

$\mathbf{g}$ being the Almansi strain measure (1.27)

$$\mathbf{g} = \mathbf{F} \cdot \mathbf{F}^T. \tag{1.254}$$

Finally, due to assumed condition of incompressibility, we have

$$\boldsymbol{\Lambda} = \mathbf{g}. \tag{1.255}$$

Assume that the parameters of elastic strains $\mathbf{F}_\alpha$, $\mathbf{d}_\alpha$ and $\boldsymbol{\varepsilon}_\alpha$ and viscous strain $\mathbf{F}_\beta$, $\mathbf{d}_\beta$ and $\boldsymbol{\varepsilon}_\beta$ are described by eqs. (1.252), (1.253), (1.254) and (1.255). Assume also that the elastic strain is governed by the classical equations of the nonlinear theory of elasticity (1.132), (1.133) and (1.134), so that the dissipative inequality for an elastic element takes the form of an equality

$$\mathbf{s}_\alpha : \mathbf{d}_\alpha - \rho \dot{F}_{*\alpha} = 0. \tag{1.256}$$

The first approach to construction of a nonlinear theory of viscoelastic material uses the constitutive equation for viscosity (1.143)

$$\mathbf{s}_\beta = \mathbf{H}\left(\mathbf{d}_\beta\right), \tag{1.257}$$

where $\mathbf{H}(...)$ denotes a deviator of an isotropic tensor function which fully agrees with (1.143).

The requirement of the second law of thermodynamics for the constitutive equation for viscosity (1.257) reduces to the inequality

$$\mathbf{H}(\mathbf{d}) : \mathbf{d} \geq 0. \tag{1.258}$$

The function $\mathbf{H}(...)$ is assumed to be chosen so that inequality (1.258) holds for any argument $\mathbf{d}$.

Dealing with a material having the Maxwell rheological model, we take (as in eqs. (1.181) and (1.183))

$$s = \mathbf{s}_\alpha = \mathbf{s}_\beta, \quad F_* = F_{*\alpha}. \tag{1.259}$$

Substitution of eqs. (1.132) and (1.257) into the latter equation leads to the first equation for the elastic and viscous kinematic parameters

$$\mathbf{s} = -2\rho \mathrm{Dev}\left(\frac{\partial F_{*\alpha}}{\partial \mathbf{\Lambda}_\alpha} \cdot \mathbf{\Lambda}_\alpha\right)^s = \mathbf{H}\left(\mathbf{d}_\beta\right). \tag{1.260}$$

Combining this equation with each of the above decompositions leads to a system of equations for elastic and viscous kinematic values. After these values have been found, the strain deviator $\mathbf{s}$ may be evaluated by means of the first equation in (1.260). In such a manner three versions of the constitutive equations for the Maxwell material are obtained. However, not each version satisfies the second law of thermodynamics. The requirement of this law, under the above restrictions, is as follows

$$\mathbf{s} : \mathbf{d} - \rho \dot{F}_* \geq 0. \tag{1.261}$$

By combining eqs. (1.261), (1.259), (1.257) and (1.256) it is easy to show that the requirement of the second law reduces to the following inequality

$$\mathbf{H}\left(\mathbf{d}_\beta\right) : \left(\mathbf{d} - \mathbf{d}_\alpha\right) \geq 0. \tag{1.262}$$

It is easy to prove that, by virtue of conditions (1.258), this inequality holds for the first method of decomposition. This result appears to be quite evident from the perspective of the method of rheological models. It is, however, more difficult to show that inequality (1.262) may be violated when other methods of decomposition of viscoelastic strain into elastic and viscous components are applied. In order to demonstrate this some counter-examples will be considered. Let us limit our consideration to the case in which the phenomenon of viscosity is described by the simplest equation of Newton, i.e.

$$\mathbf{s}_\beta = \mathbf{H}(\mathbf{d}_\beta) = 2\eta \mathbf{d}_\beta, \tag{1.263}$$

where η denotes a viscosity which is non-negative due to eq. (1.147). Inserting (1.263) into (1.262) we obtain an expression for the dissipative function

$$\Phi = 2\eta \mathbf{d}_\beta : (\mathbf{d} - \mathbf{d}_\alpha). \tag{1.264}$$

It is interesting to note that this expression does not depend upon the constitutive equation of elasticity (1.132).

Consider a uniform triaxial tension and a shear in a plane under the condition of incompressibility, i.e.

$$x_1 = X_1\varphi + X_2 s,\ x_2 = X_2\delta,\ x_3 = X_3\omega,\ \omega\delta\varphi = 1,$$

where x_k are the Cartesian coordinates of a material point in its actual configuration, while X_k are those in the initial configuration; φ, δ and ω stand for elongations, and s denotes a shear in the plane x_1x_2. The inverse relations are as follows

$$X_1 = \frac{x_1}{\varphi} - x_2\frac{s}{\varphi\delta},\ X_2 = \frac{x_2}{\delta},\ X_2 = \frac{x_3}{\omega}.$$

The deformation gradient is given by

$$\mathbf{F} = \nabla\mathbf{R} = \frac{\mathbf{i}_1\mathbf{i}_1}{\varphi} + \frac{\mathbf{i}_2\mathbf{i}_2}{\delta} + \frac{\mathbf{i}_3\mathbf{i}_3}{\omega} - \mathbf{i}_1\mathbf{i}_2 s\omega, \tag{1.265}$$

where $\mathbf{i}_k$ denotes the unit base vectors of the Cartesian coordinates, $\mathbf{R}$ is the position vector of a generic point in the initial configuration, and ∇ denotes the Hamilton operator in the actual configuration.

Assume that the gradients of elastic and viscous deformations have similar expressions

$$\mathbf{F}_\alpha = \frac{\mathbf{i}_1\mathbf{i}_1}{\varphi_\alpha} + \frac{\mathbf{i}_2\mathbf{i}_2}{\delta_\alpha} + \frac{\mathbf{i}_3\mathbf{i}_3}{\omega_\alpha} - \mathbf{i}_1\mathbf{i}_2 s_\alpha\omega_\alpha, \tag{1.266}$$

$$\mathbf{F}_\beta = \frac{\mathbf{i}_1\mathbf{i}_1}{\varphi_\beta} + \frac{\mathbf{i}_2\mathbf{i}_2}{\delta_\beta} + \frac{\mathbf{i}_3\mathbf{i}_3}{\omega_\beta} - \mathbf{i}_1\mathbf{i}_2 s_\beta\omega_\beta, \tag{1.267}$$

where the subindexes α and β refer to elastic and viscous components, respectively.

The rate strain tensors are obtained by means of eq. (1.252)

$$\mathbf{d} = \mathbf{i}_1\mathbf{i}_1\frac{\dot\varphi}{\varphi} + \mathbf{i}_2\mathbf{i}_2\frac{\dot\delta}{\delta} + \mathbf{i}_3\mathbf{i}_3\frac{\dot\omega}{\omega} + (\mathbf{i}_1\mathbf{i}_2 + \mathbf{i}_2\mathbf{i}_1)\frac{\varphi}{2\delta}\left(\frac{s}{\varphi}\right)^{\bullet},$$

$$\mathbf{d}_\alpha = \mathbf{i}_1\mathbf{i}_1 \frac{\dot{\varphi}_\alpha}{\varphi_\alpha} + \mathbf{i}_2\mathbf{i}_2 \frac{\dot{\delta}_\alpha}{\delta_\alpha} + \mathbf{i}_3\mathbf{i}_3 \frac{\dot{\omega}_\alpha}{\omega_\alpha} + (\mathbf{i}_1\mathbf{i}_2 + \mathbf{i}_2\mathbf{i}_1) \frac{\varphi_\alpha}{2\delta_\alpha} \left(\frac{s_\alpha}{\varphi_\alpha}\right)^\bullet, \tag{1.268}$$

$$\mathbf{d}_\beta = \mathbf{i}_1\mathbf{i}_1 \frac{\dot{\varphi}_\beta}{\varphi_\beta} + \mathbf{i}_2\mathbf{i}_2 \frac{\dot{\delta}_\beta}{\delta_\beta} + \mathbf{i}_3\mathbf{i}_3 \frac{\dot{\omega}_\beta}{\omega_\beta} + (\mathbf{i}_1\mathbf{i}_2 + \mathbf{i}_2\mathbf{i}_1) \frac{\varphi_\beta}{2\delta_\beta} \left(\frac{s_\beta}{\varphi_\beta}\right)^\bullet.$$

Let us proceed to the method of multiplicative decomposition (1.251). This equation yields the following kinematic relations

$$\varphi = \varphi_\alpha\varphi_\beta, \; \delta = \delta_\alpha\delta_\beta, \; s = s_\alpha\delta_\beta + \varphi_\alpha s_\beta. \tag{1.269}$$

Using eqs. (1.268) and (1.269) we arrive, after some algebra, at the following expressions for the dissipative function (1.263)

$$\begin{aligned} \Phi &= 4\eta \left\{ \left(\frac{\dot{\varphi}_\beta}{\varphi_\beta}\right)^2 + \left(\frac{\dot{\delta}_\beta}{\delta_\beta}\right)^2 + \frac{\dot{\varphi}_\beta\dot{\delta}_\beta}{\omega_\beta\delta_\beta} + \frac{1}{2}\frac{\varphi\varphi_\beta}{\delta\delta_\beta}\left[\left(\frac{s_\beta}{\varphi_\beta}\right)^\bullet\right]^2 + \right. \\ & \left. + \frac{1}{8}\frac{\varphi s_\alpha}{\delta\varphi_\alpha}\left(\frac{s_\beta}{\varphi_\beta}\right)^\bullet \left(\frac{\dot{\delta}_\beta}{\delta_\beta} - \frac{\dot{\varphi}_\beta}{\varphi_\beta}\right)\right\}. \end{aligned} \tag{1.270}$$

This quadratic form in rates is nonnegative provided that

$$s_\alpha^2 < \frac{8}{\omega_\alpha}. \tag{1.271}$$

Only if this inequality holds is the dissipative inequality satisfied. However, conversely, it may be violated. Therefore, the method of multiplicative decomposition is not unconditionally acceptable.

Consider now the second method of decomposition, namely an additive decomposition of the Almansi strain tensor. Paper [135] contains a counterexample showing that the requirement of the second law of thermodynamics may be violated when this decomposition method is used. A uniaxial tension under the conditions of incompressibility and material isotropy has been studied there. It means that one must take

$$\delta = \varphi = \frac{1}{\sqrt{\omega}}$$

in eq. (1.265). The dissipative function takes the following form

$$\Phi = 3\eta\frac{\dot{\omega}_\beta}{\omega}\left(\frac{\dot{\omega}}{\omega} - \frac{\dot{\omega}_\alpha}{\omega_\alpha}\right).$$

For the sake of simplicity we write down eq. (1.250) in the form

$$\mathrm{Dev}\varepsilon = \mathrm{Dev}\varepsilon_\alpha + \mathrm{Dev}\varepsilon_\beta\,.$$

This leads to the following expression for the dissipative function

$$\Phi = \frac{3\eta}{B(\omega_\beta)} \left\{ B(\omega) \left(\frac{\dot{\omega}}{\omega} \right)^2 - [B(\omega) + B(\omega_\alpha)] \frac{\dot{\omega}}{\omega} \frac{\dot{\omega}_\alpha}{\omega_\alpha} + B(\omega_\alpha) \left(\frac{\dot{\omega}_\alpha}{\omega_\alpha} \right)^2 \right\},$$

where

$$B(x) = \frac{1}{x} + 2x^2 > 0.$$

The determinant of this quadratic form is given by

$$\Delta = -\frac{1}{4} [B(\omega) - B(\omega_\alpha)]^2 \leq 0.$$

The determinant is seen to be non-positive! Therefore, the second law of thermodynamics may be violated when the second method of decomposition is applied.

Thus the following theorem is proved

Theorem 4: The constitutive equations for the viscoelastic material of Maxwell are thermodynamically consistent only if an additive decomposition of the rate strain tensor is applied, provided that the classic constitutive equations for elasticity and viscosity are utilized.

The main equation of the first approach to this problem is obtained by inserting (1.249) into (1.260)

$$\mathbf{s} = -2\rho \mathrm{Dev} \left(\frac{\partial F_{*\alpha}}{\partial \mathbf{\Lambda}_\alpha} \cdot \mathbf{\Lambda}_\alpha \right)^s = \mathbf{H}(\mathbf{d} - \mathbf{d}_\alpha). \qquad (1.272)$$

The second approach to the problem of decomposition of a viscoelastic strain into components implies utilizing the generalised equation of viscosity

$$\mathbf{s}_\beta = \mathbf{H}(\mathbf{d} - \mathbf{d}_\alpha). \qquad (1.273)$$

Substituting eqs. (1.273) and (1.132) into (1.259) leads to eq. (1.272) for the kinematic parameters describing elastic strain as in the previous approach. However the parameters that describe the viscous strain remain completely undetermined. Combining eqs. (1.261), (1.256), (1.259) and (1.273) we come to the following form of the second law of thermodynamics in the framework of this approach

$$\mathbf{H}(\mathbf{d} - \mathbf{d}_\alpha) : (\mathbf{d} - \mathbf{d}_\alpha) \geq 0. \qquad (1.274)$$

In accordance with conditions (1.258) it holds without any restrictions!

Hence, we have proved the following theorem, cf. [136].

Theorem 5: The constitutive equations of the Maxwell viscoelastic material are thermodynamically consistent in the framework of any decomposition method, provided that the classical constitutive equation for elasticity and the generalised equation for viscosity are taken.

It is the author's view that it is meaningless when some mechanical parameter (e.g. plastic strain) has no single-valued definition. A regularisation is needed under these circumstances. In other words, the problem is to be deemed from a more general perspective with a subsequent consideration of the limiting cases.

In order to present this more general perspective we generalise the generalised equation for viscosity (1.273) and replace it by a generalised Kelvin-Voigt equation

$$\mathbf{s}_\beta = \mathbf{H}(\mathbf{d} - \mathbf{d}_\alpha) + \mathbf{s}_h. \tag{1.275}$$

Here $\mathbf{s}_h$ denotes a stress tensor determining a kinematic hardening. Assume that this hardening is described by an equation of pure elasticity (1.132)

$$\mathbf{s}_h = -2\rho \mathrm{Dev}\left(\frac{\partial F_*(\mathbf{\Lambda}_\alpha, \mathbf{\Lambda}_\beta)}{\partial \mathbf{\Lambda}_\beta} \cdot \mathbf{\Lambda}_\beta\right)^s, \tag{1.276}$$

with the free energy being assumed to depend on two strain measures $\mathbf{\Lambda}_\alpha$ and $\mathbf{\Lambda}_\beta$. Inserting eqs. (1.275) and (1.132) into eq. (1.259) yields the main equation

$$\mathbf{s} = -2\rho \mathrm{Dev}\left(\frac{\partial F_*}{\partial \mathbf{\Lambda}_\alpha} \cdot \mathbf{\Lambda}_\alpha\right)^s = \mathbf{H}(\mathbf{d} - \mathbf{d}_\alpha) - 2\rho \mathrm{Dev}\left(\frac{\partial F_*}{\partial \mathbf{\Lambda}_\beta} \cdot \mathbf{\Lambda}_\beta\right)^s. \tag{1.277}$$

Equation (1.272) may be considered a limiting case of the latter equation when $\partial F_*/\partial \mathbf{\Lambda}_\beta \to 0$. Let us perform a thermodynamical analysis of this approach to the problem of strain decomposition. As before, the thermodynamical inequality has the form of inequality (1.261), but now we have a more complicated expression for $\rho \dot{F}_*$

$$\rho \dot{F}_* = \rho \dot{F}_*(\mathbf{\Lambda}_\alpha, \mathbf{\Lambda}_\beta) = \mathbf{s}_\alpha : \mathbf{d}_\alpha + \mathbf{s}_h : \mathbf{d}_\beta. \tag{1.278}$$

Inserting eq. (1.278), (1.275) and (1.259) into (1.261) and combining the terms yields the following inequality

$$\mathbf{H}(\mathbf{d} - \mathbf{d}_\alpha) : (\mathbf{d} - \mathbf{d}_\alpha) + \mathbf{s}_h : (\mathbf{d} - \mathbf{d}_\alpha - \mathbf{d}_\beta) \geq 0. \tag{1.279}$$

It is easy to see that the left hand side of this inequality is a linear function in tensor $\mathbf{s}_h$. The latter depends only on the strain measures $\mathbf{\Lambda}_\alpha$ and $\mathbf{\Lambda}_\beta$ and on this account it can take a value independent on the rate strain tensors $\mathbf{d}$, $\mathbf{d}_\alpha$ and $\mathbf{d}_\beta$. By virtue of the theorem on linear inequalities [179], the necessary and sufficient conditions for inequality (1.279) are as follows

$$\mathbf{H}(\mathbf{d} - \mathbf{d}_\alpha) : (\mathbf{d} - \mathbf{d}_\alpha) \geq 0, \; \mathbf{d} - \mathbf{d}_\alpha - \mathbf{d}_\beta = 0. \tag{1.280}$$

The first inequality (1.280) is satisfied identically due to assumption (1.258). The second equation (1.280) gives a single-valued definition of viscous strain.

Thus we have proved the theorem.

Theorem 6: Only additive decomposition of the strain rate of shape change onto elastic and viscous components leads to a thermodynamically consistent constitutive equation for the Maxwell viscoelastic material, provided that the classic constitutive equation for pure elasticity and the generalised equation for viscosity with a kinematic hardening are taken.

The regularisation performed leads to the conclusion that only additive decomposition of the strain rate of shape change is thermodynamically consistent! This analysis can be easily generalised for the Prandtl elastoplastic material and the Bingham viscoelastoplastic material.

The main result of the regularisation means that all methods of decomposition of strain must be rejected except the additive decomposition of the strain rate of shape change into elastic and viscoplastic or plastic components.

1.13 Linearisation of the equations of continuum mechanics

Linearised equations of continuum mechanics are admitted while studying the vast majority of engineering structures. The reason for this is that the normal functioning of these structures is possible only when there exists a fixed or a moving coordinate system relative to which the structural displacements are small in some sense. It is natural to analyse the structural behaviour under this simplifying assumption.

We therefore assume that the displacement $\mathbf{u} = \mathbf{r} - \mathbf{R}$ is small compared to a typical linear structural size L, i.e.

$$|\mathbf{u}| \ll L. \tag{1.281}$$

However, this assumption alone does not suffice for a correct construction of a linearised theory. It is necessary to accept a more rigid assumption that

$$\mathbf{u} = \alpha \mathbf{U}(\mathbf{r}, t), \tag{1.282}$$

where α is a constant small parameter and $\mathbf{U}$ is a bounded function with bounded derivatives. Apparently, inequality (1.281) holds for sufficiently small α. Moreover, the derivatives of $\mathbf{U}$ with respect to the space coordinates and time are small for small α.

Construction of a linearised theory implies a systematic retaining of terms of low order in α (or displacement $\mathbf{u}$ and its derivatives) and neglecting terms of higher order.

Consider the Almansi strain measure $\mathbf{g}$. Due to (1.11) and (1.27) we obtain

$$\mathbf{g} = \mathbf{E} - 2\varepsilon + (\nabla \mathbf{u}) \cdot (\mathbf{u} \nabla), \tag{1.283}$$

the notation ε being used to denote the linear strain tensor

$$\varepsilon = (\nabla \mathbf{u})^s . \tag{1.284}$$

We split this into spherical and deviatoric parts

$$\varepsilon = \frac{\vartheta}{3}\mathbf{E} + \mathbf{e}, \quad \mathbf{e} = \mathrm{Dev}\varepsilon, \ \vartheta = \nabla \cdot \mathbf{u} \tag{1.285}$$

and substitute it into (1.283) to obtain

$$\mathrm{g} = \mathbf{E}\left(1 - \frac{2}{3}\vartheta\right) - 2\mathbf{e} + (\nabla \mathbf{u}) \cdot (\mathbf{u}\nabla) . \tag{1.286}$$

The third invariant of this tensor is equal to

$$I_3(\mathrm{g}) = \left(1 - \frac{2}{3}\vartheta\right)^3 \left|\mathbf{E} - [2\mathbf{e} - (\nabla \mathbf{u}) \cdot (\mathbf{u}\nabla)]\left(1 - \frac{2}{3}\vartheta\right)^{-1}\right| .$$

Keeping terms up to second order in small components of tensor $\nabla \mathbf{u}$ we have

$$I_3(\mathrm{g}) = \left(1 - \frac{2}{3}\vartheta\right)^3 |1 - 2\mathbf{e} : \mathbf{e} + (\nabla \mathbf{u}) : (\mathbf{u}\nabla)| . \tag{1.287}$$

By (1.33) we obtain a linear approximation

$$\frac{\rho_0}{\rho} = [I_3(\mathrm{g})]^{-1/2} = 1 + \vartheta, \tag{1.288}$$

where ϑ is called a small dilatation.

Further, due to (1.286) and (1.287) we obtain the following equation for the strain of shape change

$$\mathbf{\Lambda} = [I_3(\mathrm{g})]^{-1/3} \mathrm{g} = \mathbf{E} - 2\mathbf{e} + \frac{2}{3}(\mathbf{e} : \mathbf{e})\mathbf{E} + \mathrm{Dev}[(\nabla \mathbf{u}) \cdot (\mathbf{u}\nabla)] , \tag{1.289}$$

in which terms up to second order are kept. Retaining the terms up to first order yields a simple expression

$$\mathbf{\Lambda} = \mathbf{E} - 2\mathbf{e}. \tag{1.290}$$

Let us proceed now to equation (1.28) which can be rewritten as follows

$$\mathbf{v} \cdot (\mathbf{E} - \nabla \mathbf{u}) = \frac{\partial \mathbf{u}}{\partial t} .$$

The second term in parentheses is small compared with the first one and thus can be neglected. This leads to a linear equation for velocity

$$\mathbf{v} = \dot{\mathbf{u}} = \frac{\partial \mathbf{u}}{\partial t} ,$$

from which the velocity is seen to be first order. Therefore, both the spin tensor $\mathbf{\Omega}$ and the strain rate of shape change $\mathbf{d}$ have the same order. Accounting for this fact, substituting (1.290) into the kinematic equation (1.39) and keeping the terms of first order yields

$$\dot{\mathbf{e}} = \mathbf{d}, \tag{1.291}$$

where the velocity may be evaluated by means of the linear equation

$$\dot{\mathbf{e}} = \frac{\partial \mathbf{e}}{\partial t} . \tag{1.292}$$

Similarly, for any first order function no distinction is made between its material derivative and its partial derivative with respect to time.

Let us next linearise the constitutive equations.

By virtue of (1.288) a linearised version of the equation for dilatation is

$$\sigma = k(\theta)\vartheta - m(\theta), \tag{1.293}$$

the thermal strain m/k being considered to be a value of the same order as the dilatation ϑ. Under volumetric deformations, due to (1.119) and (1.121) free energy and entropy to second order are given by

$$\rho_0 F_0 = \frac{k(\theta)}{2}\vartheta^2 - m(\theta)\vartheta + n(\theta), \tag{1.294}$$

$$\rho_0 S_0 = -\frac{\vartheta^2}{2}\frac{dk(\theta)}{d\theta} + \vartheta\frac{dm(\theta)}{d\theta} - \frac{dn(\theta)}{d\theta}. \tag{1.295}$$

Linearisation of the constitutive equation of elastic material (1.137) for a shape distortion is carried out by means of linear equations (1.290) and (1.288). Assuming that free energy F_* is an analytic function of the invariants a and b of tensor $\mathbf{\Lambda}$ and keeping the first order terms only, leads to the following result

$$\mathbf{s} = 2G\mathbf{e}, \tag{1.296}$$

where the shear modulus is given by

$$G = G(\theta) = \rho_0 \left(2\frac{\partial F_*}{\partial a} + 4\frac{\partial F_*}{\partial b} \right) |_{\mathbf{\Lambda}=\mathbf{E}} \cdot \tag{1.297}$$

Let us obtain an equation for free energy (1.136) up to the second order. Substituting (1.289) into the equation for invariant a (1.135) yields

$$a = \mathbf{\Lambda} : \mathbf{E} = 3 + 2\mathbf{e} : \mathbf{e}. \tag{1.298}$$

Next,

$$\begin{aligned} 2b - \frac{1}{3}a^2 &= \mathbf{\Lambda} : \mathbf{\Lambda} - \frac{1}{3}(\mathbf{\Lambda} : \mathbf{E})(\mathbf{E} : \mathbf{\Lambda}) = \\ &= (\mathrm{Dev}\mathbf{\Lambda}) : (\mathrm{Dev}\ \mathbf{\Lambda}) = 4\mathbf{e} : \mathbf{e}. \end{aligned}$$

Hence, with an accuracy up to second order, we have

$$b = \frac{3}{2} + 4\mathbf{e} : \mathbf{e}. \tag{1.299}$$

The free energy is assumed to allow expansion in series in terms of invariants a and b

$$F_* (a, b, \theta) = F_*|_{\mathbf{\Lambda}=\mathbf{E}} + \frac{\partial F_*}{\partial a}|_{\mathbf{\Lambda}=\mathbf{E}} (a - 3) + \frac{\partial F_*}{\partial b}|_{\mathbf{\Lambda}=\mathbf{E}} \left(b - \frac{3}{2}\right) + \ldots$$

Substituting a and b due to (1.298) and (1.299) and taking into account (1.114) we obtain

$$\rho_0 F_* = G\mathbf{e} : \mathbf{e}, \tag{1.300}$$

where G is the above shear modulus.

Comparing (1.296) and (1.300) we arrive at the equation

$$\mathbf{s} = \frac{\partial \rho_0 F_*}{\partial \mathbf{e}}, \tag{1.301}$$

which is analogous of (1.132).

Specific internal entropy of the shape distortion is due to (1.134) and has the form

$$S_* = -\mathbf{e} : \mathbf{e}\frac{\partial G}{\partial \theta}. \tag{1.302}$$

Let us proceed to linearisation of the equation for viscous material (1.143). The strain rate of shape change **d** should be considered small and one should keep only first order terms on the right hand side of (1.143), assuming η and φ analytical functions of the invariants of tensor **d**. This results in the Newton equation (1.146), which due to (1.291) takes the form

$$\mathbf{s} = 2\eta\dot{\mathbf{e}}, \quad \eta = \eta(\theta). \tag{1.303}$$

Specific free energy and specific internal entropy equate to zero, as in the general case of viscous material.

Finally, we address the constitutive equations for a plastic material. They have non-analytic nonlinearities, e.g. $N[\mathbf{f}(\mathbf{d})]$ in the denominator in (1.157). For this reason the equations for a plastic material cannot be presented in a linear form. They can only be simplified by accounting for the smallness of the components of tensor $\mathbf{d}$. Let us assume that function $\mathbf{f}(\mathbf{d})$ in (1.156) is an analytical function of $\mathbf{d}$. The linearisation then results in (1.164) in which a positive factor is omitted since, because of the norm property, it will not appear in the resulting equation (1.157). Aiming for maximum simplicity of the resulting equations one usually takes the shear stress intensity as a stress deviator norm. This leads to the St. Venant-Mises equations (1.172), which due to (1.291) are given by

$$\left[\begin{array}{l} \dot{\mathbf{e}} = 0, \ \ \tau \leq \tau_s, \\ \dot{\mathbf{e}} \neq 0, \ \ \mathbf{s} = \tau_s \dot{\mathbf{e}}/v, \end{array}\right. \tag{1.304}$$

where

$$\tau = \sqrt{\frac{1}{2}\mathbf{s} : \mathbf{s}}, \ \ v = \sqrt{\frac{1}{2}\dot{\mathbf{e}} : \dot{\mathbf{e}}}, \ \ \tau_s = \tau_s(\theta). \tag{1.305}$$

A non-analytic nonlinearity $\dot{\mathbf{e}}/v$ is kept in eq. (1.304).

Note also that the equations for connection of rheological elements in series are simplified. Due to (1.182)

$$\mathbf{d} = \mathbf{d}_\alpha + \mathbf{d}_\beta \tag{1.306}$$

and linearised expression (1.291), we have

$$\dot{\mathbf{e}} = \dot{\mathbf{e}}_\alpha + \dot{\mathbf{e}}_\beta. \tag{1.307}$$

Because of the trivial initial conditions for $\mathbf{e}$, $\mathbf{e}_\alpha$ and $\mathbf{e}_\beta$ integration of (1.307) gives

$$\mathbf{e} = \mathbf{e}_\alpha + \mathbf{e}_\beta.$$

We can state that the strain deviator of a connection in series is a sum of the strain deviators of the elements. This allows the axioms of a connection in series to be written in the form

$$\mathbf{s} = \mathbf{s}_\alpha = \mathbf{s}_\beta, \ \ \mathbf{e} = \mathbf{e}_\alpha + \mathbf{e}_\beta. \tag{1.308}$$

By analogy, the axioms of a parallel connection turn out to take the form

$$\mathbf{s} = \mathbf{s}_\alpha + \mathbf{s}_\beta, \ \ \mathbf{e} = \mathbf{e}_\alpha = \mathbf{e}_\beta. \tag{1.309}$$

In conclusion, we can write down the dynamics equation (1.56). Keeping terms of first order leads to the result

$$\nabla \cdot \boldsymbol{\tau} + \rho(\mathbf{K} - \ddot{\mathbf{u}}) = 0 \tag{1.310}$$

where $\rho = \rho_0$ and

$$\ddot{\mathbf{u}} = \frac{\partial^2 \mathbf{u}(\mathbf{r}, t)}{\partial t^2}.$$

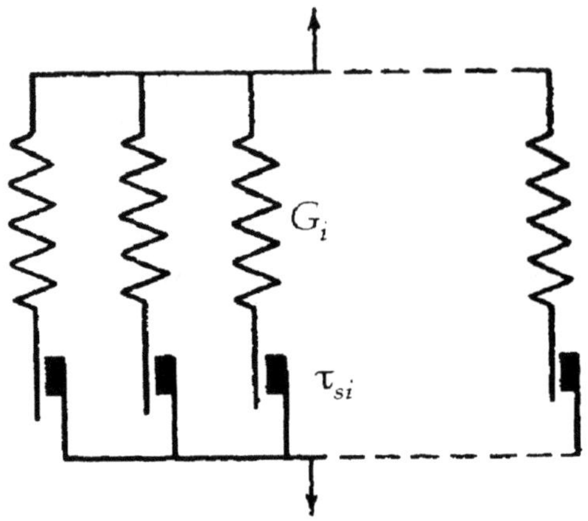

FIGURE 1.11. The generalised Prandtl rheological model.

Note that linearisation allows us to ignore distinctions between the initial and current coordinates of the material points **R** and **r** in arguments of the functions in the linearised equations. In particular, it allows the boundary conditions to be set on the undeformed body surface.

1.14 Theories of small microplastic deformations

In the present section we derive constitutive equations for some elastoplastic materials, whose rheological models consist of a large number of various elastic and plastic elements. Linearised equation of continuum mechanics are used to this aim.

1.14.1 The generalized Prandtl material

The rheological model for this material is shown in Fig. 1.11 and consists of the Prandtl elements connected in parallel. In accordance with the conditions of the parallel connection we have

$$\mathbf{s} = \sum_{i=1}^{N} \mathbf{s}_i, \tag{1.311}$$

where $\mathbf{s}_i$ stands for the stress deviator of the i-th Prandtl element.

Let $\mathbf{e}_i$ denote plastic strain of the i-th Prandtl element. Due to the axioms of parallel connection of the fundamental rheological elements, the strain of the i-th elastic element is

$$\mathbf{e} - \mathbf{e}_i \ .$$

Hence, by virtue of (1.296) deviator $\mathbf{s}_i$ is given by

$$\mathbf{s}_i = 2G_i(\mathbf{e} - \mathbf{e}_i), \tag{1.312}$$

where G_i is the shear modulus which corresponds to the i-th elastic element.

Due to (1.304) the constitutive equation for the i-th plastic element is as follows

$$\left[\begin{array}{ll} \dot{\mathbf{e}}_i = 0, & \tau_i \leq \tau_{si}, \\ \dot{\mathbf{e}}_i \neq 0, & \mathbf{s}_i = \tau_{si}\dot{\mathbf{e}}_i/v_i, \end{array}\right. \tag{1.313}$$

where

$$\tau_i = \sqrt{\frac{1}{2}\mathbf{s}_i\colon \mathbf{s}_i}, \quad v_i = \sqrt{\frac{1}{2}\dot{\mathbf{e}}_i\colon \dot{\mathbf{e}}_i} \tag{1.314}$$

and τ_{si} stands for the yield stress of the i-th plastic element.

We eliminate the internal forces $\mathbf{s}_i$ from the equations. Substituting (1.312) into (1.313) yields the following equations for plastic strain in the i-th Prandtl element

$$\left[\begin{array}{ll} \dot{\mathbf{e}}_i = 0, & \sqrt{\frac{1}{2}(\mathbf{e} - \mathbf{e}_i)\colon(\mathbf{e} - \mathbf{e}_i)} \leq h_i, \\ \dot{\mathbf{e}}_i \neq 0, & h_i\dot{\mathbf{e}}_i/v_i = \mathbf{e} - \mathbf{e}_i, \end{array}\right. \tag{1.315}$$

where h_i is a non-dimensional yield stress of the i-th Prandtl element

$$h_i = \frac{\tau_{si}}{2G_i}. \tag{1.316}$$

Substituting (1.312) into (1.311), we obtain an expression for the stress deviator

$$\mathbf{s} = \sum_{i=1}^{N} 2G_i(\mathbf{e} - \mathbf{e}_i). \tag{1.317}$$

The system of equations (1.317) and (1.315) represents the constitutive equations for the generalised Prandtl material. Given dependence of the strain deviator on time $\mathbf{e} = \mathbf{e}(t)$, plastic strain $\mathbf{e}_i$ should be found from (1.315) and then stress deviator $\mathbf{s}$ from (1.317).

The case in which $N \to \infty$ and $G_i \to 0$ $(i = 1, 2, 3, ...N)$ is of considerable interest for applications. This is the case of continuously distributed yield stresses over the model elements. We restrict our consideration to an isothermal case, that is the yield stress of each element does not vary in time. We arrange the Prandtl elements in the sum (1.317) in ascending order of nondimensional yield stress h and put

$$G_i = F(h_i)\,\Delta h_i,$$

where $F(h)$ is a non-negative function. Let

$$\mathbf{e}_{hi} = \mathbf{e}_i \tag{1.318}$$

denote plastic strain in a Prandtl element with a nondimensional yield stress $h = h_i$. Substituting two latter equations into (1.317), we arrive at the sum

$$\mathbf{s} = \sum_{i=1}^{N} 2F(h_i)(\mathbf{e} - \mathbf{e}_{hi})\,\Delta h_i,$$

which can be considered as a Riemann sum of the following integral

$$\mathbf{s} = \int_0^\infty 2F(h)(\mathbf{e} - \mathbf{e}_h)\,dh. \tag{1.319}$$

In other words, equation (1.319) is a limiting form of eq. (1.317) for $N \to \infty$. To determine plastic strain $\mathbf{e}_h$ in (1.319) one must solve eq. (1.315) which due to (1.318) takes the form

$$\left[\begin{array}{ll} \dot{\mathbf{e}}_h = 0, & \sqrt{\frac{1}{2}(\mathbf{e} - \mathbf{e}_h):(\mathbf{e} - \mathbf{e}_h)} \le h, \\ \dot{\mathbf{e}}_h \ne 0, & h\dot{\mathbf{e}}_h / v_h = \mathbf{e} - \mathbf{e}_h. \end{array}\right. \tag{1.320}$$

where the intensity of shear strain rate v_h is given by

$$v_h = \sqrt{\frac{1}{2}\dot{\mathbf{e}}_h : \dot{\mathbf{e}}_h}.$$

Equations (1.319) and (1.320) form a system of constitutive equations for the model with a continuous spectrum of yield stresses.

Let us introduce the notation

$$G = \int_0^\infty F(h)\,dh, \quad p(h) = \frac{F(h)}{G}, \tag{1.321}$$

and rewrite (1.319) in the form

$$\mathbf{s} = 2G\left[\mathbf{e} - \int_0^\infty \mathbf{e}_h p(h)\,dh\right]. \tag{1.322}$$

As seen from definition (1.321), function $p(h)$ satisfies the following conditions

$$p(h) \ge 0, \quad \int_0^\infty p(h)\,dh = 1 \tag{1.323}$$

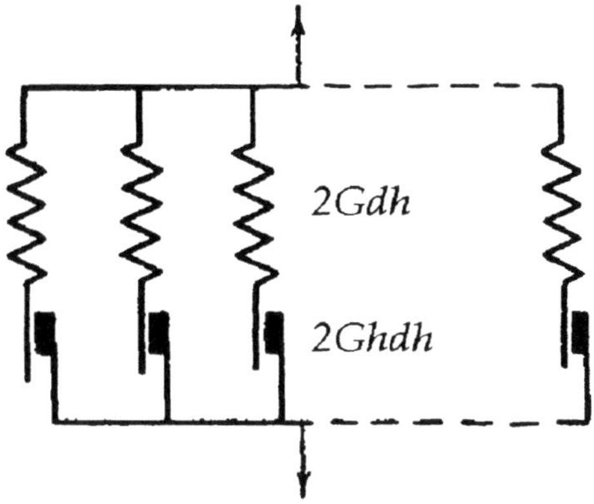

FIGURE 1.12. The Ishlinsky rheological model.

and thus can be treated as a density of the distribution of yield stresses of the Prandtl elements in the rheological model with an infinite number of elements. Provided that all plastic strains vanish, i.e. $\mathbf{e}_h = 0$, equation (1.322) gives

$$\mathbf{s} = 2G\mathbf{e},$$

that is, G is the shear modulus.

Equations (1.320) and (1.322) with a continuous yield stress distribution are visualised by the rheological model shown in Fig. 1.12. This model consists of an infinite number of ideal elastoplastic elements (the Prandtl elements). In each Prandtl element a "spring" of rigidity $2Gdh$ and an ideal plastic element with the yield stress $2Ghdh$ are connected in series. A "force" in a typical element h is given by

$$2G(\mathbf{e} - \mathbf{e}_h)\, dh.$$

We multiply this "force" in element h by the probability $p(h)$ of the element h and sum the products. The result is the full "force"

$$\mathbf{s} = \int_0^\infty 2G(\mathbf{e} - \mathbf{e}_h)\, p(h)\, dh.$$

Taking into account the property of $p(h)$, eq. (1.323), we obtain (1.322). The equation for plastic strain $\mathbf{e}_h$ is obtained from the condition of equality of "forces" in elastic and plastic elements in the element h. It is easy to see that this is given by (1.320).

1.14.2 The Novozhilov-Kadashevich model

Consider the rheological model depicted in Fig. 1.13. The model consists

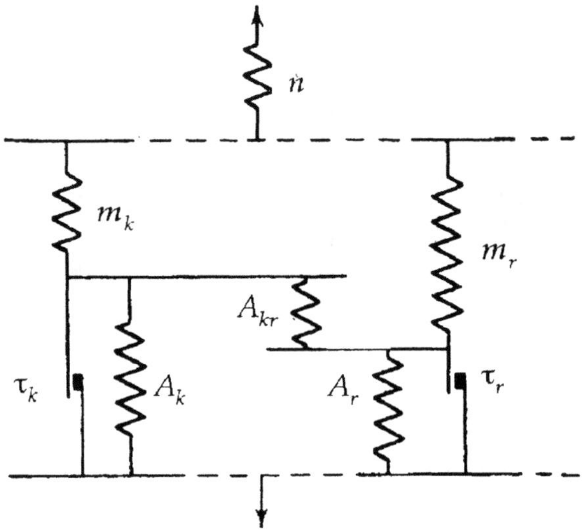

FIGURE 1.13. Rheological model for the Novozhilov-Kadashevich material.

of two parts in series, namely an elastic element of rigidity n and a system of parallel and interacting arms. Only two arms and an elastic element $A_{kr} = A_{rk}$ visualising their interaction are shown in Fig. 1.13. Each arm corresponds to a Prandtl rheological model which is completed by an elastic element A_k determining linear kinematic hardening.

We introduce notation which is very close to that used in [69]. Let $\boldsymbol{\sigma}$ be the value of the stress deviator which corresponds to the force in element n, $\boldsymbol{\sigma}_k$ - to that in element m_k, $\boldsymbol{\tau}_k$ - to that in the ideal plastic element with the yields stress $\boldsymbol{\tau}_{sk}$. Let the difference of the two latter tensors be equal to $\mathbf{s}_k$. Then

$$\boldsymbol{\sigma}_k = \boldsymbol{\tau}_k + \mathbf{s}_k. \tag{1.324}$$

It is clear that $\mathbf{s}_k$ denotes the sum of the force in element A_k and the forces in all interacting elements acting on the arm k. Let $\mathbf{e}$, $\boldsymbol{\varepsilon}$ and $\boldsymbol{\varepsilon}_k$ denote the deviator of the total strain, the strain deviator of the system of the parallel arms, and the strain deviator of the k-th plastic elements, respectively.

Let us now obtain the system of constitutive equations.

The equation of deformation for elastic elements n and m_k are given by

$$\boldsymbol{\sigma} = n\left(\mathbf{e} - \boldsymbol{\varepsilon}\right), \quad \boldsymbol{\sigma}_k = m_k\left(\boldsymbol{\varepsilon} - \boldsymbol{\varepsilon}_k\right). \tag{1.325}$$

By means of analogous equations for elastic elements A_k and A_{kr} we obtain

$$\mathbf{s}_k = A_k\boldsymbol{\varepsilon}_k + \sum_r A_{kr}\left(\boldsymbol{\varepsilon}_k - \boldsymbol{\varepsilon}_r\right). \tag{1.326}$$

The equation for deformation of an ideal plastic element $\boldsymbol{\tau}_{sk}$ is as follows

$$\left[\begin{array}{l} \dot{\boldsymbol{\varepsilon}}_k = 0, \ \sqrt{\frac{1}{2}\boldsymbol{\tau}_k : \boldsymbol{\tau}_k} \leq \tau_{sk}, \\ \dot{\boldsymbol{\varepsilon}}_k \neq 0, \ \ \boldsymbol{\tau}_k = \tau_{sk}\dot{\boldsymbol{\varepsilon}}_k / \sqrt{\frac{1}{2}\dot{\boldsymbol{\varepsilon}}_k : \dot{\boldsymbol{\varepsilon}}_k}, \end{array}\right. \tag{1.327}$$

where τ_{sk} denotes the yield stress of k-th plastic element.

Summation of the forces in the arms m_k gives the last equation

$$\boldsymbol{\sigma} = \sum_{k=1}^{N} \boldsymbol{\sigma}_k, \tag{1.328}$$

where N is the number of plastic elements in the model.

Let us show that if all values of m_k coincide, i.e. $m_k = m$, the present system of equations allows us to obtain equation of [69]. To this aim, we introduce mean stresses and mean plastic strain, cf. [69]

$$\langle \boldsymbol{\sigma}_k \rangle = \frac{1}{N} \sum_{k=1}^{N} \boldsymbol{\sigma}_k, \ \ \langle \boldsymbol{\varepsilon}_k \rangle = \frac{1}{N} \sum_{k=1}^{N} \boldsymbol{\varepsilon}_k. \tag{1.329}$$

By adding the right and left hand sides of the second equations in (1.325) it is easy to obtain

$$\langle \boldsymbol{\sigma}_k \rangle = m\left(\boldsymbol{\varepsilon} - \langle \boldsymbol{\varepsilon}_k \rangle\right).$$

Substracting the second equation in (1.325) we find

$$\langle \boldsymbol{\sigma}_k \rangle - \boldsymbol{\sigma}_k = m\left(\boldsymbol{\varepsilon}_k - \langle \boldsymbol{\varepsilon}_k \rangle\right). \tag{1.330}$$

Equations (1.324), (1.327) and (1.330) coincide exactly with those in [69]. Equation (1.326) differs from the following equation in [69]

$$\mathbf{s}_k = \sum_{r=1}^{N} C_{kr}\boldsymbol{\varepsilon}_k, \ \ (C_{kr} = C_{rk}) \tag{1.331}$$

only in that

$$C_{kr} = \left\{ \begin{array}{l} -A_{kr}, \ k \neq r, \\ A_k + \sum\limits_{l=1}^{N} A_{kl}, \ \ k = r. \end{array}\right.$$

The theories considered in this section are termed theories of microplastic deformations. The name is due to the fact that these theories present a phenomenological description of heterogeneous polycrystal materials.

It has been repeatedly pointed out, cf. [3], [68], [114], [89] and [90], that there is a need for theories of plasticity which take into account microplastic properties of real heterogeneous materials. The above theory of the elasto-plastic material with the rheological model in Fig. 1.13 has been proposed by Novozhilov and Kadashevich [66], [69] and [70], and is the most complicated theory of microplasticity. The theory of the generalised Prandtl material is a simplified case ($A_k = A_{kr} = 0, n \to \infty$) of this. The latter is essentially a simpler version, keeping however all typical features of the microplasticity theory. The "forces" in some arms of the model may be qualitatively interpreted as microstresses in grains of a polycrystal aggregate. Apparently, they have a decisive influence on the deformation process and hold memories of previous loadings.

Equations for the generalised Prandtl material with a discrete spectrum of yield stresses are obtained by Besseling [12]. The idea of a model with a continuous spectrum of yield stresses belongs to Ishlinsky who first studied a one-dimensional version in his paper [54]. Therefore, it is advisable to call the rheological model of Fig. 1.12 *the Ishlinsky model* and the material with this rheological model - *the Ishlinsky material.* Equations for the Ishlinsky material in a particular three-dimensional state of stress are contained in papers by Novozhilov and Kadashevich [69] and others. These equations were later derived in paper [129] by the present author.

2

Plasticity theory and internal friction in materials

2.1 Models of elastoplastic materials in the theory of internal friction

The vast majority of both experimental and theoretical papers on the amplitude-dependent internal friction in metals is devoted to the analysis of harmonic or near-harmonic deformation laws. The main problem encountered with the theoretical analysis of non-harmonic motions, which are present in various applications, is that there exist no analytic expressions for inelastic forces for arbitrary time-varying deformations. It is worth mentioning that existing expressions for this force, cf. [140] and [144], deal exclusively with a harmonic deformation law. However, it is unclear how these expressions should be modified to make them applicable for arbitrary deformation.

Based on the performed experimental studies, Davidenkov (1938) made a hypothesis [28] that internal friction at considerable stresses is an effect of microplastic deformations. He also indicated that internal friction must be studied by means of the Mises-Hencky plasticity theory, cf. [65]. This rational idea, however, was applied only for cyclic deformations in a uniaxial state of stress, and for a particular loading curve. As a result, the famous formula for a hysteretic loop was proposed. Accordingly, the energy loss in a material during an oscillation is a power function of the amplitude of deformation or stress.

Nowadays the fundamental role of plasticity theory in the applied theory of energy dissipation is explicitly or implicitly assumed. Consequently, the

Davidenkov formula [28] and its generalisations have been intensively used by Pisarenko [144] and his colleagues from Kiev. The same formula has been used as the basis for other, more simple applied theories of internal friction; among them, the Panovko theory [140] is the most widespread.

In 1960 Sorokin applied the plasticity theory equations to the analysis of internal friction for an uniaxial state of stress and a harmonic deformation, cf. [171].

Hence, plasticity theory is a background for the nowadays popular equations of energy dissipation at intensive stresses. The ideas of plasticity theory are systematically used for the analysis of harmonic and non-harmonic vibrations in the present book.

The main reason for using this approach becomes apparent when we realise that one can generalise it to the case of three-dimensional stresses. This problem is very important in the applied theory of energy dissipation, and was pointed out in [171] and subsequently confirmed by a great number of approaches to the problem of harmonic vibrations. It has been proposed in the literature that this generalisation can be made by means of the methods of linear viscoelasticity [171], the superposition principle [144], the hypothesis that the volumetric energy dissipation during a cycle of an oscillation is modelled by a power function of the amplitude value of specific potential energy [115] and, finally, the methods of plasticity theory [144]. However, only the latter can be correctly generalised to non-harmonic motions.

Let us consider the question of selecting a version of plasticity theory suitable for description of the internal friction. Following the work by Davidenkov, amplitude-dependent internal friction presents an effect of microplastic strains. Microplastic strains are those strains which occur at any stress level, even at stresses below the macroscopic yield stress. From this perspective, application of the theories of microplastic strains by Novozhilov and Kadashevich is suitable. The theory of the Ishlinsky elastoplastic material which is the simplest theory of microplasticity is systematically used in this book.

In this chapter, a one-dimensional variant is considered and its applicability to the description of some of the simplest effects of amplitude-dependent internal friction is proved.

2.2 One-dimensional theory of microplasticity

In this section we follow the main idea of the paper [69] by Ishlinsky. Consider an uniaxial stress state in which the material is described by the rheological model shown in Fig. 2.1. The model consists of an infinite number of the Prandtl elements. In each Prandtl element a spring of rigidity Edh is connected in series with an element of dry friction $Ehdh$. The stiffnesses of

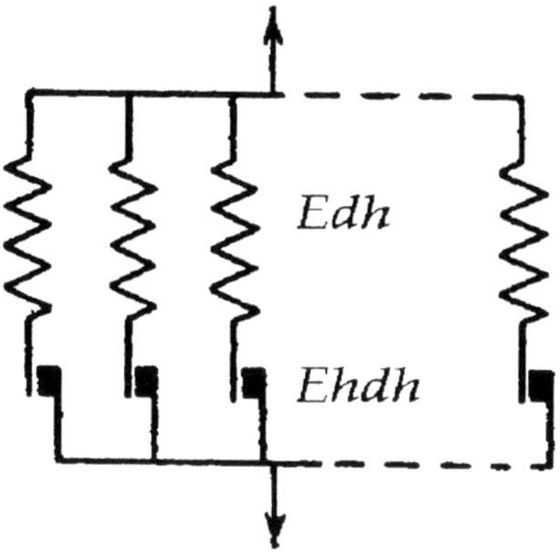

FIGURE 2.1. Rheological model for one-dimensional theory of microplasticity.

all springs are assumed to coincide, while the non-dimensional yield stress h varies such that it is continuously distributed and has a probability density $p(h)$.

The deformation law of a Prandtl element h is given by

$$\begin{aligned} d\sigma &= E(\varepsilon - \varepsilon_h)\,dh, \\ E(\varepsilon - \varepsilon_h)\,dh &= Ehdh\,\mathrm{sgn}\dot{\varepsilon}_h, \end{aligned} \tag{2.1}$$

where $d\sigma$ is stress in an arbitrary Prandtl element h, ε is the total strain in any Prandtl element and ε_h is the plastic strain in a Prandtl element h. It should be mentioned that in (2.1) the function $\mathrm{sgn}\dot{\varepsilon}_h$ is understood as follows: $\mathrm{sgn}\dot{\varepsilon}_h = 1$ if $\dot{\varepsilon}_h > 0$ and $\mathrm{sgn}\dot{\varepsilon}_h = -1$ if $\dot{\varepsilon}_h < 0$. We assume that if the time derivative of the plastic deformation vanishes, i.e. $\dot{\varepsilon}_h = 0$, function $\mathrm{sgn}\dot{\varepsilon}_h$ is between -1 and 1 and is prescribed by the second equation in (2.1). This definition of sgn allows the system of equations (2.1) to be suitable for the processes of loading and unloading any Prandtl element both in the presence and absence of plastic deformations.

Multiplying the force in a Prandtl element h by its probability we obtain after summation over all elements

$$\sigma = E\int_0^\infty (\varepsilon - \varepsilon_h)\,p(h)\,dh, \quad \varepsilon - \varepsilon_h = h\,\mathrm{sgn}\dot{\varepsilon}_h, \tag{2.2}$$

or

$$\sigma = E\varepsilon - E\int_0^\infty \varepsilon_h p(h)\,dh, \quad h\,\mathrm{sgn}\dot{\varepsilon}_h + \varepsilon_h = \varepsilon. \tag{2.3}$$

In this expression for stress the first term represents an elastic force which obeys Hooke's law, while the second term describes the dissipation effect.

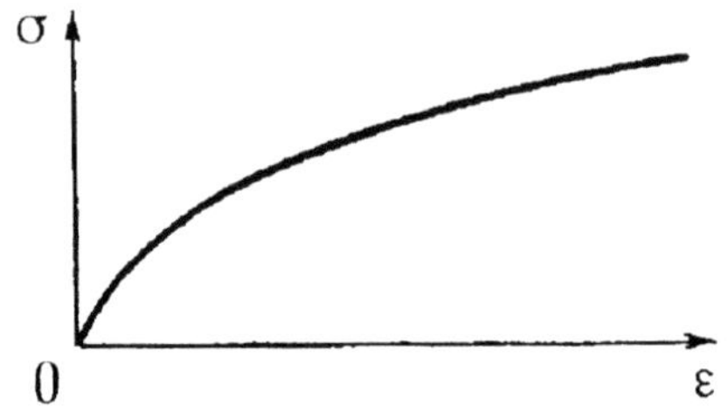

FIGURE 2.2. A typical loading diagram.

The second equation in (2.3) serves for determining the plastic strain in an arbitrary Prandtl element h.

The system of equations (2.3) is suitable for describing any time-dependent processes of loading and unloading.

For example, consider a monotonic loading, i.e. $\varepsilon > 0$ and $\dot{\varepsilon} > 0$. Assuming that the initial values of plastic strain in all Prandtl elements vanish, the second equation (2.3) yields

$$\begin{aligned} \varepsilon_h &= 0, \qquad & h > \varepsilon, \\ \varepsilon_h &= \varepsilon - h, \qquad & h \leq \varepsilon. \end{aligned} \tag{2.4}$$

Substituting this into the first equation (2.3) we obtain an expression for the stress

$$\sigma = E \left[\varepsilon - \int_0^{\varepsilon} (\varepsilon - h)\, p(h)\, dh \right]. \tag{2.5}$$

This equation was obtained and used for analysis of the model behaviour in [54], [171] and [58]. The following equations

$$\frac{d\sigma}{d\varepsilon} = E \int_{\varepsilon}^{\infty} p(h)\, dh, \tag{2.6}$$

$$\frac{d^2\sigma}{d\varepsilon^2} = -Ep(\varepsilon). \tag{2.7}$$

are obtained in [54] by means of differentiating (2.5). Equation (2.7) indicates a physical meaning of density $p(h)$ and shows that the loading curve is convex upwards ($p \geq 0$). Condition (2.6) indicates a monotonity of the loading curve. A typical loading diagram which satisfies these conditions is shown in Fig. 2.2. The above restrictions imposed on the loading curve are not very rigid and are quite acceptable from a physical point of view. Indeed, for an overwhelming majority of structural materials the loading curve exhibits this type of behaviour at small strains. However, only when the strains are small can one study the effect of internal friction. Otherwise intensive plastic deformations take place.

Let us introduce the following notation

$$\Phi(\varepsilon) = E\left[\varepsilon - \int_0^{\varepsilon} (\varepsilon - h)\, p(h)\, dh\right]. \tag{2.8}$$

Thus, during the loading we have

$$\sigma = \Phi(\varepsilon).$$

Let a strain ε_0 be attained. This means that the following stress and plastic strain

$$\sigma_0 = \Phi(\varepsilon_0), \tag{2.9}$$

$$\varepsilon_{h0} = \left[\begin{array}{ll} 0, & h > \varepsilon_0, \\ \varepsilon_0 - h, & h < \varepsilon_0. \end{array}\right. \tag{2.10}$$

are attained.

We proceed now to an unloading process. Let $\varepsilon_0 - \varepsilon = \varepsilon_1 > 0$, $\dot{\varepsilon} < 0$ and $\varepsilon_1 < 2\varepsilon_0$. The difference

$$\varepsilon - \varepsilon_{h0} = \varepsilon - \varepsilon_0 + h = h - \varepsilon_1 > 0$$

decreases at unloading, i.e. when ε_1 increases.

If

$$|\varepsilon - \varepsilon_{h0}| < h,$$

which is equivalent to the following equality

$$h > \frac{\varepsilon_1}{2}$$

no plastic deformation occurs, i.e.

$$\varepsilon_h = \varepsilon_{h0} = \left[\begin{array}{ll} \varepsilon_0 - h, & \varepsilon_1/2 < h < \varepsilon_0, \\ 0, & h > \varepsilon_0. \end{array}\right.$$

If

$$|\varepsilon - \varepsilon_{h0}| > h$$

or

$$h < \frac{\varepsilon_1}{2}$$

then according to the second equation in (2.3) we obtain

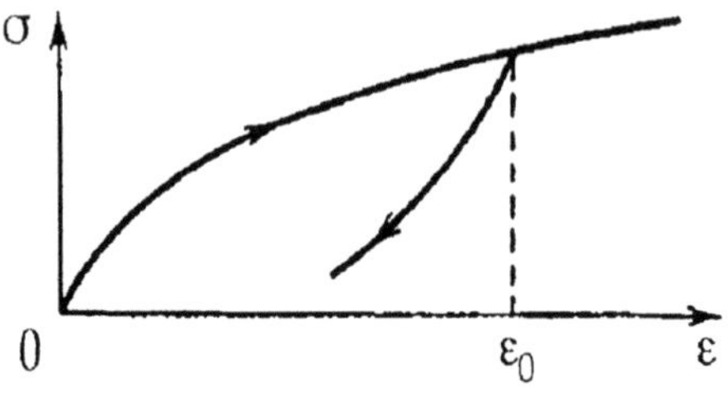

FIGURE 2.3. An unloading curve.

$$\varepsilon_h = \varepsilon + h = h + \varepsilon_0 - \varepsilon_1.$$

Therefore, we have

$$\varepsilon_h = \begin{bmatrix} h + \varepsilon_0 - \varepsilon_1, & 0 < h < \varepsilon_1/2, \\ \varepsilon_0 - h, & \varepsilon_1/2 < h < \varepsilon_0, \\ 0, & h > \varepsilon_0. \end{bmatrix} \tag{2.11}$$

Substituting these expressions into the first equation in (2.3) we obtain the following equation for the unloading curve

$$\sigma = E\left[\varepsilon - \int_0^{\varepsilon_1/2} (h + \varepsilon_0 - \varepsilon_1)\, p(h)\, dh - \int_{\varepsilon_1/2}^{\varepsilon_0} (\varepsilon_0 - h)\, p(h)\, dh\right]. \tag{2.12}$$

After some simple transformations we arrive at the formula

$$\sigma = \sigma_0 - 2\Phi\left(\frac{\varepsilon_1}{2}\right), \tag{2.13}$$

which expresses *the Masing principle* in plasticity theory, cf. [109]. Accordingly, the unloading curve can be obtained from the loading curve by means of the following coordinate transformation

$$\sigma_* = \frac{\sigma_0 - \sigma}{2}, \quad \varepsilon_* = \frac{\varepsilon_1}{2} = \frac{\varepsilon_0 - \varepsilon}{2}.$$

A typical unloading curve is depicted in Fig. 2.3.

The Masing principle is known to be inadequate in describing intensive deformations of materials, see [109]. The question therefore arises as to whether or not this model of internal friction should be rejected? The answer to this is a categoric no! Firstly, only very small strains, namely of the order of elastic strains, are of interest in the theory of internal friction. Indeed, in case of very small plastic deformations tests confirm the Masing principle. Secondly, an indirect confirmation of the validity of the Masing

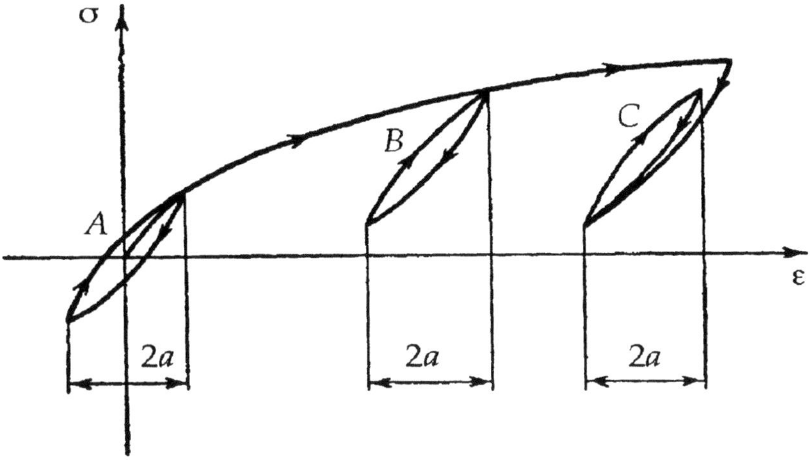

FIGURE 2.4. Some hysteresis loops.

principle for small deformations, due to an effect of internal friction, is described below.

As follows from (2.13), the shape of the unloading curve does not depend upon the strain attained at the loading. This means that the shape of the hysteresis loops A, B and C depicted in Fig. 2.4 coincides. Hence, the dynamic properties of material under cyclic deformation must not depend upon an additional static load. This effect is well known to experts in internal friction, and was pointed out in [32]. Tests expose a very weak dependence of the dissipative properties of many materials on additional static load, cf. [171] and [143] if the latter does not exceed admissible bounds. There exist, however, materials for which the influence of the static load on the energy dissipation is considerable, see [105], [146] and [82].

The one-dimensional version of the elastoplastic material appeared several times in the mechanical literature. For example, Jenkins applied it in 1922, see reference [109]. However, it was Masing in 1923, who used this model to validate the principle named after him in the theory of plasticity, [109]. In 1930 Timoshenko recommended that it should be used in problems of hysteretic friction [177]. In 1944 Ishlinsky gave an analytic background of the model [54], whilst in 1953 Afanasiev [1] applied it to the problem of fatigue strength. In 1960 Sorokin [171] used it in the applied theory of internal friction. The above model was later considered in [32], [59], [58], [64], [101] and [187]. A "discrete" particular case, the so-called model of bilinear hysteresis [57], [11], [24], [26], [62], [77] etc., is widely used in various theoretical investigations.

It is worth noting that in all the above cited papers the rheological model was used only for plotting the loading and unloading diagrams. The differential form of deformation equations (2.3) has not yet been used. However, it considerably simplifies the analysis of cyclic deformation by using the methods of nonlinear mechanics and is a suitable starting point for analysis of random deformations.

2.3 Harmonic deformation of material

Consider the behaviour of a material under a harmonic deformation in time, i.e.

$$\varepsilon = a \cos \omega t, \tag{2.14}$$

where a denotes the strain and ω is a frequency, $\omega > 0$. Obtaining the stress harmonic of the same frequency is of practical interest. In order to find it, we use the method of harmonic linearization [149]. According to this method we assume that the plastic strain in each Prandtl element is due to a harmonic law and carry out a harmonic linearisation of the nonlinear function only in (2.3), to get

$$\operatorname{sgn}\dot{\varepsilon}_h \approx \frac{n}{b_h \omega}\dot{\varepsilon}_h, \quad n = \frac{4}{\pi}, \tag{2.15}$$

where b_h is the amplitude of strain ε_h and ω is the deformation frequency.

Linearization of the second nonlinear equation in (2.3) leads to the following equation

$$\frac{n}{b_h \omega}\dot{\varepsilon}_h + \varepsilon_h = \varepsilon. \tag{2.16}$$

It should be noted that the latter equation is valid only when $b_h > 0$. If $b_h = 0$, which means absence of the plastic strain in this arm, the following equation

$$\varepsilon_h = 0 \tag{2.17}$$

must be used instead of (2.16). After ε_h has been determined, the stress is obtained from the first linear equation (2.3)

$$\sigma = E\varepsilon - E\int_0^\infty \varepsilon_h p(h)\, dh. \tag{2.18}$$

Since equations (2.16)-(2.18) are linear, we can use the general complex exponential form for the variables

$$\sigma = Ae^{i\omega t}, \; \varepsilon = Be^{i\omega t}, \; \varepsilon_h = B_h e^{i\omega t}, \tag{2.19}$$

where A, B and B_h are complex values. It goes without saying that only the real parts of expressions (2.19) have a physical meaning. The amplitude of the total strain a and the amplitude of the plastic strain b_h are absolute values of the corresponding complex functions ε and ε_h, respectively

$$a = |\varepsilon|, \quad b_h = |\varepsilon_h|. \tag{2.20}$$

Substituting (2.19) into (2.16) yields

$$\varepsilon_h = \left(1 + i\frac{nh}{b_h}\right)^{-1} \varepsilon. \tag{2.21}$$

Evaluating the absolute value of both sides of the latter equation results in an equation for b_h

$$b_h = \left[1 + \left(\frac{nh}{b_h}\right)^2\right]^{-1/2} a. \tag{2.22}$$

One finds easily that

$$b_h = \left[a^2 - (nh)^2\right]^{1/2}. \tag{2.23}$$

This expression makes sense only if $h < a/n$, otherwise it is meaningless. This means that (2.21), and hence (2.16) has no solution if $h \geq a/n$. In this case, one must use (2.17) which yields $b_h = 0$.

Thus, we have

$$b_h = \left[\begin{array}{ll} \left[a^2 - (nh)^2\right]^{1/2} & 0 < h < a/n, \\ 0, & h \geq a/n. \end{array}\right. \tag{2.24}$$

Substituting this dual expression into eqs. (2.21) and (2.17), respectively, we obtain

$$\varepsilon_h = \left[\begin{array}{ll} \left[1 - (nh/a)^2 - inh/a\sqrt{1 - (nh/a)^2}\right] \varepsilon, & 0 < h < a/n, \\ 0, & h \geq a/n. \end{array}\right. \tag{2.25}$$

Finally, substitution of the latter equation into equation (2.18) for stress leads to the following result

$$\sigma = E_c \varepsilon, \quad E_c = E \left[1 - \int\limits_0^1 \left(1 - \eta^2 - i\eta\sqrt{1 - \eta^2}\right) p\left(\frac{a\eta}{n}\right) \frac{a}{n} d\eta\right]. \tag{2.26}$$

In what follows, E_c is referred to as *the complex Young's modulus*. As follows from (2.26) the complex Young's modulus does not depend on the deformation frequency for any density $p(h)$. This property of the model corresponds to the law of internal friction observed by most metals and structural materials. In this sense, the model considered here successfully describes internal friction under intensive stresses.

2.4 The choice of the yield stress distribution density $p(h)$. Conclusions

We bring now another argument in favour of the considered model. Let us pose the following question: what distribution $p(h)$ appears to be the most suitable from a physical perspective? If the yield stress in each Prandtl element were infinite, then the material would be ideally elastic and have no damping. Existence of finite values of h may be considered a manifestation of the fact that the material has defects. There is much evidence to suggest that these defects have the same physical nature as the defects in the theories of fatigue and brittle failure [17]. For this reason, we adopt the distribution function used in these theories, so that the distribution function of defects for small values of h is as follows

$$F(h) = Hh^{\alpha}, \tag{2.27}$$

where H and α are some positive constants. Differentiating eq. (2.27) with respect to h one obtains the following probability density

$$p(h) = \alpha H h^{\alpha-1}. \tag{2.28}$$

The structure of equation (2.8) suggests that the dependence

$$p = p(h)$$

influences the stress value only over a small range of strains. Thus, information contained in eq. (2.28) suffices for analysis of (2.8). Substituting distribution (2.28) into (2.8) and evaluating the integral yields the following closed form expression for loading curve

$$\sigma = \Phi(\varepsilon) = E\left(\varepsilon - \frac{H}{\alpha+1}\varepsilon^{\alpha+1}\right).$$

This equation differs from that for the loading curve by Davidenkov [28] in notation only. Substitution of the latter result into the Masing formula, eq. (2.13), gives an equation for the "descending branch" in the Davidenkov formula for a hysteresis loop. An equation for the "ascending branch" can be obtained by analogy.

The above should be considered as a plausible substantiation for the Davidenkov formula for a hysteresis loop.

Substitution of distribution (2.28) into (2.26) and evaluation of the integrals leads to the following expression for the complex Young's modulus

$$E_c = E\left(1 - ra^{\alpha} + iga^{\alpha}\right), \tag{2.29}$$

where

$$r = \frac{2H}{2+\alpha} n^{-\alpha}, \quad g = \frac{\alpha H}{2} n^{-\alpha} \mathrm{B}\left(\frac{\alpha+1}{2}, \frac{3}{2}\right), \tag{2.30}$$

B being the Eulerian beta-function.

Equation (2.29) indicates a power law dependence of the resistance force amplitude $Ega^{\alpha+1}$ on the strain amplitude. The same power law is recommended by Panovko [140] and gives agreement with numerous tests of internal friction. A list of some of these tests can be found in [140], [142] and [143]. However, contrary to the recommendation of [140] the amplitude of the elastic force $E(1 - ra^{\alpha})\, a$ depends on the strain amplitude nonlinearly, namely in the form of [142], [143] and [144]. In [140] the amplitude of the elastic force was equated to Ea. It should be kept in mind that the indicated difference often turns out to be insignificant and may be neglected, as the nonlinear term in the equation for the elastic force is of the same order as the inelastic resistance, and thus is small. There exist, however, some situations in which the above difference cannot be neglected. Such a case is when parameter α is close to zero. Indeed, expression (2.30) says that for small α the parameter of the force of inelastic resistance g can be small, while the nonlinearity parameter r in the expression for the elastic force is finite. Altogether, it allows us to say that the nonlinearity of the elastic component of internal friction is soft. The effect of a soft nonlinearity of elastic component of internal friction was observed in a test with a concrete beam, [170]. Another manifestation of this effect is an amplitude dependence of the "modulus defect" observed in a series of experimental studies and reported in [150].

It is worth noting that, in principle, the internal friction model considered does not admit existence of a linear friction since the very friction force vanishes ($g \to 0$) with vanishing α, $\alpha \to 0$. This fact has been validated by available test data, for example [28], [142], [143], [144] and [170], in which the resistance force was a power law with positive α for the materials studied.

For real materials, however, dependence of the dissipative properties on amplitude is a power law only when the stresses are not too large. For large stresses this dependence is observed to show considerable deviation from a power law, cf. [171]. However, for the theory presented here the power law was theoretically derived only for small stresses when approximation (2.27) is valid! Therefore general expression (2.26) contains essentially more general dependences that can be used to approximate the natural laws.

The general conclusion is as follows. The model of the Ishlinsky elasto-plastic material is an abstract model of a hypothetical material with the following specific properties.

1. Superimposing a statical load does not influence the energy dissipation in the material under cyclic deformation.

2. Energy dissipation in material under cyclic deformation does not depend upon the deformation frequency.

3. Dependence of the energy dissipation in the material on the strain amplitude is a power function for stresses which are not too large. The model admits a generalisation of this dependence in case of a broad range of stress.

4. Energy dissipation is nonlinear in principle. "Linear" dissipation is impossible in this model.

5. In the cases in which energy dissipation in a material is considerable, the elastic component of the material reaction is essentially nonlinear.

Each of these five properties is satisfactorily confirmed by available experimental data on the energy-absorbing properties of materials under harmonic deformations. The author does not, however, believe that all real materials behave as the model suggests or requires. There will always exist materials whose behaviour turns out to be more complex and contradicts at least one of the above properties. However, this means that the model considered is not suitable for this particular material and another, more sophisticated model must be used in its place. The model presented is considered useful as it reflects typical properties of a considerable group of structural materials and admits theoretical analysis for polyharmonic and random deformations. It also admits a generalisation to the cases of three-dimensional stress fields. All these aspects will be demonstrated in what follows.

2.5 Forced vibration of an oscillator with nonlinear internal friction

The above properties of internal friction are to some extent known in mechanics. The fifth property, however, requires further justification. For this reason, we consider some simple examples which emphasize the real value of this property.

Consider the forced vibration of an oscillator with internal friction driven by a harmonic force, see Fig. 2.5. The mass of the deformable element is assumed to be negligible. The governing equation for the oscillator is

$$m\ddot{w} + F\sigma\{\varepsilon\} = p_0 \cos \omega t, \tag{2.31}$$

where m is the oscillator mass, F is the cross-sectional area of the elastic element, p_0 is the amplitude of the harmonic load, σ is the axial stress and ε is the strain. It is apparent that the displacement w is expressed in terms of strain as follows

$$w = L\varepsilon, \tag{2.32}$$

where L denotes the length of the deformable element.

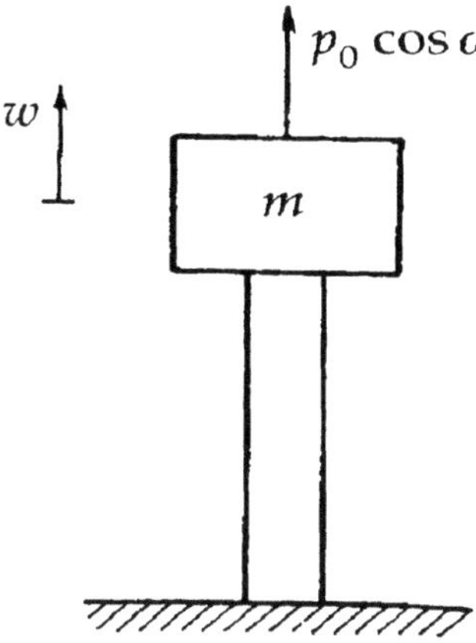

FIGURE 2.5. A harmonically driven oscillator.

Assume that the stress σ is expressed in terms of the strain ε by means of (2.3). Linearising the nonlinearity $\sigma\{\varepsilon\}$ and using the complex exponential form for the variables yields

$$m\ddot{\varepsilon} + c_c\varepsilon = p_* e^{i\omega t}, \tag{2.33}$$

where c_c denotes a complex stiffness and p_* denotes the amplitude of the generalised force. These are given by

$$c_c = \frac{E_c F}{L}, \quad p_* = \frac{p_0}{L}. \tag{2.34}$$

A steady-state solution of (2.33) is

$$\varepsilon = \frac{p_* e^{i\omega t}}{c_c - m\omega^2}. \tag{2.35}$$

Let us denote the complex stiffness as follows

$$c_c = u + iv, \tag{2.36}$$

where, due to (2.34) and (2.26), the real and imaginary parts are given by

$$\begin{aligned} u &= c\left[1 - \int_0^1 (1-\eta^2)\, p\left(\frac{a\eta}{n}\right)\frac{a}{n}d\eta\right], \\ v &= c\int_0^1 \eta\sqrt{1-\eta^2}\, p\left(\frac{a\eta}{n}\right)\frac{a}{n}d\eta, \\ c &= EF/L. \end{aligned} \tag{2.37}$$

Substituting (2.36) into (2.35) and evaluating the absolute value of both sides of the resulting equation we obtain the following equation for the strain amplitude

$$a^2 = \frac{p_*^2}{\left(u - m\omega^2\right)^2 + v^2}. \tag{2.38}$$

Solving it for frequency we arrive at the following equation for the amplitude-frequency response function

$$m\omega^2 = u(a) \pm \sqrt{\frac{p_*^2}{a^2} - v^2(a)}\ . \tag{2.39}$$

The resonance amplitude is seen to be determined by the following equation

$$\frac{p_*}{a} = v(a), \tag{2.40}$$

while the resonance frequency is given by

$$\omega = \sqrt{\frac{u(a)}{m}}\ . \tag{2.41}$$

Equations (2.40) and (2.41) are valid for any monotonic decreasing function $v = v(a)$. The imaginary part of the complex rigidity determines the resonance amplitude, while the real part gives the resonance frequency. We see from eqs. (2.37) and (2.41) that the decrease in the resonance frequency relative to its value without damping is of the order of the *energy dissipation factor* $v(a)$. Hence, any change of the load amplitude must result in a change of the resonance amplitude and thus, by virtue of (2.41), a considerable change of the resonance amplitude. This effect has been observed by Sorokin [170] when testing a concrete beam, where a resonance frequency change of 7% was observed when increasing the amplitude of the load by a factor of 15.

The question of the resonance frequency shift for a general dependence $c_c = c_c(a)$ was discussed by Sorokin in [171] where u and v were taken in the form

$$u = c\frac{1-\gamma^2/4}{1+\gamma^2/4},\ v = c\frac{\gamma}{1+\gamma^2/4} \tag{2.42}$$

and $\gamma = \gamma(a)$. As γ is usually very small, we have approximately

$$u = c - \frac{v^2}{2c},\ v = c\gamma. \tag{2.43}$$

Hence, in this theory the resonance frequency decreases with the second order in v, and there will be negligible change for small v.

The same question of the resonance frequency shift was considered in the theories by Pisarenko and Panovko, both theories having used the hysteresis

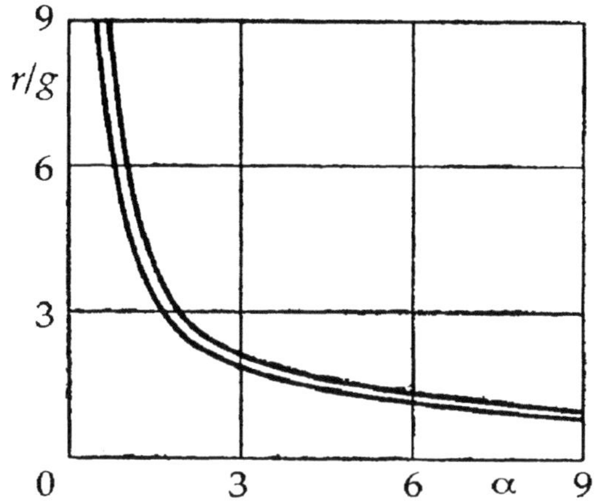

FIGURE 2.6. Comparison of the present theory (lower curve) with the Pisarenko theory (upper curve).

loop proposed by Davidenkov. In order to compare the present theory with the theories by Pisarenko and Panovko, we use an equation for the complex Young's modulus (2.29). This leads to the following expressions for the coefficients

$$u = c\,(1 - ra^{\alpha})\,,\; v = cga^{\alpha}. \tag{2.44}$$

One can easily find a direct relation between u and v, i.e.

$$u = c - \frac{r}{g}\frac{v}{c}. \tag{2.45}$$

The ratio r/g due to (2.30) is equal to

$$\frac{r}{g} = 4\left[\alpha\,(2+\alpha)\,\mathrm{B}\left(\frac{\alpha+1}{2},\frac{3}{2}\right)\right]^{-1} \tag{2.46}$$

The lower curve in Fig. 2.6 shows the ratio r/g versus α.

The Pisarenko theory leads to expressions which are similar to (2.44). The only difference lies in the expressions for the coefficients r and g which is due to various methods of linearisation. In [143] one finds the following equation for the ratio r/g

$$\frac{r}{g} = 2^{2n}\,\frac{n}{n-1}\,\frac{\left[\Gamma\left(n+\frac{1}{2}\right)\right]^2}{\Gamma\,(2n+1)},\; n = \alpha + 1. \tag{2.47}$$

This ratio versus α is presented in Fig. 2.6 by the upper curve. A comparison of the curves shows that they are in good agreement. Hence, for a power dependence of the hysteresis loss on the strain amplitude, the theory proposed and the Pisarenko theory give nearly identical results.

In the Panovko theory [140] it is assumed that

$$u = c,\ v = cga^{\alpha}, \tag{2.48}$$

so the latter theory does not cause any shift of the resonance frequency. This discrepancy between the Panovko theory and the Pisarenko theory has been discussed in [182]. A typical amplitude-frequency response function in this paper shows that the resonance frequency shift is usually negligible (about 0.7%). The view of the author is that the resonance frequency shift for the majority of applied problems is very small, so in actuality the Pisarenko theory, the Panovko theory and the present theory give very similar results.

2.6 Free vibration of an oscillator with nonlinear internal friction

The equation for free vibration of such an oscillator is obtained from (2.31) by equating the external force to zero. This gives

$$m\ddot{w} + F\sigma\{\varepsilon\} = 0. \tag{2.49}$$

A weakly decaying oscillating process is sought in the following form

$$\varepsilon = a(t)\cos\varphi(t). \tag{2.50}$$

The instantaneous frequency $\dot{\varphi}$ plays the part of frequency and eq. (2.50) allows for slow variations of the instantaneous frequency.

While studying damped oscillations in mechanical systems with internal damping one usually assumes that the reaction of the deformable element in a weakly decaying process depends upon the current value of the strain amplitude as if the process were not decaying. That is, we take a to be const and apply the harmonic linearization of nonlinearity (2.3) as has already been done in Section 2.3. Its solution is sought in the general complex exponential form of (2.50)

$$\varepsilon = a(t)\,e^{i\varphi(t)}. \tag{2.51}$$

As a result we obtain the following equation for ε

$$m\ddot{\varepsilon} + c_c\varepsilon = 0, \tag{2.52}$$

where the complex rigidity c_c is given by (2.34).

Substituting the assumed solution (2.51) into (2.52) and separating the real and imaginary parts we obtain, by virtue of (2.36), the following equations for amplitude and phase of the decaying process

$$\ddot{a} - a(\dot{\varphi})^2 = -\frac{au(a)}{m}, \tag{2.53}$$

$$\frac{1}{a}\left(a^2\dot{\varphi}\right)^{\bullet} = -\frac{av(a)}{m}. \tag{2.54}$$

As a weakly decaying process is sought, we may neglect the first term in (2.53) in comparison with the other terms, to obtain

$$\dot{\varphi} = \sqrt{\frac{u(a)}{m}}. \tag{2.55}$$

Substitution of (2.55) into (2.54) yields the following equation for a

$$\left(a^2\sqrt{\frac{u(a)}{m}}\right)^{\bullet} = -\frac{a^2 v(a)}{m}. \tag{2.56}$$

Retaining in eqs. (2.55) and (2.56) only asymptotically dominant terms in plastic strain (i.e. terms of the first order in v) we obtain closed form expressions for the instantaneous frequency

$$\dot{\varphi} = \sqrt{\frac{c}{m}} \tag{2.57}$$

and the envelope of damped vibration

$$\dot{a} = -\frac{av(a)}{2\sqrt{cm}}. \tag{2.58}$$

Equation (2.58) cannot be solved for an arbitrary function v. But it may be used to find the logarithmic decrement of vibration

$$\Delta = \ln\frac{a_n}{a_{n+1}}, \tag{2.59}$$

where a_n and a_{n+1} are two consecutive maxima (or minima) of ε. Assuming that the variation in amplitude a is small during the time interval between two maxima we obtain by means of eqs. (2.57) and (2.58)

$$\Delta = \pi\frac{v(a)}{c}. \tag{2.60}$$

This is an equation which holds in general. For the Davidenkov hysteresis loop we obtain a power law for the logarithmic decrement of vibration

$$\Delta = \pi g a^{\alpha+1}, \tag{2.61}$$

and the following equation for the envelope of decaying vibration

$$\dot{a} = -\frac{ga^{\alpha+1}}{2}\sqrt{\frac{c}{m}}. \tag{2.62}$$

The solution to this equation is

$$a = a_0 \left(1 + \frac{\alpha g a_0^\alpha}{2} \sqrt{\frac{c}{m}}\, t \right)^{-1/\alpha}, \tag{2.63}$$

where a_0 is the strain amplitude at the initial instant of time, i.e. at $t = 0$.

As in the proposed theory α is always positive, the following conclusions are valid:

1. Vibration decay in systems with an amplitude-dependent internal friction never obeys an exponential law.
2. The vibrations never die out in a finite time.

3

Three-dimensional cyclic deformations of elastoplastic materials

3.1 Constitutive equations for elastoplastic materials

In what follows, the equations for the Ishlinsky material are used in isothermal case. Consequently,

1. The dilatation ϑ depends linearly on the mean normal stress

$$\sigma = k\vartheta. \tag{3.1}$$

This is eq. (1.293) for $m(\theta_0) = 0$ where θ_0 is a fixed temperature.

2. The relation between the stress deviator and the strain deviator is given by the rheological model of Fig. 1.12, and the governing equations are

$$\mathbf{s} = 2G\left[\mathbf{e} - \int_0^\infty \mathbf{e}_h p(h)\, dh\right]. \tag{3.2}$$

$$\left[\begin{array}{ll} \dot{\mathbf{e}}_h = 0, & \sqrt{\frac{1}{2}(\mathbf{e}-\mathbf{e}_h):(\mathbf{e}-\mathbf{e}_h)} \le h, \\ \dot{\mathbf{e}}_h \ne 0, & h\dot{\mathbf{e}}_h/v_h = \mathbf{e} - \mathbf{e}_h. \end{array}\right. \tag{3.3}$$

Thus, the material properties are determined by two constants, namely the dilatation modulus k and the shear modulus G, as well as the probability density function of the nondimensional yield stress $p(h)$.

The system of equations (3.3) permits the following interpretation which simplifies the solution process.

Let the Mises condition

$$\sqrt{\frac{1}{2}(\mathbf{e}-\mathbf{e}_h):(\mathbf{e}-\mathbf{e}_h)} = h \tag{3.4}$$

hold. By means of direct substitution one can prove that any solution ($\dot{\mathbf{e}}_h \neq 0$) of the second line in (3.3)

$$\frac{h}{v_h}\dot{\mathbf{e}}_h = \mathbf{e}-\mathbf{e}_h, \tag{3.5}$$

subject to certain initial conditions satisfies the restriction (3.4). It follows from (3.3) that (3.4) may be realised at $\dot{\mathbf{e}}_h = 0$ as well. But this case must be considered as a limiting case for solutions of (3.5). If

$$\sqrt{\frac{1}{2}(\mathbf{e}-\mathbf{e}_h):(\mathbf{e}-\mathbf{e}_h)} < h, \tag{3.6}$$

equation (3.3) reduces to

$$\dot{\mathbf{e}}_h = 0, \tag{3.7}$$

in which there is no plastic strain. Altogether, it means that equations (3.3) can be replaced by the following system

$$\left[\begin{array}{ll} \sqrt{\frac{1}{2}(\mathbf{e}-\mathbf{e}_h):(\mathbf{e}-\mathbf{e}_h)} = h, & h\dot{\mathbf{e}}_h/v_h + \mathbf{e}_h = \mathbf{e}, \\ \sqrt{\frac{1}{2}(\mathbf{e}-\mathbf{e}_h):(\mathbf{e}-\mathbf{e}_h)} < h, & \dot{\mathbf{e}}_h = 0. \end{array}\right. \tag{3.8}$$

Thus, if (3.4) holds then plastic strain $\mathbf{e}_h$ is determined by (3.5), while if (3.6) holds then $\mathbf{e}_h$ must be obtained from (3.7). Any solution of eq. (3.5) automatically satisfies its applicability condition, i.e. (3.4), and therefore does not satisfy condition (3.6). Thus, if (3.5) has a solution, the latter is the solution of system (3.3). When the initial conditions do not allow eq. (3.5) to have a solution, we must consider (3.6), the only alternative of (3.4), and make use of (3.7).

In conclusion, we note that in spite of the seeming complexity, the differential equations (3.2) and (3.3) turn out to be very useful and convenient for analysis of harmonic and especially polyharmonic and random deformation.

Another possible applications of the plasticity theory equations to the analysis of vibrational energy dissipation have been reported in the literature. Equations for small elastoplastic deformations of material which obey the "hypothesis of the universal curve" are derived and recommended for application in [183], [184], [144], [145] and [154]. The hysteresis loop form was substantiated only for harmonic deformation and it remains unclear how to generalize the equations to the case of non-harmonic deformation. The theory of plasticity proposed is free of this considerable shortcoming and allows us to analyse both harmonic and non-harmonic deformations.

3.2 Simple deformation

The content of this Section is based on paper [129].

A state of material is called *the natural state of material* provided that

$$\mathbf{s} = 0, \quad \mathbf{e} = 0, \quad \mathbf{e}_h = 0. \tag{3.9}$$

The concept of the natural state is usually attributed to the whole body. The above definition may be considered as a definition of natural state at a point. Being applied to materials under consideration this definition is expedient since it excludes the very possibility of internal "microstresses" at the point. In the light of the given definition, the natural state of a body is referred to the state in which each point of the body is in the natural state.

A deformation from the natural state at which the strain deviator varies in time as follows

$$\mathbf{e} = \mathbf{e}_0 \lambda(t), \quad \lambda(0) = 0, \tag{3.10}$$

is termed *simple deformation*, cf. [52]. Here $\mathbf{e}_0$ is some constant tensor and λ is a scalar function of time.

Simple deformation under the condition that $\dot{\lambda}$ does not change its sign in called *simple loading*. Because of the trivial initial condition $\lambda(0) = 0$ the parameter λ keeps its sign during any simple loading, the sign of λ coinciding with the sign of $\dot{\lambda}$.

A deformation which follows after a simple loading and obeys the law $\mathbf{e} = \mathbf{e}_0 \lambda(t)$ with the sign of $\dot{\lambda}$ opposite to that of simple loading, is termed *simple unloading.*

Let us introduce the following parameter

$$\gamma = \gamma_0 \lambda(t), \quad \gamma_0 = +\sqrt{\frac{1}{2}\mathbf{e}_0 : \mathbf{e}_0}. \tag{3.11}$$

Its absolute value coincides with the shear strain intensity, but it has the sign of λ. Parameter γ is useful for analyzing deformations with alternating sign.

Consider a simple loading.

Due to the initial condition (3.9) for $\mathbf{e}_h$ the second line in (3.8) yields

$$\mathbf{e}_h = 0, \quad |\gamma| \leq h. \tag{3.12}$$

If $|\gamma| > h$, the first line of (3.8), i.e.

$$\frac{h}{v_h} \dot{\mathbf{e}}_h + \mathbf{e}_h = \mathbf{e}_0 \lambda(t) \tag{3.13}$$

must be used.

Solution of the tensor equation (3.13) is sought in the form

$$\mathbf{e}_h = \mathbf{e}_0 \left(A + B\lambda\right), \tag{3.14}$$

where A and B are unknown constants. Then

$$\dot{\mathbf{e}}_h = \mathbf{e}_0 B\dot{\lambda}, \;\; v_h = \left|B\dot{\lambda}\right| \gamma_0 \,. \tag{3.15}$$

Substituting (3.14) and (3.15) into (3.13) yields the following equation for A and B

$$\frac{hB\dot{\lambda}}{\gamma_0 \left|B\dot{\lambda}\right|}\mathbf{e}_0 + \mathbf{e}_0 \left(A + B\lambda\right) = \mathbf{e}_0\lambda. \tag{3.16}$$

From here we find

$$B = 1, \; A = -\frac{h}{\gamma_0}\operatorname{sgn}\dot{\lambda}. \tag{3.17}$$

As pointed out above, the sign of λ coincides with sign of $\dot{\lambda}$ for a simple loading. Hence, an expression for $\mathbf{e}_h$ is

$$\mathbf{e}_h = \mathbf{e}_0\lambda \left(1 - \frac{h}{|\gamma|}\right), \; |\gamma| \geq h. \tag{3.18}$$

For $|\gamma| = h$, the latter expression must give the same value of plastic deformation as (3.12). A direct proof shows that this is the case.

By combining eqs. (3.12) and (3.18) we obtain

$$\mathbf{e}_h = \left[\begin{array}{ll} \mathbf{e}_0\lambda \left(1 - h/\left|\gamma\right|\right), & |\gamma| \geq h, \\ 0, & |\gamma| \leq h. \end{array} \right. \tag{3.19}$$

Substituting these equations into (3.2) we obtain an expression for the stress deviator

$$\mathbf{s} = 2G\mathbf{e}_0\lambda \left[1 - \int\limits_0^{|\gamma|} \left(1 - \frac{h}{|\gamma|}\right) p\left(h\right) dh\right]. \tag{3.20}$$

We introduce now the following scalar value

$$\tau = 2G\gamma_0\lambda \left[1 - \int\limits_0^{|\gamma|} \left(1 - \frac{h}{|\gamma|}\right) p\left(h\right) dh\right]. \tag{3.21}$$

It is easy to understand that its absolute value is equal to the shear stress intensity introduced in Section 1.9 and is evaluated for the stress deviator by means of (1.165).

With the help of definition (3.11), eq. (3.21) may be rewritten as follows

$$\tau = \Phi(\gamma), \tag{3.22}$$

where

$$\Phi(\gamma) = 2G\gamma \left[1 - \int_0^{|\gamma|} \left(1 - \frac{h}{|\gamma|} \right) p(h)\, dh \right]. \tag{3.23}$$

Using (3.21) we transform (3.20) to a simpler form

$$\mathbf{s} = \frac{\tau}{\gamma} \mathbf{e}. \tag{3.24}$$

Equations (3.24) and (3.22) are easily recognized as equations of the deformation theory of plasticity of a hardening material which obeys "the hypothesis of the universal curve", cf. [65] and [79]. It should be noted that these equations are a consequence of the assumed rheological model and condition of simple loading. However, they are not valid for a complex loading.

Let us draw the loading curve in the $\gamma\tau$ plane. By differentiating (3.22) and (3.23) for $\gamma > 0$ we obtain

$$\frac{d\tau}{d\gamma} = 2G \int_\gamma^\infty p(h)\, dh \geq 0, \tag{3.25}$$

$$\frac{d^2\tau}{d\gamma^2} = -2Gp(\gamma) \leq 0. \tag{3.26}$$

These equations show that for $\gamma > 0$ the loading curve is a monotonic convex upwards curve. The loading curve for negative λ is easily plotted because (3.23) is an odd function of γ. A typical loading curve marked by LL is depicted in Fig. 3.1.

We proceed now to consider simple unloading.

Assume that $\lambda = 1$ corresponds to the loading end, i.e. the following values

$$\mathbf{e} = \mathbf{e}_0,\ \gamma = \gamma_0,\ \tau = \tau_0 = \Phi(\gamma_0),$$

$$\mathbf{e}_h = \mathbf{e}_{h0} = \left[\begin{array}{ll} \mathbf{e}_0 (1 - h/\gamma_0), & \gamma_0 \geq h, \\ 0, & \gamma_0 \leq h \end{array} \right. \tag{3.27}$$

are attained. By means of the difference

$$\mathbf{e} - \mathbf{e}_{h0} = \left[\begin{array}{ll} \mathbf{e}_0 (\lambda - 1 + h/\gamma_0), & h \leq \gamma_0, \\ \mathbf{e}_0 \lambda, & h \geq \gamma_0 \end{array} \right. \tag{3.28}$$

we evaluate the following scalar value

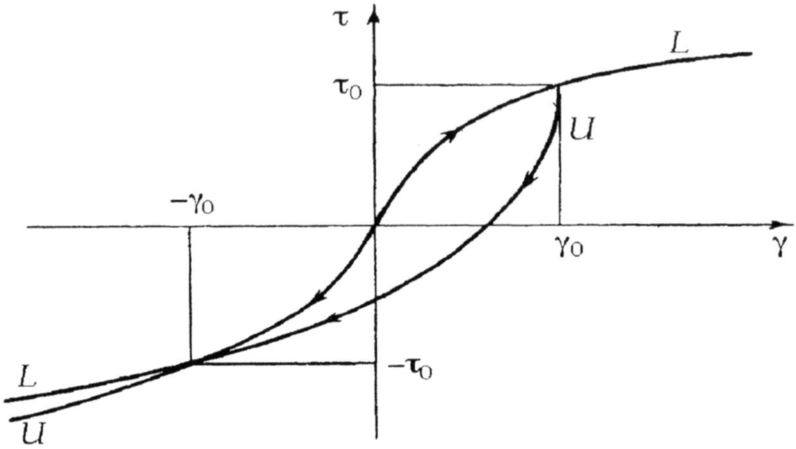

FIGURE 3.1. Simple loading (curve LL) and simple unloading (curve UU).

$$\sqrt{\frac{1}{2}(\mathbf{e}-\mathbf{e}_{h0}):(\mathbf{e}-\mathbf{e}_{h0})} = \left[\begin{array}{ll} \gamma_0 \left|\lambda - 1 + h/\gamma_0\right|, & h \le \gamma_0, \\ \gamma_0 \left|\lambda\right|, & h \ge \gamma_0. \end{array} \right. \tag{3.29}$$

Let

$$-1 \le \lambda \le 1, \tag{3.30}$$

that corresponds to the following interval for γ

$$-\gamma_0 \le \gamma \le \gamma_0. \tag{3.31}$$

By virtue of (3.30) we obtain the following set of inequalities

$$\sqrt{\frac{1}{2}(\mathbf{e}-\mathbf{e}_{h0}):(\mathbf{e}-\mathbf{e}_{h0})} = \left[\begin{array}{ll} \gamma_0 \left|\lambda - 1 + h/\gamma_0\right| \ge h, & 0 \le h \le \gamma_1/2, \\ \gamma_0 \left|\lambda - 1 + h/\gamma_0\right| \le h, & \gamma_1/2 \le h \le \gamma_0, \\ \gamma_0 \left|\lambda\right| \le h, & h \ge \gamma_0, \end{array} \right. \tag{3.32}$$

where

$$\gamma_1 = \gamma_0 (1 - \lambda). \tag{3.33}$$

Inequalities (3.32) indicate that for $h \ge \gamma_1/2$ the second line of (3.8) must be used. Due to initial conditions (3.27) we obtain

$$\mathbf{e}_h = \mathbf{e}_{h0} = \left[\begin{array}{ll} \mathbf{e}_0 (1 - h/\gamma_0), & \gamma_1/2 \le h \le \gamma_0, \\ 0, & h \ge \gamma_0. \end{array} \right. \tag{3.34}$$

If $h < \gamma_1/2$, then the first equation (3.8) must be integrated. Its solution is sought in the form (3.14) with the only difference that now $\dot{\lambda} < 0$. By virtue of (3.14) and (3.17) we have

$$\mathbf{e}_h = \mathbf{e}_0 \left(\lambda + \frac{h}{\gamma_0}\right), \quad 0 \leq h \leq \frac{\gamma_1}{2}. \tag{3.35}$$

For $h = \gamma_1/2$, or because of (3.33), for

$$\lambda = 1 - \frac{2h}{\gamma_0}$$

we obtain

$$\mathbf{e}_h = \mathbf{e}_0 \left(1 - \frac{h}{\gamma_0}\right),$$

as predicted by (3.34).

Summarising,

$$\mathbf{e}_h = \left[\begin{array}{ll} \mathbf{e}_0 \left(\lambda + h/\gamma_0\right), & 0 \leq h \leq \gamma_1/2, \\ \mathbf{e}_0 \left(1 - h/\gamma_0\right), & \gamma_1/2 \leq h \leq \gamma_0, \\ 0, & h \geq \gamma_0. \end{array}\right. \tag{3.36}$$

Substituting (3.36) into (3.2) leads to the following expression for the stress deviator

$$\mathbf{s} = 2G\mathbf{e}_0 \left[\lambda - \int\limits_0^{\gamma_1/2} \left(\lambda + \frac{h}{\gamma_0}\right) p(h)\, dh - \int\limits_{\gamma_1/2}^{\gamma_0} \left(1 - \frac{h}{\gamma_0}\right) p(h)\, dh\right]. \tag{3.37}$$

We introduce now the following parameter

$$\tau = 2G\gamma_0 \left[\lambda - \int\limits_0^{\gamma_1/2} \left(\lambda + \frac{h}{\gamma_0}\right) p(h)\, dh - \int\limits_{\gamma_1/2}^{\gamma_0} \left(1 - \frac{h}{\gamma_0}\right) p(h)\, dh\right] \tag{3.38}$$

and note that its absolute value coincides with the shear stress intensity evaluated by (3.37).

Equation (3.37) may be rewritten in the form

$$\mathbf{s} = \frac{\tau}{\gamma}\mathbf{e}, \tag{3.39}$$

which is conventional in the theory of small elastoplastic deformations.

Transformation of (3.38) leads to the following result

$$\tau = \tau_0 - 2\Phi\left(\frac{\gamma_0 - \gamma}{2}\right), \tag{3.40}$$

which is the Masing principle.

The unloading curve is denoted by UU and depicted in Fig. 3.1. However, it should be mentioned that the unloading follows this line only for γ prescribed by the interval (3.31).

Consider now the question of a subsequent deformation $\gamma < -\gamma_0$. In this case

$$\lambda < -1 \tag{3.41}$$

and (3.29) becomes

$$\sqrt{\frac{1}{2}(\mathbf{e} - \mathbf{e}_{h0}) : (\mathbf{e} - \mathbf{e}_{h0})} = \left[\begin{array}{ll} \gamma_0 \left|\lambda - 1 + h/\gamma_0\right| \geq h, & 0 \leq h \leq \gamma_0, \\ \gamma_0 \left|\lambda\right| \geq h, & \gamma_0 \leq h \leq |\gamma|, \\ \gamma_0 \left|\lambda\right| \leq h, & h \geq |\gamma|. \end{array}\right. \tag{3.42}$$

The above gives the following values of plastic strains

$$\mathbf{e}_h = \left[\begin{array}{ll} \mathbf{e}_0 (\lambda + h/\gamma_0), & 0 \leq h \leq |\gamma|, \\ 0, & h \geq |\gamma|. \end{array}\right. \tag{3.43}$$

With the aid of (3.43) we find

$$\mathbf{s} = 2G\mathbf{e}_0\lambda \left[1 - \int_0^{|\gamma|} \left(1 - \frac{h}{|\gamma|}\right) p(h)\, dh\right]. \tag{3.44}$$

This expression exactly coincides with (3.20) which describes loading for negative γ. Hence, for $\gamma < -\gamma_0$

$$\tau = -\Phi(|\gamma|), \tag{3.45}$$

that is, the deformation follows the loading curve LL, cf. Fig. 3.1. For $\gamma = -\gamma_0$, the value of τ and its two derivatives on the curve (3.45) are given by

$$\tau = -\Phi(\gamma_0), \; \frac{d\tau}{d\gamma} = \Phi'(\gamma_0), \; \frac{d^2\tau}{d\gamma^2} = -\Phi''(\gamma_0), \tag{3.46}$$

where $\Phi'(\gamma)$ and $\Phi''(\gamma)$ denote the first and the second derivatives of the function $\Phi(\gamma)$ with respect to its argument.

For the same value of γ, i.e. $\gamma = -\gamma_0$, we obtain the following values for τ and its derivative for the unloading curve (3.40)

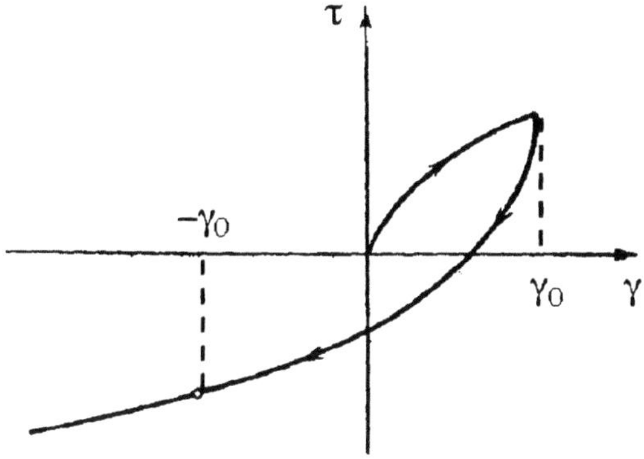

FIGURE 3.2. Diagram of a loading and a subsequent unloading.

$$\tau = -\Phi(\gamma_0), \frac{d\tau}{d\gamma} = \Phi'(\gamma_0), \frac{d^2\tau}{d\gamma^2} = -\frac{1}{2}\Phi''(\gamma_0). \tag{3.47}$$

Comparing (3.46) and (3.47) we come to the conclusion that for $\gamma = -\gamma_0$ the loading and unloading curves have the same slope and intersect, and the curvature of the loading curve (3.45) is twice as much as that of the unloading curve.

Figure 3.2 shows a diagram of loading up to $\gamma = \gamma_0$ and a subsequent unloading into the region $\gamma < -\gamma_0$, which is drawn by means of the above properties of loading and unloading.

3.3 Simple cyclic deformation after a simple deformation

In the previous Section we studied simple loading and simple unloading. For these processes the following equality

$$\mathbf{s} = \frac{\tau}{\gamma}\mathbf{e} \tag{3.48}$$

holds, where for the loading curve LL

$$\tau = \Phi(\gamma), \tag{3.49}$$

and for the unloading curve UU

$$\tau = \tau_0 - 2\Phi\left(\frac{\gamma_0 - \gamma}{2}\right). \tag{3.50}$$

Apparently, equations (3.48), (3.49) and (3.50) hold for an arbitrarily simple deformation provided that τ_0 and γ_0 denote the unloading starting

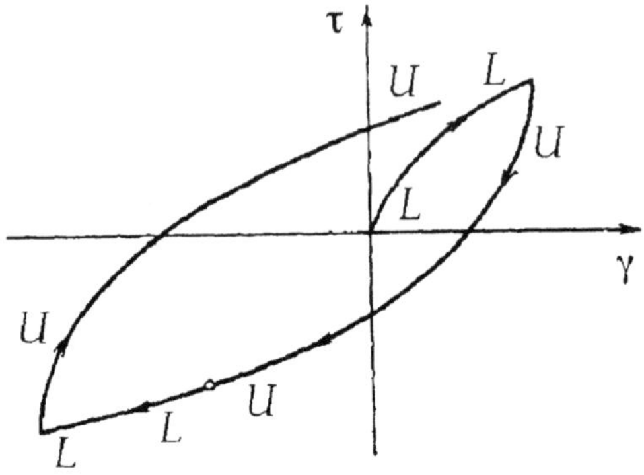

FIGURE 3.3. Arbitrary deformation as a combination of loading and unloading curves.

values of τ and γ, respectively. It follows from this fact that any deformation can be represented as a combination of loading and subsequent unloadings. A typical deformation diagram is shown in Fig. 3.3.

Figure 3.4 shows two types of deformation, namely a simple deformation with a subsequent cyclic deformation (A) and a cyclic deformation (B). The principle of construction of both hysteretic loops is the same, i.e. they consist of unloading curves (3.50). Therefore, the hysteresis loops A and B are absolutely equal. Denoting by $\tau_{cyclic} = \tau_B$ and $\gamma_{cyclic} = \gamma_B$ the values of τ and γ for the cyclic deformation B we may write the following equations for cycle A

$$\tau_A = \tau_{st} + \tau_B, \tag{3.51}$$

$$\gamma_A = \gamma_{st} + \gamma_B, \tag{3.52}$$

with τ_{st} and γ_{st} denoting static values of intensity of shear stresses and shear strains, respectively.

By virtue of (3.52) and (3.49) parameter λ may be represented in the form

$$\lambda_A = \lambda_{st} + \lambda_B. \tag{3.53}$$

Then, (3.10) yields

$$\mathbf{e}_A = \mathbf{e}_{st} + \mathbf{e}_B. \tag{3.54}$$

It remains to transform (3.48) for cycle A. Equations (3.9), (3.10) and (3.48) give

$$\mathbf{s}_A = \tau_A \frac{\mathbf{e}_A}{\gamma_A} = \tau_A \frac{\mathbf{e}_0}{\gamma_0}. \tag{3.55}$$

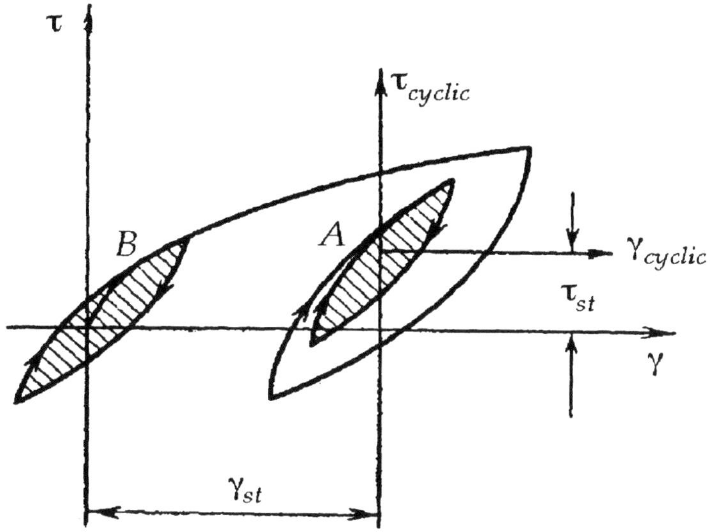

FIGURE 3.4. Cyclic deformation (B) and a simple deformation with a subsequent cyclic deformation (A).

Substituting for τ_A from (3.51) into the latter equation and simplifying, we obtain

$$\mathbf{s}_A = \mathbf{s}_{st} + \mathbf{s}_B, \tag{3.56}$$

where

$$\mathbf{s}_{st} = \tau_{st}\frac{\mathbf{e}_0}{\gamma_0} = \tau_{st}\frac{\mathbf{e}_{st}}{\gamma_{st}}, \tag{3.57}$$

$$\mathbf{s}_B = \tau_B\frac{\mathbf{e}_0}{\gamma_0} = \tau_B\frac{\mathbf{e}_B}{\gamma_B}. \tag{3.58}$$

It follows from (3.54) and (3.56) that the material behaviour under a simple cyclic deformation does not depend upon superposition of an additional static load, the latter being attained by a simple deformation.

3.4 The method of harmonic linearisation for tensor equations

In the applied theory of vibration, it is difficult to work with the equations for elastoplastic materials because they are nonlinear. Exact solutions are obtained only for some simple mechanical systems and for a few simple hysteresis loops. One has to use the approaches of nonlinear mechanics in order to obtain an effective solution of this challenging problem. The method of harmonic linearisation, e.g. [16] and [149], appears to be the most convenient for the analysis of single-frequency vibrational processes and can be applied for scalar and vector nonlinear equations. However, in continuum mechanics, the relationship between the variables is of tensor character. Application of the method of harmonic linearisation to tensor equations is quite specific. Below we explain the method of harmonic linearisation for a nonlinear isotropic tensor operator

$$\mathbf{z}(t) = \mathbf{z}\{\mathbf{x}(\tau)\}, \ t_0 < \tau < t, \tag{3.59}$$

where $\mathbf{z}$ is an isotropic tensor operator depending on the time-history of $\mathbf{x}$. In what follows, we assume that $\mathbf{z}$ and $\mathbf{x}$ are symmetric tensors.

Tensor relations of type (3.59) occur frequently in the mechanics of continuous media and appear for constitutive equations linking stress and strain in isotropic materials. The equations for the Ishlinsky elastoplastic material, see Section 3.1, are a particular case of (3.59). In Section 3.1 the deviators are linked and we will assume for simplicity that $\mathbf{z}$ and $\mathbf{x}$ in (3.59) are deviators, i.e. tensors with zero first main invariants.

In accordance with the method of harmonic linearisation we assume a harmonic law for $\mathbf{x}$

$$\mathbf{x} = \mathbf{x}_0 + \mathbf{x}_1, \tag{3.60}$$

where $\mathbf{x}_0$ is a tensor with constant components, i.e. $\mathbf{x}_0 = \text{const}$ and $\mathbf{x}_1$ is a tensor with vibrating components, i.e. in a fixed orthogonal coordinate system its components are given by

$$(\mathbf{x}_1)_{ij} = \mathbf{X}_{ij} \cos\left(\omega t - \varphi_{ij}\right). \tag{3.61}$$

Here $\mathbf{X}_{ij}$ are components of the amplitude tensor $\mathbf{X}$, φ_{ij} are phases and ω is frequency.

We approximate a nonlinear isotropic tensor operator by a linear isotropic tensor function of $\mathbf{x}_0$, $\mathbf{x}_1$ and $\dot{\mathbf{x}}_1$, i.e.

$$\mathbf{z} \approx p\mathbf{x}_0 + q\mathbf{x}_1 + r\dot{\mathbf{x}}_1, \tag{3.62}$$

where p, q and r are scalar constant factors.

It is worth noting that there is no need for tensor $\mathbf{E}s$ to appear in approximation (3.62) since only deviators were assumed to be considered and tensor $\mathbf{E}s$ has a nonzero first invariant.

The linearisation factors p, q and r are obtained from condition of minimum of the mean square of the error provided by approximation (3.62)

$$R = \int_0^T (\mathbf{z} - p\mathbf{x}_0 - q\mathbf{x}_1 - r\dot{\mathbf{x}}_1) : (\mathbf{z} - p\mathbf{x}_0 - q\mathbf{x}_1 - r\dot{\mathbf{x}}_1)\, dt = \min \tag{3.63}$$

under the assumption of harmonic motion (3.60) and (3.61).

The condition of minimising R with respect to the parameters p, q and r yields

$$\begin{aligned} \frac{\partial R}{\partial p} &= 0, \\ \frac{\partial R}{\partial q} &= 0, \\ \frac{\partial R}{\partial r} &= 0. \end{aligned} \tag{3.64}$$

Solving these equations for p, q and r we obtain

$$p = \frac{\int_0^T \mathbf{z}\left[\mathbf{x}(\tau)\right] : \mathbf{x}_0 dt}{\int_0^T \mathbf{x}_0 : \mathbf{x}_0 dt}, \tag{3.65}$$

$$q = \frac{\int_0^T \mathbf{z}\left[\mathbf{x}(\tau)\right] : \mathbf{x}_1 dt}{\int_0^T \mathbf{x}_1 : \mathbf{x}_1 dt}, \tag{3.66}$$

$$r = \frac{\int_0^T \mathbf{z}\left[\mathbf{x}(\tau)\right] : \dot{\mathbf{x}}_1 dt}{\int_0^T \dot{\mathbf{x}}_1 : \dot{\mathbf{x}}_1 dt}. \tag{3.67}$$

When evaluating the latter integrals the argument of $\mathbf{z}$ is given by (3.60) and (3.61).

Note that the linearisation factors depend upon the components of the constant tensor $\mathbf{x}_0$, the components X_{ij} of the amplitude tensor $\mathbf{x}_1$, phases φ_{ij} and frequency ω.

3.5 Harmonic linearisation of the equation for the elastoplastic material

In the present Section we consider a simple deformation due to a harmonic law

$$\mathbf{e} = \mathbf{E}\cos\left(\omega t - \varphi\right), \tag{3.68}$$

where $\mathbf{E}$ denotes an amplitude tensor, ω is frequency and φ is phase.

When dealing with the theory of internal friction one is mainly interested in obtaining the stress deviator components of the same frequency ω. With this in view, the most appropriate approaches are the methods of nonlinear mechanics [16] and, in particular, the method of harmonic linearisation [149], see Section 3.4.

The constitutive equations for the elastoplastic material are given by eqs. (3.2) and (3.3). Let $\dot{\mathbf{e}}_h \neq 0$, then the only nonlinearity is the following isotropic tensor function

$$\dot{\mathbf{e}}_h / v_h, \tag{3.69}$$

where

$$v_h = \sqrt{\frac{1}{2}\dot{\mathbf{e}}_h \colon \dot{\mathbf{e}}_h}.$$

According to the method of harmonic linearisation we assume that the plastic strain tensor $\mathbf{e}_h$ (by analogy with $\mathbf{e}$) obeys a harmonic law

$$\mathbf{e}_h = \mathbf{E}_h \cos\left(\omega t - \varphi_h\right). \tag{3.70}$$

We approximate the nonlinear isotropic tensor function (3.69) by a linear isotropic tensor function

$$\frac{\dot{\mathbf{e}}_h}{v_h} \approx k_h \dot{\mathbf{e}}_h, \tag{3.71}$$

where k_h is a scalar constant.

The linearisation factor k_h is obtained by means of formula (3.66)

$$k_h = \frac{\int_0^T v_h^{-1} \dot{\mathbf{e}}_h : \dot{\mathbf{e}}_h dt}{\int_0^T \dot{\mathbf{e}}_h : \dot{\mathbf{e}}_h dt} = \frac{\int_0^T v_h dt}{\int_0^T v_h^2 dt}. \tag{3.72}$$

By virtue of (3.70)

$$v_h^2 = \omega^2 \Gamma_h^2 \sin^2\left(\omega t - \varphi_h\right), \tag{3.73}$$

with

$$\Gamma_h = \sqrt{\frac{1}{2}\mathbf{E}_h \colon \mathbf{E}_h} \tag{3.74}$$

denoting the time-maximum value of the shear strain intensity.

Substituting (3.73) into (3.72) and evaluating the integrals leads to the following expression for the linearisation factor

$$k_h = \frac{n}{\omega \Gamma_h}, \tag{3.75}$$

where

$$n = 4/\pi. \tag{3.76}$$

Substituting (3.75) into (3.71) and then into the system (3.2) and (3.5) we obtain a linearised variant of the constitutive equations

$$\begin{aligned} &\dot{\mathbf{e}}_h nh/\omega\Gamma_h + \mathbf{e}_h = \mathbf{e}, \\ &\mathbf{s} = 2G\left[\mathbf{e} - \int_0^\infty \mathbf{e}_h p(h)\, dh\right]. \end{aligned} \tag{3.77}$$

Linearity of these equations allows the variables to be written in a general complex exponential form

$$\mathbf{e} = \mathbf{E}e^{i(\omega t - \varphi)}, \mathbf{e}_h = \mathbf{E}_h e^{i(\omega t - \varphi_h)}, \mathbf{s} = \mathbf{S}e^{i(\omega t - \psi)}, \tag{3.78}$$

where $\mathbf{E}$, $\mathbf{E}_h$ and $\mathbf{S}$ are real amplitude tensors and ω is frequency, $\omega > 0$. It should be mentioned that only real parts of (3.78) have physical meaning. The amplitude values of the intensities are easily expressed over the complex values as follows

$$T = \sqrt{\frac{1}{2}\mathbf{s} : \mathbf{s}^*}, \tag{3.79}$$

$$\Gamma_h = \sqrt{\frac{1}{2}\mathbf{e}_h : \mathbf{e}_h^*}, \tag{3.80}$$

$$\Gamma = \sqrt{\frac{1}{2}\mathbf{e} : \mathbf{e}^*}, \tag{3.81}$$

where the asterisk denotes the complex conjugate. The amplitude of the shear strain intensity is introduced by (3.81) and is positive. Substituting (3.78) into the first equation of (3.77) we obtain

$$\mathbf{e} = \mathbf{e}_h\left(1 + inh/\Gamma_h\right). \tag{3.82}$$

Evaluating for the amplitude value of the shear strain intensity by means of (3.81) and (3.82) we arrive at an equation for Γ_h

$$\Gamma^2 = \Gamma_h^2\left[1 + \left(nh/\Gamma_h\right)^2\right],$$

from which we find

$$\Gamma_h = \sqrt{\Gamma^2 - (nh)^2}. \tag{3.83}$$

Substituting this into (3.82) and solving for $\mathbf{e}_h$ yields

$$\mathbf{e}_h = \mathbf{e}\left[1 - \left(\frac{nh}{\Gamma}\right)^2 - i\frac{nh}{\Gamma}\sqrt{1 - \left(\frac{nh}{\Gamma}\right)^2}\right]. \tag{3.84}$$

This solution is valid only for $\Gamma > nh$ since only under this assumption does a real value of amplitude (3.83) exist. If $\Gamma < nh$ then, due to Section 3.1, we must take eq. (3.7) from which under the trivial initial conditions, we obtain $\mathbf{e}_h = 0$. Thus

$$\mathbf{e}_h = \left[\begin{array}{ll} \mathbf{e}\left[1 - (nh/\Gamma)^2 - inh/\Gamma\sqrt{1 - (nh/\Gamma)^2}\right], & h \leq \Gamma/n, \\ 0, & h > \Gamma/n. \end{array}\right. \tag{3.85}$$

Substituting this into the second equation (3.77) yields the following equation for the stress deviator

$$\mathbf{s} = 2G_c\mathbf{e}, \tag{3.86}$$

where

$$G_c = G\left[1 - \int_0^1 \left(1 - \eta^2 - i\eta\sqrt{1 - \eta^2}\right) p\left(\frac{\Gamma}{n}\eta\right)\frac{\Gamma}{n}d\eta\right] \tag{3.87}$$

is referred to as a *complex shear modulus.*

Formula (3.87) generalises eq. (2.26) for the three-dimensional case, in which the strain amplitude is now replaced by the amplitude of shear strain intensity. This is quite natural since it was the model of the elastoplastic material which was chosen to describe the internal friction within the material. As above, G_c turns out to be independent of the deformation frequency.

Distribution (2.28) was shown in Section 2.4 to be of crucial importance for the theory of internal friction. Substituting probability density $p(h)$ due to (2.28) into (3.87) we come to the following expression for the complex shear modulus

$$G_c = G\left(1 - r\Gamma^\alpha + ig\Gamma^\alpha\right), \tag{3.88}$$

where r and g, by analogy with (2.30) are given by

$$r = \frac{2H}{2+\alpha}n^{-\alpha}, \quad g = \frac{\alpha H}{2}n^{-\alpha}\mathrm{B}\left(\frac{\alpha+1}{2}, \frac{3}{2}\right), \quad \omega > 0. \tag{3.89}$$

As in eq. (2.29) a power dependence of the energy dissipation factor $g\Gamma^\alpha$ on the amplitude of shear strain intensity Γ is obtained.

3.6 The correspondence principle

It turns out that in the theory of internal friction one can formulate a principle which is analogous to the known correspondence principle in the theory of viscoelasticity. Strictly speaking, the correspondence principle for elastoplasticity is less general and is analogous with elasticity theory only in regard to the equations' derivation, but not their solution.

Let us conclude the results of this Chapter so far.

Equation (3.1) and the set of equations (3.2) and (3.3) were taken as constitutive equations for the material. Equation (3.1) characterises the material behaviour under volumetric deformations, whilst eqs. (3.2) and (3.3) describe the material behaviour under shape distortion. By means of the method of harmonic linearisation and complex analysis, the latter two equations were transformed to the form (3.86)

$$\mathbf{s} = 2G_c\mathbf{e}, \tag{3.90}$$

with G_c being a complex shear modulus given by (3.87) or (3.88).

Equation (3.1) is linear, that is, there is no need to linearise it. The complex form of this equation

$$\sigma = k\vartheta \tag{3.91}$$

does not change this.

Formally, the constitutive equations (3.90) and (3.91) coincide with the corresponding equations of the linear theory of elasticity [99]. The only difference is the complex shear modulus. Thus, Hooke's law [99] for complex moduli is valid in terms of k and G_c. We only write down expressions for the complex Young's modulus and the complex Poisson's ratio, cf. [99]

$$\frac{1}{E_c} = \frac{1}{3G_c} + \frac{1}{9k}, \; \frac{\nu_c}{E_c} = \frac{1}{6G_c} - \frac{1}{9k}, \tag{3.92}$$

as they are very popular in elasticity theory.

To simplify the future transformations, we rewrite the complex shear modulus in the form

$$G_c = G\,(1+\beta)\,, \tag{3.93}$$

where due to (3.87)

$$\beta\,(\Gamma) = -\int\limits_0^1 \left(1-\eta^2 - i\eta\sqrt{1-\eta^2}\right) p\left(\frac{\Gamma}{n}\eta\right)\frac{\Gamma}{n}d\eta. \tag{3.94}$$

Substituting (3.93) into (3.92) and simplifying yields

$$E_c = \frac{E(1+\beta)}{1+\dfrac{1-2\nu}{3}\beta}, \tag{3.95}$$

$$\nu_c = \frac{\nu - \dfrac{1-2\nu}{3}\beta}{1+\dfrac{1-2\nu}{3}\beta}; \tag{3.96}$$

with E and ν denoting Young's modulus and Poisson's ratio in the classical theory of elasticity, respectively.

In these equations, factor β characterises the effect of plastic deformations. Assuming that this effect is small, i.e.

$$|\beta| \ll 1, \tag{3.97}$$

and retaining terms of the first order in β only, we obtain from (3.95) and (3.96)

$$E_c = E\left[1+\frac{2}{3}(1+\nu)\beta\right], \tag{3.98}$$

$$\nu_c = \nu - \frac{1-2\nu}{3}(1+\nu)\beta. \tag{3.99}$$

By virtue of (3.94) β depends on the amplitude value of shear stress intensity Γ

$$\beta = \beta(\Gamma). \tag{3.100}$$

Sometimes it is more convenient to have a dependence of the complex moduli on the amplitude of the shear stress intensity T. In order to derive this equation, we take (3.90) from which, due to (3.79) and (3.93), we obtain

$$T = 2G\Gamma|1+\beta|. \tag{3.101}$$

This simplifies by means of (3.97) and becomes

$$T = 2G\Gamma. \tag{3.102}$$

There is no necessity to retain terms of first order in β since this would cause the appearance of terms of higher order than those retained in eqs. (3.93), (3.98) and (3.99).

By means of (3.102), eq. (3.100) may be rewritten in the following form

$$\beta = \beta\left(\frac{T}{2G}\right). \tag{3.103}$$

Distribution (2.28) is important for the theory of internal friction in materials. In this particular case, due to (3.88)

$$\beta = l\Gamma^{\alpha} \tag{3.104}$$

or

$$\beta = l\left(\frac{T}{2G}\right)^{\alpha}, \tag{3.105}$$

where

$$l = -r + ig. \tag{3.106}$$

Concluding the Section we formulate a simple way of obtaining a boundary value problem for elastoplastic bodies in case of single-frequency vibrations. To this end, it is sufficient to write down the differential equations and the boundary conditions for the dynamic theory of elasticity, replace their moduli by their complex values and transform to a general complex exponential form. We obtain then the following equations

$$\begin{aligned} &\nabla \cdot \boldsymbol{\tau} + \rho\left(\mathbf{K} - \ddot{\mathbf{u}}\right) = 0, \\ &\mathbf{s} = 2G_c\mathbf{e}, \ \boldsymbol{\tau} = \sigma\mathbf{E} + \mathbf{s}, \\ &\sigma = k\vartheta, \ \boldsymbol{\varepsilon} = \tfrac{\vartheta}{3}\mathbf{E} + \mathbf{e}, \\ &\boldsymbol{\varepsilon} = \left(\nabla\mathbf{u}\right)^s \end{aligned} \tag{3.107}$$

and the boundary conditions

$$\begin{aligned} &\text{on}\, O_1, \quad \mathbf{u} = \mathbf{U}, \\ &\text{on}\, O_2, \quad \boldsymbol{\tau}_n = \mathbf{p}, \end{aligned} \tag{3.108}$$

where O_1 and O_2 are parts of the body surface O ($O = O_1 + O_2$).

This is the correspondence principle.

Let us point out some difficulties associated with solving such a boundary value problem, eqs. (3.107) and (3.108). The main complication is that the shear modulus G_c depends on the value of the shear stress intensity T and the latter depending upon the spatial coordinate, the dependence of which is unknown *a priori*. Therefore, in order to solve the boundary value problem, eqs. (3.107) and (3.108), it is necessary to have a solution of the dynamic problem of elasticity theory for arbitrary heterogeneous body. This solution does not exist in general case. A considerable amount of simplification can be achieved when the stress state can be sufficiently accurately approximated by a homogeneous one. A series of simple problem of this sort will be considered in the next Section. Some simplification can be achieved in the problem of uniaxial wave propagation which will be studied below. However, the overwhelming majority of applied problems requires application of the Galerkin method which will be considered in Section 4.1.

3.7 Energy dissipation in materials in some particular stress states

3.7.1 Stress state of pure shear

Consider the stress state of a homogeneous pure shear, say in plane xy. Under cyclic deformation with a shear stress intensity T_{xy} we have

$$T = T_{xy}, \tag{3.109}$$

and hence

$$G_c = G\left[1 + \beta\left(\frac{T_{xy}}{2G}\right)\right]. \tag{3.110}$$

It is worthwhile remembering that the complex shear modulus (3.110) links the complex stress τ_{xy} and the complex shear γ_{xy} as follows

$$\tau_{xy} = G_c\gamma_{xy}. \tag{3.111}$$

A stress state of pure shear occurs in a thin-walled circular cantilever tube with a massive disc attached at its free end. Neglecting the tube mass we obtain a differential equation for free vibrations of the disc

$$I\ddot{\varphi} + \frac{FR^2}{L}G_c\varphi = 0, \tag{3.112}$$

where I is the moment of inertia of the disc, F is the cross-sectional area of the tube, L is its length, R is its radius and φ is an angle of rotation of the disc.

Equation (3.112) is analogous to eq. (2.52). By means of this analogy, it is easy to determine the decrement of torsional vibration

$$\Delta = \pi \operatorname{Im}\frac{G_c}{G} = D(T), \tag{3.113}$$

where

$$D(T) = \pi \operatorname{Im}(G_c/G). \tag{3.114}$$

3.7.2 Uniaxial stress state

In the case of an uniaxial homogeneous stress state with the amplitude of the normal stress, given by Σ_x, we have

$$T = \frac{\Sigma_x}{\sqrt{3}}, \tag{3.115}$$

and therefore

$$E_c = E\left[1 + \frac{2}{3}(1+\nu)\beta\left(\frac{T}{2G}\right)\right] = E\left[1 + \frac{2}{3}(1+\nu)\beta\left(\frac{\Sigma_x}{2G\sqrt{3}}\right)\right]. \tag{3.116}$$

This state of stress occurs in the case of a longitudinal vibration of a massive weight attached to a weightless rod. This problem has been already studied in Sections 2.5 and 2.6. The decrement of longitudinal vibration is

$$\Delta = \pi \operatorname{Im} \frac{E_c}{E} = \frac{2}{3}(1+\nu) D(T). \tag{3.117}$$

A comparison of (3.117) and (3.113) shows that for the same T the decrement of torsional vibration is greater than that of longitudinal vibration.

A homogeneous uniaxial state of stress takes place for example under axisymmetric vibration of a circular ring. So, expression (3.117) is relevant to this problem.

3.7.3 Isotropic biaxial stress state

Let us proceed to a homogeneous isotropic biaxial state of stress. The stress tensor is

$$\boldsymbol{\tau} = (\mathbf{ii} + \mathbf{jj})\,\sigma_x, \; \sigma_x = \Sigma_x e^{i(\omega t - \psi)}, \tag{3.118}$$

where $\mathbf{i}$ and $\mathbf{j}$ are unit base vectors of axes x and y, respectively, and Σ_x is the stress amplitude.

Equations (3.118) and (3.79) yield

$$T = \frac{\Sigma_x}{\sqrt{3}}. \tag{3.119}$$

Let us link complex stresses and complex strains. A relation is given by equations similar to those in elasticity theory under the condition that $\sigma_y = \sigma_x$

$$\begin{aligned} \varepsilon_x &= \frac{1}{E_c}(\sigma_x - \nu_c \sigma_x), \\ \varepsilon_y &= \frac{1}{E_c}(\sigma_x - \nu_c \sigma_x). \end{aligned} \tag{3.120}$$

The strains are seen to equal each other

$$\varepsilon_x = \varepsilon_y,$$

and are determined by

$$\sigma_x = \frac{E_c}{1-\nu_c}\varepsilon_x . \tag{3.121}$$

By virtue of (3.98) and (3.99) the following equation

$$\frac{E_c}{1-\nu_c} = \frac{E}{1-\nu}\left[1+\frac{(1+\nu)}{3(1-\nu)}\beta\left(\frac{T}{2G}\right)\right] \tag{3.122}$$

holds asymptotically for small β.

The state of stress under consideration takes place in a thin-walled spherical shell under polar-symmetrical vibrations. Due to the correspondence principle its equation is as follows

$$\frac{\rho R^2}{2}\ddot{u} + \frac{E_c}{1-\nu_c}u = 0, \tag{3.123}$$

where u is the radial displacement, ρ is the mass density of the shell and R is its radius.

Equation (3.123) yields the following decrement of the shell vibration

$$\Delta = \frac{(1+\nu)}{3(1-\nu)}D(T) . \tag{3.124}$$

It is seen to be lower than that for torsional and longitudinal vibrations.

3.7.4 *Deformation without shape distortion*

Finally, note the fact that the energy dissipation is absent at any deformation without shape distortion. This discredits the proposed theory to some extent. Indeed, in this case $\Gamma = T = 0$, and due to (3.94), (3.93), (3.98) and (3.99) all moduli turn out to be equal to those in the theory of ideally elastic bodies.

4

Single-frequency vibrations in elastoplastic bodies

4.1 The Galerkin method in problems of vibrations of elastoplastic bodies

In what follows we offer an approach which is a direct application of the Galerkin method in the form of a single term approximation to the system of equations for an elastoplastic body with the subsequent use of the method of harmonic linearisation. The vibration mode of the elastic body is assumed to be given. This assumption is crucial and is generally accepted in the analysis of nonlinear systems with distributed parameters.

Consider the body depicted in Fig. 4.1. The dynamics of this body are

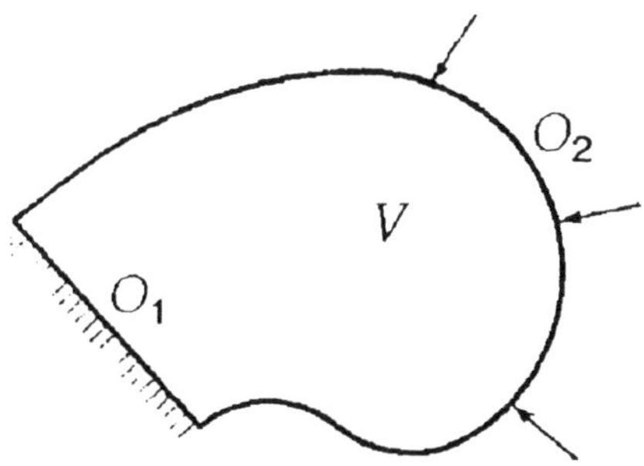

FIGURE 4.1. Schematics of an elastoplastic body under consideration.

governed by equation (1.310)

$$\nabla \cdot \boldsymbol{\tau} + \rho\left(\mathbf{K} - \ddot{\mathbf{u}}\right) = 0 \tag{4.1}$$

which holds for the volume V occupied by the body.

Let part of the body surface O_1 be fixed, i.e. we have the boundary condition

$$\mathbf{u} = 0. \tag{4.2}$$

Assume that the rest of the body surface O_2 experiences an external surface load $\mathbf{p}$, such that

$$\mathbf{p} - \boldsymbol{\tau}_n = 0, \ \boldsymbol{\tau}_n = \mathbf{n} \cdot \boldsymbol{\tau}. \tag{4.3}$$

Finally, note that the strain is due to (1.284)

$$\boldsymbol{\varepsilon} = \left(\nabla \mathbf{u}\right)^s, \tag{4.4}$$

whilst the stresses are related to strains by means of a complex chain of equations (1.58), (1.285), and (3.1), (3.2) and (3.3). An exact form of this relation is not essential at the moment, for this reason we write it in an operational form

$$\boldsymbol{\tau} = \boldsymbol{\tau}\left\{\boldsymbol{\varepsilon}\right\}, \tag{4.5}$$

with $\boldsymbol{\tau}\left\{\boldsymbol{\varepsilon}\right\}$ being a tensor operator of $\boldsymbol{\varepsilon}$.

If we substitute (4.4) into (4.5) and put the result in (4.1) and (4.3), we obtain the only vector equation for the vector field $\mathbf{u}$ and the boundary conditions (4.2) and (4.3). These correspond to the following variational equation

$$\int\limits_V \left(\nabla \cdot \boldsymbol{\tau} + \rho\mathbf{K} - \rho\ddot{\mathbf{u}}\right) \cdot \delta\mathbf{u} dV + \int\limits_O \left(\mathbf{p} - \boldsymbol{\tau}_n\right) \cdot \delta\mathbf{u} dO = 0, \tag{4.6}$$

where $\delta\mathbf{u}$ is a variation of $\mathbf{u}$ which is arbitrary in the volume V and on the surface O_2 and vanishes on the surface O_1, cf. (4.2).

By means of the formula

$$\left(\nabla \cdot \boldsymbol{\tau}\right) \cdot \delta\mathbf{u} = \boldsymbol{\nabla} \cdot \left(\boldsymbol{\tau} \cdot \delta\mathbf{u}\right) - \boldsymbol{\tau} : \delta\boldsymbol{\varepsilon} \tag{4.7}$$

and the Ostrogradsky-Gauss integral theorem

$$\int\limits_V \nabla \cdot \mathbf{b} dV = \int\limits_O \mathbf{n} \cdot \mathbf{b} dO, \tag{4.8}$$

with $\mathbf{n}$ being a unit vector normal and $\mathbf{b}$ being a vector field, we transform the variational equation (4.6) to the form

$$\int\limits_{O_2} \mathbf{p} \cdot \delta\mathbf{u} dO + \int\limits_V \rho\mathbf{K} \cdot \delta\mathbf{u} dV - \int\limits_V \left(\boldsymbol{\tau} : \delta\boldsymbol{\varepsilon} + \rho\ddot{\mathbf{u}} \cdot \delta\mathbf{u}\right) dV = 0. \tag{4.9}$$

This equation is of a convenient form for the forthcoming analysis.

Confining ourselves to the analysis of resonant and near-resonant vibrations we look for a solution of the variational equation in the form

$$\mathbf{u} = \mathbf{u}_0 q\left(t\right), \tag{4.10}$$

where $q(t)$ stands for an unknown function of time and $\mathbf{u}_0$ denotes a normal mode of the elastic vibration in the body, the mode being obtained by ignoring the plastic properties of the body material.

Considering only those variations which belong to the class of functions (4.10) yields

$$\delta\mathbf{u} = \mathbf{u}_0\delta q,\ \delta\boldsymbol{\varepsilon} = \boldsymbol{\varepsilon}_0\delta q,\ \boldsymbol{\varepsilon}_0 = \left(\nabla\mathbf{u}_0\right)^s. \tag{4.11}$$

Substituting them into (4.9) and taking into account that δq is arbitrary, we obtain the following equation for q

$$\int\limits_{O_2} \mathbf{p}\cdot\mathbf{u}_0 dO + \int\limits_V \left(\rho\mathbf{K}\cdot\mathbf{u}_0 - \boldsymbol{\tau} : \boldsymbol{\varepsilon}_0 - \rho\ddot{\mathbf{u}}\cdot\mathbf{u}_0\right) dV = 0. \tag{4.12}$$

This equation contains an acceleration term which is linear in the unknown function $q(t)$

$$\ddot{\mathbf{u}} = \mathbf{u}_0\ddot{q} \tag{4.13}$$

and a stress tensor $\boldsymbol{\tau}$ which in nonlinear in q. For this reason, equation (4.12) is nonlinear in q.

The method used here to reduce partial differential equations to ordinary differential equations is a standard tool of nonlinear mechanics, see [108]. In mathematical physics this technique is called the Kantorovich approach [76].

Let us proceed now to (4.12). Restricting ourselves to consideration of harmonic or near-harmonic motions we perform a harmonic linearisation of the nonlinearities. The only nonlinearity is the dependence $\boldsymbol{\tau}\left\{\boldsymbol{\varepsilon}\right\}$ which has already been linearised in Section 3.5. By means of the general complex exponential form for the variables we obtain due to (3.90) and (3.91)

$$\sigma = k\vartheta,\ \mathbf{s} = 2G_c\mathbf{e}, \tag{4.14}$$

where G_c denotes a complex shear modulus.

Decomposition (1.58) renders

$$\boldsymbol{\tau} = \mathbf{E}k\vartheta + 2G_c\mathbf{e}. \tag{4.15}$$

Accounting of $\mathbf{u}$, due to (4.10), we may write

$$\mathbf{e} = \mathbf{e}_0 q, \vartheta = \vartheta_0 q. \tag{4.16}$$

Substituting (4.16) into (4.15), then inserting the result in (4.12) we obtain a linearised equation for q

$$m\ddot{q} + c_c q = p = p_0 e^{i\omega t}. \tag{4.17}$$

The following notation is introduced

$$p = \int_{O_2} \mathbf{p} \cdot \mathbf{u}_0 dO + \int_V \rho \mathbf{K} \cdot \mathbf{u}_0 dV, \tag{4.18}$$

$$m = \int_V \rho \mathbf{u}_0^2 dV, \tag{4.19}$$

$$c_c = \int_V \left(k\vartheta_0^2 + 4G_c\Gamma_0^2\right) dV, \tag{4.20}$$

where

$$\Gamma_0 = \sqrt{\frac{1}{2}\mathbf{e}_0 : \mathbf{e}_0}. \tag{4.21}$$

Factors m, c_c and p are the "equivalent" mass, complex rigidity and external force, respectively. It is easy to see that the form of eq. (4.17) coincides with that of eq. (2.33). Thus, the solution strategy is the same as in Sections 2.5 and 2.6.

Let us consider the structure of the expression for the complex rigidity c_c. It contains the complex shear modulus which depends either on the amplitude value of shear strain intensity or shear stress intensity. The latter are given by

$$\Gamma = \Gamma_0 a, \; T = T_0 a, \tag{4.22}$$

where a is the amplitude of q.

Substituting (3.93) due to (3.103) into (4.20) and eliminating the kinematic parameters yields

$$c_c = \int_V \left[\frac{\sigma_0^2}{k} + \frac{T_0^2}{G} + \frac{T_0^2}{G}\beta\left(\frac{T_0 a}{2G}\right)\right] dV. \tag{4.23}$$

For a power dependence β due to (3.105), we have

$$c_c = \int_V \left[\frac{\sigma_0^2}{k} + \frac{T_0^2}{G} + 4Gla^\alpha\left(\frac{T_0}{2G}\right)^{2+\alpha}\right] dV. \tag{4.24}$$

Let us now make use of the analogy between eqs. (4.17) and (2.33). Assuming, as in Section 2.5, that

$$c_c = u + iv, \tag{4.25}$$

we find an equation for the amplitude-frequency response function

$$a^2 = \frac{p_0^2}{\left(u - m\omega^2\right)^2 + v^2}, \tag{4.26}$$

an equation for the envelope of free decaying vibration (cf. Section 2.6)

$$\dot{a} = -\frac{av(a)}{2\sqrt{cm}} \tag{4.27}$$

and a closed form expression for the vibration decrement

$$\Delta = \pi \frac{v}{c}. \tag{4.28}$$

Here

$$c = \int_V \left[\frac{\sigma_0^2}{k} + \frac{T_0^2}{G}\right] dV, \tag{4.29}$$

represents rigidity in the case of pure elastic vibrations.

Multiplying the numerator and the denominator of (4.28) by a^2 and using (3.114) yields the following equation for the decrement, cf. [129]

$$\Delta = \left\{ \int_V \frac{1}{G} \left[\frac{3(1-2\nu)}{2(1+\nu)}\Sigma^2 + T^2\right] dV \right\}^{-1} \int_V \left[\frac{T^2}{G} D(T)\right] dV. \tag{4.30}$$

Here ν is the Poisson ratio and

$$\Sigma = \sigma_0 a \tag{4.31}$$

stands for the amplitude of the mean normal stress.

Note that the assumption of the uniformity of the stress state has not been used. Thus, the equations obtained are suitable for vibration analysis both in homogeneous and heterogeneous bodies.

A very simple equation for vibration decrement is obtained in case of a homogeneous deformable volume with a uniform stress state throughout. The volume integrals are easily evaluated, to give

$$\Delta = \left[1 + \frac{3(1-2\nu)}{2(1+\nu)} \frac{\Sigma^2}{T^2}\right]^{-1} D(T). \tag{4.32}$$

It follows from (4.32) that the vibration decrement maximum is achieved when $\Sigma = 0$, dependence of Δ on T being similar for various states of stress.

The opinion that the material layers near the body surface contribute more to the decrement than the internal layers is expressed in the literature. An explanation of this phenomenon is that the plastic properties of the surface layers differ considerably from those of the body interior. This difference may be caused by additional action on the surface layers caused by mechanical and thermal treatment. Equation (4.30) may also be used in this case. Assume that the plastic properties of the body material are inhomogeneous such that a thin layer near the surface of thickness t is described by a function $D_1(T)$, while the rest of the body is described by a function $D(T)$. Assuming that t is very small and neglecting variation of the stress state within the surface layer we arrive at the following equation for the vibration decrement

$$\Delta = \frac{\int_V \frac{T^2}{G} D(T)\, dV + t \int_O \frac{T^2}{G} \left[D_1(T) - D(T)\right] dO}{\int_V \frac{1}{G} \left[\frac{3(1-2\nu)}{2(1+\nu)} \Sigma^2 + T^2\right] dV}. \tag{4.33}$$

In the limiting case in which the interior of the material does not contribute to the energy dissipation, i.e. $D(T) = 0$, equation (4.33) takes the form

$$\Delta = \frac{t \int_O \frac{T^2}{G} D_1(T)\, dO}{\int_V \frac{1}{G} \left[\frac{3(1-2\nu)}{2(1+\nu)} \Sigma^2 + T^2\right] dV}. \tag{4.34}$$

In the next Section we discuss some experimental data which suggests that both terms in the numerator (4.33) may be comparable.

4.2 Analysis of experimental data on the energy dissipation of material

Below we consider some particular states of stress and compare their theoretical vibration decrements with experimental ones.

An experiment with a cantilever thin-walled tube specimen carrying a massive disc at its end is described in [143]. The disc weight considerably exceeds the specimen weight, so a uniform state of stress takes place in the tube due to the first mode of torsional and longitudinal vibration.

The general formula (4.30) will be used to evaluate the vibration decrement. The integrals in this formula must be evaluated over the entire volume of the body, i.e. over the specimen volume and the disc volume. The disc, however, is assumed to be a rigid body which is formally attained by letting $G \to \infty$ and $k \to \infty$ in the volume occupied by the disc. This ensures that the integral over the disc volume vanishes. Hence, the decrement value is given by (4.30) in which we integrate only over the volume of the deformable part of the body, i.e. over the specimen volume.

Let us now proceed to a theoretical analysis of the test.

A state of pure shear with a shear stress amplitude τ occurs in a thin-walled tube specimen when the system vibrates according to its first torsional normal mode, i.e.

$$T = T_t = \tau, \ \ \Sigma = 0. \tag{4.35}$$

Substituting these value into (4.30) we obtain the following equation for the decrement of torsional vibration

$$\Delta_t = \frac{1}{V} \int_V D(T)\, dV. \tag{4.36}$$

Here, and in what follows, we suppose that the elastic characteristics of material G and ν are constant in the specimen, while the plastic characteristics defined by function $D(T)$ vary.

A uniaxial stress state with a normal stress amplitude σ occurs in a thin-walled tube specimen when it vibrates according to its first longitudinal normal mode, i.e.

$$T = T_l = \frac{\sigma}{\sqrt{3}}, \ \ \Sigma = \frac{\sigma}{3} = \frac{T}{\sqrt{3}}. \tag{4.37}$$

With the help of (4.32) we obtain the decrement of longitudinal vibration

$$\Delta_l = \frac{2(1+\nu)}{3} \frac{1}{V} \int_V D(T)\, dV. \tag{4.38}$$

Comparison of (4.36) and (4.38) shows that the decrement of longitudinal vibration is as much as 87% of the decrement of torsional vibration for the same values of shear stress intensity T and Poisson's ratio ν. In other words, they approximately coincide regardless of the character of plastic heterogeneity in the specimen. Therefore, approximately the same value of the decrements of torsional and longitudinal vibration is observed provided that the shear stress intensities coincide, i.e.

$$\frac{\tau}{\sigma} = \frac{1}{\sqrt{3}} \approx 0.58. \tag{4.39}$$

This relationship more or less fits experimental data collected in [143] and [159]. It should be noted that experimental data may considerably deviate from (4.39). One of the main reasons for this deviation is material anisotropy caused by thermomechanical treatment of the specimen which is not taken into account in the proposed theory. These deviations have been discussed in [113]. Another possible reason is that the specimen is not a thin-walled system, i.e. the stress state is actually non-uniform, the reason for this is given in [5].

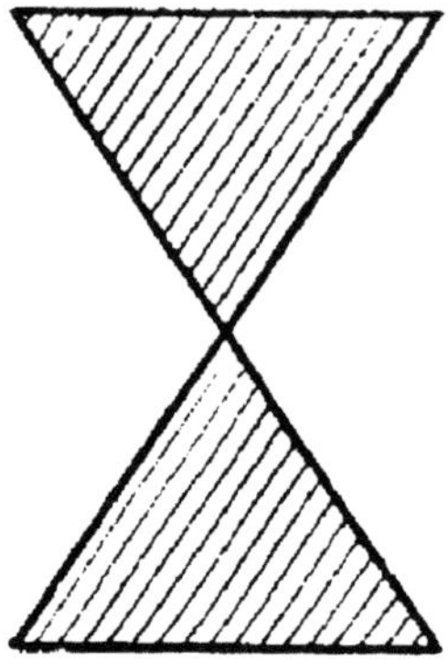

FIGURE 4.2. Cross-section of a prismatic specimen.

The same objective, i.e. a comparison of the damping ability of materials under pure shear and in a uniaxial stress state, has been pursued experimentally by Khilchevsky [80] who carried out work on specimens with a heterogeneous state of stress. The decay of torsional vibrations of a solid circular specimen and the decay of bending vibrations of a prismatic specimen with the cross-section shown in Fig. 4.2 (the bending plane is vertical) have been compared there. The test conditions ensured that the stresses did not change along the specimens.

For torsional vibrations, the stress in a circular specimen is given by

$$T = \tau\rho, \tag{4.40}$$

where ρ is a nondimensional radius, $0 < \rho < 1$, and τ is the maximum amplitude of the shear stress.

Assuming that the plastic properties of the specimen vary, but depend only on the radial coordinate we obtain by means of (4.30) the following value of the decrement of torsional vibration

$$\Delta_t = 4\int_0^1 \rho^3 D\left(\tau\rho\right) d\rho. \tag{4.41}$$

For bending vibrations of a specimen which has the cross-section shown in Fig. 4.2, the stresses are as follows

$$T = \frac{\sigma\zeta}{\sqrt{3}}, \quad \Sigma = \frac{T}{\sqrt{3}}. \tag{4.42}$$

Here σ denotes the maximum amplitude of the bending stress and ζ stands for a non-dimensional vertical coordinate, $-1 < \zeta < 1$. The cross-sectional width is given by

$$b = b_0\left|\zeta\right|, \tag{4.43}$$

where b_0 is the maximum width. Assuming that the plastic properties of the material depend on the vertical coordinate only, we obtain an expression for the logarithmic decrement

$$\Delta_b = \frac{8(1+\nu)}{3} \int_0^1 \zeta^3 D\left(\frac{\sigma\zeta}{\sqrt{3}}\right) d\zeta. \tag{4.44}$$

Experiments [80] have shown close agreement with these decrements if condition (4.39) holds. Comparison of (4.41) and (4.44) indicates that this is possible provided that the dependencies of $D(T)$ on variable ρ in the first case and on ζ in the second case coincide.

An experimental comparison of the decrement of polar-symmetrical vibrations in a thin-walled spherical shell and the decrement of vibration of a plane specimen under the condition of pure bending has been performed in [84]. In these cases both specimens were made of the same material and a bi-axial isotropic stress state took place in the shell.

Hence

$$T = \frac{\sigma_1}{\sqrt{3}}, \quad \Sigma = \frac{2\sigma_1}{3} = \frac{2T}{\sqrt{3}}, \tag{4.45}$$

where σ_1 denotes the amplitude of normal stress.

Assuming that the plastic properties of the material is heterogeneous over the shell thickness we obtain, by means of (4.30), the following expression for the decrement of the shell vibration

$$\Delta_s = \frac{1+\nu}{3(1-\nu)} \frac{1}{h} \int_{-h/2}^{h/2} D\left(\frac{\sigma_1}{\sqrt{3}}\right) dz. \tag{4.46}$$

When a plane specimen vibrates in a pure bending stress state, the normal stress varies linearly over the thickness, i.e.

$$T = \frac{\sigma_2}{\sqrt{3}}\left|\frac{2z}{h}\right|, \quad \Sigma = \frac{\sigma_2}{3}\left|\frac{2z}{h}\right| = \frac{T}{\sqrt{3}}, \tag{4.47}$$

where σ_2 is the maximum amplitude of normal stress.

There exist many reasons to suggest that the plastic properties of a plate vary over the plate thickness in the same way as in the shell. Equation (4.30) then yields the following value of the vibration decrement of the plane specimen

$$\Delta_p = \frac{2(1+\nu)}{h} \int_{-h/2}^{h/2} \left(\frac{2z}{h}\right)^2 D\left(\frac{\sigma_2}{\sqrt{3}}\left|\frac{2z}{h}\right|\right) dz. \tag{4.48}$$

It seems impossible to compare (4.46) and (4.48) in the general case. With this in mind, let us consider some reasonable limiting cases. When the plastic properties of the material are uniform and $D(T)$ is a linear function

$$D = AT, \tag{4.49}$$

we obtain

$$\Delta_s = \frac{1+\nu}{3(1-\nu)} D\left(\frac{\sigma_1}{\sqrt{3}}\right), \quad \Delta_p = \frac{1+\nu}{2} D\left(\frac{\sigma_2}{\sqrt{3}}\right). \tag{4.50}$$

These values are approximately equal for the same amplitudes $\sigma_1 = \sigma_2$ and $\nu = 0.3$. In another limiting case in which $D(T) \neq 0$ only in thin surface layers of thickness t we have

$$\Delta_s = \frac{1+\nu}{3(1-\nu)} \frac{2t}{h} D\left(\frac{\sigma_1}{\sqrt{3}}\right), \quad \Delta_p = 4(1+\nu)\frac{t}{h} D\left(\frac{\sigma_2}{\sqrt{3}}\right). \tag{4.51}$$

For the same amplitudes $\sigma_1 = \sigma_2$ and $\nu = 0.3$, the decrement of the bending vibration of the strip is about $6(1-\nu) = 4.2$ times greater than the decrement of the shell vibration.

A test [84] has shown that the decrement of bending vibration is twice as much as that of shell vibration which lies between the above two limiting values. This comparison confirms the general conclusion of [5] that the surface layers provide an essential contribution to the total value of decay.

Let us next consider this problem in detail. Let the energy dissipation in a thin layer near the surface of thickness t be described by a function $\gamma D(T)$ (γ is a constant factor) and the energy dissipation in the rest of the body be described by a function $D(T)$. Let $D(T)$ be given by (4.49). The values of the decrements for a shell and a plane strip are as follows ($t/h \ll 1$)

$$\Delta_s = \frac{1+\nu}{3(1-\nu)} D\left(\frac{\sigma_1}{\sqrt{3}}\right)\left[1 + \frac{2t}{h}(\gamma - 1)\right], \tag{4.52}$$

$$\Delta_p = \frac{1+\nu}{2} D\left(\frac{\sigma_2}{\sqrt{3}}\right)\left[1 + \frac{8t}{h}(\gamma - 1)\right]. \tag{4.53}$$

In each of these equations the second value in the square brackets characterises the relative contribution of heavily damped surface layers to the total value of the energy dissipation. Let us denote

$$y = \frac{2t}{h}(\gamma - 1) \tag{4.54}$$

and evaluate the decrement ratio for $\sigma_1 = \sigma_2$. The result is

$$\frac{\Delta_p}{\Delta_s} = \frac{3(1-\nu)}{2} \frac{1+4y}{1+y}. \tag{4.55}$$

This decrement ratio is known from experimental test results. We can then easily obtain that

$$y = \frac{\Delta_p/\Delta_s - 3(1-\nu)/2}{6(1-\nu) - \Delta_p/\Delta_s}. \qquad (4.56)$$

From experimental results [84] we have $\Delta_p/\Delta_s = 2$. If we take ν=0.3, then from (4.56) we obtain

$$y = 0.43.$$

We note here that the value of y characterises the relative contribution of heavily damped thin surface layers to the vibration decrement of the shell. This value turns out to be considerable, namely 43%. The contribution of the surface layers to the vibration decrement of the plane strip is even more significant, namely $4y$, which corresponds to 172%.

Hence, the contribution of the surface layers of the material is considerable. However, it is not considered to be a decisive factor in these cases.

The value of y obtained allows us to estimate the thickness t of the surface layer with a higher damping. Let, for example, $\gamma = 5$, i.e. the damping property of the surface layer be 5 times higher than that of the interior. Thus, eq. (4.54) yields

$$t/h = 0.054$$

which is quite plausible.

Let us proceed to [148]. In this paper single-frequency bending-torsional vibrations in a system consisting of a cantilever circular rod with an eccentrically attached massive disc have been investigated experimentally.

The theoretical description of this experiment is as follows. The mass of the rod may be neglected as it is small compared to the disc mass. In cylindrical coordinates r, θ, z (with z being the axial coordinate) the stress distribution is given by

$$\tau_{z\theta} = \tau\rho, \quad \sigma_z = \sigma\zeta\rho\cos\theta, \qquad (4.57)$$

where τ is the amplitude of shear stress on the specimen surface, σ is the maximum amplitude of the normal stress due to bending at the clamped end, ρ is a nondimensional radius ($0 \le \rho \le 1$), and ζ is a nondimensional axial coordinate ($0 \le \zeta \le 1$). The other stresses are equal to zero.

The amplitude of the shear stress intensity and amplitude of the mean normal stress are as follows

$$T = \rho\sqrt{T_t^2 + T_b^2\zeta^2\cos^2\theta}, \; \Sigma = T_b\zeta\rho\,|\cos\theta|\,, \qquad (4.58)$$

where

$$T_t = \tau, \quad T_b = \sigma/\sqrt{3}, \qquad (4.59)$$

denote maximum amplitudes of the shear stress intensity due to torsion and bending, respectively.

Substituting (4.58) into (4.30) yields the vibration decrement

$$\Delta = \frac{8\int\limits_0^1\int\limits_0^1\int\limits_0^{\pi/2} \rho^3\left(1+\beta^2\zeta^2\cos^2\theta\right) D\left(T_t\rho\sqrt{1+\beta^2\zeta^2\cos^2\theta}\right) d\theta d\rho d\zeta}{\pi\left[1+\beta^2/4\left(1+\nu\right)\right]}, \tag{4.60}$$

where β is the following ratio

$$\beta = T_b/T_t. \tag{4.61}$$

When $T_b \to 0$ one obtains the decrement of torsional vibration Δ_t as given by eq. (4.41).

When $T_t \to 0$ one arrives at the following equation for the decrement of bending vibration

$$\Delta_b = \frac{32\left(1+\nu\right)}{\pi}\int\limits_0^1\int\limits_0^1\int\limits_0^{\pi/2} \rho^3\zeta^2\cos^2\theta D\left(T_b\rho\zeta\cos\theta\right) d\theta d\rho d\zeta. \tag{4.62}$$

The values of these decrements may be found provided that the dependence $D = D\left(T\right)$ is known for a particular material. One often assumes that

$$D = gT^{\alpha}.$$

The two integrals in (4.60) can then be evaluated, to give

$$\Delta = \frac{\Delta_t}{\beta}\left[1+\frac{\beta^2}{4\left(1+\nu\right)}\right]^{-1}\int\limits_0^{\beta}\left(1+\frac{x^2}{2}\right)^{1+\alpha/2}\frac{P_{1+\alpha/2}\left(z\right)}{z^{1+\alpha/2}}dx. \tag{4.63}$$

where

$$z = \left(1+\frac{x^2}{2}\right)\left(1+x^2\right)^{-1/2}, \tag{4.64}$$

$P_{\nu}\left(z\right)$ is the Legendre function of the first kind, and Δ_t is the decrement of the torsional vibration given by

$$\Delta_t = 4T_t^{\alpha}\int\limits_0^1 \rho^{3+\alpha} g\left(\rho\right) d\rho. \tag{4.65}$$

The decrement of the bending vibration takes the form

$$\Delta_b = \frac{16\left(1+\nu\right)}{\pi\left(3+\alpha\right)} B\left(\frac{3+\alpha}{2},\frac{1}{2}\right)\int\limits_0^1 \rho^{3+\alpha} g\left(\rho\right) d\rho T_b^{\alpha}. \tag{4.66}$$

In deriving these equation it has been taken that the plastic properties of the material depend on the radius, namely $g = g\left(\rho\right)$, but $\alpha =$ const.

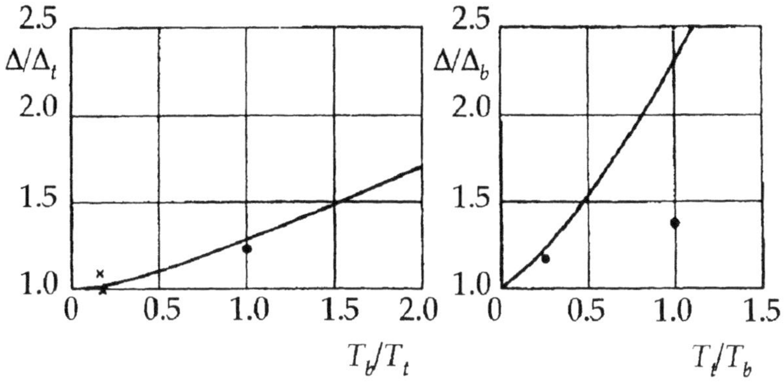

FIGURE 4.3. Ratio of logarithmic decrements versus ratio of shear stress intensities.

Results of the calculations arising from (4.63), (4.65) and (4.66) with $\alpha = 1$ and $\nu = 0.3$ are shown in Fig. 4.3. One sees that increasing any stress (normal or shear) leads to an increase in the decrement of vibration. Exactly this effect was observed in test [148], this data is denoted by points in Fig. 4.3.

Experimental data on bending-torsional vibrations of a clamped-clamped solid circular bar with a massive disc in the middle have been reported in [81]. The disc was attached eccentrically, and bending-torsional mode of vibration was excited.

The theoretical analysis for this experiment is as follows.

In the cylindrical coordinates r, θ, z (the axis z is the axial coordinate of the bar) the distribution of the extreme stresses are given by

$$\tau_{z\theta} = \tau\rho, \quad \sigma_z = \sigma\rho\,(1 - |\zeta|)\cos\theta. \tag{4.67}$$

We keep the same notation as above with the only difference being that the non-dimensional axial coordinate ζ varies in the interval $[-2, 2]$, $\zeta = 0$ denoting the disc plane. The amplitude values of the intensity of shear stresses and mean normal stress due to (4.67) are

$$\begin{aligned} T &= \rho\sqrt{T_t^2 + T_b^2\,(1 - |\zeta|)^2\cos^2\theta}, \\ \Sigma &= T_b\rho\,|1 - |\zeta||\,|\cos\theta|\,, \end{aligned} \tag{4.68}$$

where

$$T_t = \tau, \quad T_b = \sigma/\sqrt{3}. \tag{4.69}$$

Substituting (4.68) in (4.30) and simplifying leads to the expression for the vibration decrement (4.60). Hence, in this case we can use the equations derived for the analysis of the previous experiment and Fig. 4.3. In the tests of [81] only one case was considered, namely when the ratio of the stresses was equal to 0.23, which gave $T_b/T_t = 0.13$. The experimental results for materials with $\alpha \approx 1$ are denoted by crosses in Fig. 4.3. For this case it can be seen that the theoretical analysis fits the experimental data reasonably well.

4.3 Vibration of a simple torsional system

Up until now, only the problem of the nonlinear internal friction in bars with circular solid and ring cross-sections in torsion has been reported in the literature. In the present section we offer a solution for a bar with an arbitrary cross-section by means of the model of elastoplastic materials.

Consider a simple torsional system. Let a massive disc be attached to the free end of a cantilever cylindrical bar. We assume that no eccentricity occurs, that is, only pure torsional vibrations takes place. Let the bar length be L which is assumed to be much greater than the transverse dimension. This assumption is needed so that a solution of the classical problem of torsion can be applied.

Denote the angle of the disc rotation by φ, then the displacement field is given by

$$\mathbf{u} = \mathbf{e}_\varphi r \varphi, \tag{4.70}$$

where r and φ denote cylindrical coordinates and $\mathbf{e}_\varphi$ stands for the unit base vector in the circumferential direction.

Assume that external loads act on the disc only, i.e. the lateral surface of the bar is traction-free and the bar volume is free of loads. Under this condition the relative angle of twist turns out to be constant over the bar and is equal to

$$\varphi' = \varphi/L. \tag{4.71}$$

The following stresses are known to appear in the bar

$$\tau_{zx} = G\frac{\varphi}{L}\frac{\partial \Phi}{\partial y}, \quad \tau_{zy} = -G\frac{\varphi}{L}\frac{\partial \Phi}{\partial x}, \tag{4.72}$$

where Φ denotes the stress function of torsion. In the domain Ω occupied by the bar cross-section, Φ satisfies Poisson's equation

$$\Delta \Phi = -2, \tag{4.73}$$

whilst on the domain boundary its derivative in the tangential direction vanishes.

Equations (4.70) and (4.72) indicate that the angle φ may be considered a generalised coordinate q. Then in the domain occupied by the disc

$$\mathbf{u}_0 = \mathbf{e}_\varphi r. \tag{4.74}$$

In the domain occupied by the bar the stresses are

$$(\tau_{zx})_0 = \frac{G}{L}\frac{\partial \Phi}{\partial y}, \ (\tau_{zy})_0 = -\frac{G}{L}\frac{\partial \Phi}{\partial x}. \tag{4.75}$$

This renders the following intensities of shear stress and mean normal stress

$$T_0 = \frac{G}{L}\left|\nabla \Phi\right|, \ \sigma_0 = 0. \tag{4.76}$$

By means of the equations of Section 4.1 we obtain

$$m = \int\limits_{V_d} \rho r^2 dV = I, \tag{4.77}$$

$$p = \int\limits_{O_d} p_\varphi r dO + \int\limits_{V_d} K_\varphi r dV = M. \tag{4.78}$$

We integrate only over the surface and the volume of the disc since the bar is assumed to have no mass density ($\rho = 0$) and no load acts on its surface and in its volume. As follows from (4.77) the factor m is equal to the moment of inertia I about the z axis. As in (4.78), p_φ and K_φ are the tangential components of the external surface and volumetric loads, and M is equal to the torque applied to the disc. Equation (4.17) then takes the form

$$I\ddot{\varphi} + c_c \varphi = M. \tag{4.79}$$

Hence, c_c may be interpreted as a *complex torsional rigidity of a bar* of length L. An expression for this is obtained by substituting of the shear stress intensity and mean normal stress, due to (4.76), into (4.24)

$$c_c = c\left[1 + m\beta\left(\frac{T_*}{2G}\right)\right], \tag{4.80}$$

where

$$c = \frac{G}{L}\int\limits_{\Omega} \left|\nabla \Phi\right|^2 d\Omega, \tag{4.81}$$

$$m = \frac{\int\limits_{\Omega} \left|\nabla \Phi\right|^{2+\alpha} d\Omega}{\left|\nabla \Phi\right|_*^{\alpha} \int\limits_{\Omega} \left|\nabla \Phi\right|^2 d\Omega} \tag{4.82}$$

and the integration domain is the cross-sectional area of the bar. The value of T_* is the maximum value of the shear stress intensity

$$T_* = G\left|\nabla \Phi\right|_* \frac{a}{L} \tag{4.83}$$

and $\beta(\ldots)$ is a material characteristic due to (3.105).

The decrement of torsional vibration is given by (4.28) and takes the form

$$\Delta = mD\left(T_*\right), \tag{4.84}$$

where $D\left(T_*\right)$ is given by (3.114).

Comparing the latter result with (3.113) shows that the factor m characterises the difference in the decrement of the torsional vibration of the bar and the decrement of vibration in the state of uniform pure shear, both cases having the same maximum intensity of shear stress.

Equation (4.81) gives the static torsional rigidity of the bar of length L. In the literature one can find the following parameter

$$\Xi = cL = G\int_{\Omega} |\nabla\Phi|^2 \, d\Omega, \tag{4.85}$$

which is the torsional rigidity of the bar per unit length. In what follows, this is referred to as the torsional rigidity of the bar. The values of Ξ for various cross-sections can be found in [2].

The value of m have been computed only for bars with circular and ring cross-sections since the integral evaluations are elementary. The results of evaluations for some other cross-sections are given in the next Section.

Below we will use the so-called complex torsional rigidity introduced by analogy with (4.80)

$$\Xi_c = \Xi\left[1 + m\beta\left(\frac{T_*}{2G}\right)\right]. \tag{4.86}$$

Extracting from (4.86) the amplitude of torsion

$$\left|\gamma'\right| = a/L, \tag{4.87}$$

we obtain the value of the complex torsional rigidity

$$\Xi_c = \Xi\left[1 + m\left|\gamma'\right|^{\alpha}\beta\left(\frac{|\nabla\Phi|_*}{2}\right)\right] \tag{4.88}$$

for a power function $\beta\left(T/2G\right)$.

4.4 Examples of estimation of the complex torsional rigidity

4.4.1 Elliptical cross-section

Let the contour of the bar cross-section be an ellipse

$$\frac{x^2}{a^2} + \frac{y^2}{b^2} = 1,$$

where a and b are the semi-axes.

In other parts of this book we denote the amplitude of a time-dependent function by a. However in this Section we use a traditional denotation a for an ellipse semi-axis. It should not lead to any confusion since we do not deal with the amplitude a in this Section.

For an elliptic cross-section the stress function of torsion is known to be given by

$$\Phi = \frac{a^2 b^2}{a^2 + b^2} \left(1 - \frac{x^2}{a^2} - \frac{y^2}{b^2} \right). \tag{4.89}$$

The square of the absolute value of the gradient is

$$|\nabla \Phi|^2 = 4 \left(\frac{a^2 b^2}{a^2 + b^2} \right)^2 \left(\frac{x^2}{a^4} + \frac{y^2}{b^4} \right). \tag{4.90}$$

Substituting this into (4.85) and using the new integration variables ρ and φ defined by

$$x = a\rho \cos \varphi, \ y = b\rho \sin \varphi, \tag{4.91}$$

yields

$$\Xi = 4G \left(\frac{a^2 b^2}{a^2 + b^2} \right)^2 \int_0^1 \int_0^{2\pi} \rho^2 \left(\frac{\cos^2 \varphi}{a^2} + \frac{\sin^2 \varphi}{b^2} \right) ab\rho d\varphi d\rho. \tag{4.92}$$

The integrals over ρ and φ are easy to evaluate and yield the following result

$$\Xi = G \frac{\pi a^3 b^3}{a^2 + b^2}. \tag{4.93}$$

Evaluation of m is less trivial. Substituting (4.90) in (4.82), transforming to the new variables (4.91) and accounting for integral (4.92) gives

$$m = \frac{4a^2 b^{2+\alpha}}{\pi (a^2 + b^2)} \int_0^1 \int_0^{2\pi} \left(\frac{\cos^2 \varphi}{a^2} + \frac{\sin^2 \varphi}{b^2} \right)^{1+\alpha/2} \rho^{3+\alpha} d\varphi d\rho.$$

The integral over ρ is elementary, while for an integration over φ one has to transform the expression in the parentheses as follows

$$\frac{\cos^2 \varphi}{a^2} + \frac{\sin^2 \varphi}{b^2} = \frac{1}{2} \left(\frac{1}{a^2} + \frac{1}{b^2} \right) - \frac{1}{2} \left(\frac{1}{b^2} - \frac{1}{a^2} \right) \cos 2\varphi.$$

This integral can be represented in the form of the first Laplace integral [93]

$$P_\nu (z) = \frac{1}{\pi} \int_0^\pi \left[z + \sqrt{z^2 - 1} \cos \psi \right]^\nu d\psi, \tag{4.94}$$

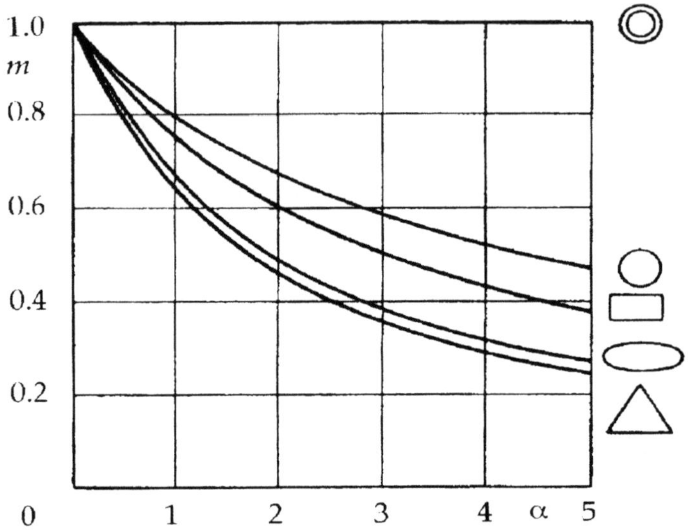

FIGURE 4.4. Parameter m versus α for some cross-sections.

where $P_\nu(z)$ is the Legendre function of the first kind. The latter is a real-valued function of z only for $z > 1$. The result of the integration is as follows

$$m = \frac{1}{1+\alpha/4}\left(\frac{a}{b}\right)^{\alpha/2}\frac{2ab}{a^2+b^2}P_{1+\alpha/2}\left(\frac{a^2+b^2}{2ab}\right). \tag{4.95}$$

In the case of a circular cross-section, i.e. $a = b$, (4.95) becomes

$$m = (1+\alpha/4)^{-1}. \tag{4.96}$$

Equations (4.41) and (4.84) lead to the same result for a power function β. Function (4.96) is drawn in Fig. 4.4 and denoted by a circle.

For a very "flat" ellipse such that $a/b \to \infty$ eq. (4.95) becomes

$$m = \frac{2}{\pi}\frac{1}{1+\alpha/4}B\left(\frac{3+\alpha}{2}, \frac{1}{2}\right), \tag{4.97}$$

where B stands for the Eulerian beta-function. Dependence (4.97) is also shown in Fig. 4.4 and is denoted by an ellipse.

Numerical results indicate that for $a/b > 1.5$ the dependence $m(\alpha)$ given by eq. (4.95) slightly differs from the dependence (4.97), so the latter can be used for any ellipse for which $a/b > 1.5$. For $a/b < 1.5$ the dependences (4.95) lie between the dependences (4.96) and (4.97). As these curves depicting these dependences are very close, the dependencies (4.95) are not shown in Fig. 4.4.

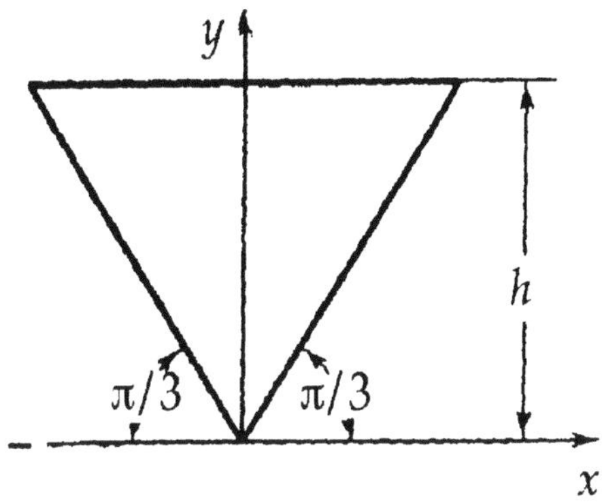

FIGURE 4.5. Coordinate system for the triangular cross-section.

4.4.2 Rectangular cross-section

Let the bar cross-section be a high aspect ratio rectangle with the sides $2a$ and $2b$. An approximated stress function for this cross-section is

$$\Phi = b^2 - y^2. \tag{4.98}$$

Substitution of (4.98) into (4.82) leads to the following result

$$m = (1 + \alpha/3)^{-1}. \tag{4.99}$$

This dependence is shown in Fig. 4.4 and denoted by a rectangle.

4.4.3 Ring cross-section

If the cross-section is a thin-walled ring, the stresses are approximately equally distributed over the cross-section, and thus $\nabla\Phi$ = const. In this case (4.82) yields m=1, which forms the upper line in Fig. 4.4.

4.4.4 Triangular cross-section

Let the cross-section be an equilateral triangle as shown in Fig. 4.5. It is known that in the coordinate system of Fig. 4.5, the stress function is given by [94]

$$\Phi = \frac{1}{2h}\left(y^2 - 3x^2\right)(h - y), \tag{4.100}$$

where h is the height of the triangle.

Substitution of the latter equation into (4.85) and evaluation of the integral leads to the known result [94]

$$\Xi = \frac{Gh^4}{15\sqrt{3}} = 0.0376Gh^4. \tag{4.101}$$

Substitution of Φ from (4.100) into (4.82) yields an integral which cannot be evaluated in closed form. The result of numerical integration of this is shown in Fig. 4.4 by the lower curve. Figure 4.4 allows a comparison of the decrements of torsional vibration in bars with various cross-sections but the same maximal stresses.

4.5 Torsional vibrations of elastoplastic rods

Consider a cylindrical bar or a bar with a slowly varying cross-section. Let the centre of gravity of each cross-section lie on the z axis.

Let γ denote the angle of rotation of a generic cross-section along z when the bar vibrates in a particular mode

$$\gamma = \gamma(z),$$

where $\gamma(z)$ is assumed to be given. Provided that $\gamma(z)$ is a slowly varying function of its argument the stresses in the cross-sections may be approximated by the formulae

$$(\tau_{zx})_0 = G\gamma' \frac{\partial \Phi}{\partial y}, \ (\tau_{zy})_0 = -G\gamma' \frac{\partial \Phi}{\partial x}, \tag{4.102}$$

where the stress function Φ is taken for the cross-section z, and the prime stands for the derivative with respect to z.

In the problem of torsion the displacement in the plane of the cross-section is given by

$$\mathbf{u}_0 = \mathbf{e}_\varphi \gamma r. \tag{4.103}$$

The warping usually has a higher order of smallness and for this reason is neglected.

Let us proceed now to evaluate the coefficients of the basic equation of dynamics.

It follows from (4.102) that

$$T_0 = G\left|\gamma'\right| |\nabla\Phi|, \ \sigma_0 = 0. \tag{4.104}$$

Substituting these into (4.24) and separating the integration over the cross-section from that over the bar length, yields

$$c_t = \int_L \left(\gamma'\right)^2 \Xi_c dz, \tag{4.105}$$

where Ξ_c denotes the complex torsional stiffness of the bar given by (4.88). In this case the stiffness has the following form

$$\Xi_c = \Xi\left[1 + l_t\left(\gamma' a\right)^\alpha\right], \tag{4.106}$$

where

$$l_t = m\beta\left(\frac{|\nabla\Phi|_*}{2}\right).$$

Substituting (4.106) into (4.105) gives

$$c_c = c\left[1 + l_* a^\alpha\right], \tag{4.107}$$

where

$$c = \int\limits_L \Xi\left(\gamma'\right)^2 dz, \tag{4.108}$$

$$l_* = \frac{1}{c}\int\limits_L \Xi l_t \left(\gamma'\right)^{2+\alpha} dz. \tag{4.109}$$

Inserting (4.103) into (4.18) and (4.19) gives the other parameters

$$m = \int\limits_L I\gamma^2 dz, \tag{4.110}$$

$$p = \int\limits_L M\gamma dz, \tag{4.111}$$

where

$$I = \int\limits_\Omega \rho r^2 d\Omega, \tag{4.112}$$

$$M = \int\limits_\Omega K_\varphi r d\Omega + \oint\limits_\lambda p_\varphi r d\lambda. \tag{4.113}$$

The integral over Ω is an area integral over the cross-section, while the integral over λ is an integral over the cross-sectional contour. It has been assumed for simplicity that both ends of the bar are either free or clamped. The physical meaning of I is the moment of inertia about the z axis per unit length, and M denotes an external torque per unit length.

The coefficients of the basic equation of dynamics are thus determined. The further analysis is similar to that of Section 4.1.

4.6 Vibrations of elastoplastic plates

Consider a thin plate. Let the orthogonal coordinate system xyz be chosen in such a way that the xy axes contain the mid-plane and the z axis is normal to the mid-plane.

Let $W(x, y)$ denote the plate deflection due to a vibration mode. As shown in plate theory [176] the displacements of a plate point in the xy plane are as follows

$$u = -z\frac{\partial W}{\partial x}, \quad u = -z\frac{\partial W}{\partial y}. \tag{4.114}$$

The principal stresses in the theory of thin plate bending are given by

$$\begin{aligned} \sigma_x &= -\frac{zE}{1-\nu^2}\left(\frac{\partial^2 W}{\partial x^2} + \nu\frac{\partial^2 W}{\partial y^2}\right), \\ \sigma_y &= -\frac{zE}{1-\nu^2}\left(\frac{\partial^2 W}{\partial y^2} + \nu\frac{\partial^2 W}{\partial x^2}\right), \\ \tau_{xy} &= -\frac{zE}{1-\nu^2}(1-\nu)\frac{\partial^2 W}{\partial x \partial y}, \end{aligned} \tag{4.115}$$

cf. [176], where E and ν are the Young's modulus and the Poisson's ratio, respectively. The secondary stresses are at least of one order higher than the principal ones. With this in view, we do not consider them.

By means of (4.115) we obtain the mean tensile stress

$$\sigma_0 = -\frac{1}{3}\frac{zE}{1-\nu^2}(1+\nu)\Delta W \tag{4.116}$$

and the components of the stress deviator

$$\begin{aligned} s_x &= -\frac{zE}{1-\nu^2}\left(\frac{\partial^2 W}{\partial x^2} + \nu\frac{\partial^2 W}{\partial y^2} - \frac{1+\nu}{3}\Delta W\right), \\ s_y &= -\frac{zE}{1-\nu^2}\left(\frac{\partial^2 W}{\partial y^2} + \nu\frac{\partial^2 W}{\partial x^2} - \frac{1+\nu}{3}\Delta W\right), \\ s_z &= -\frac{zE}{1-\nu^2}\left(-\frac{1+\nu}{3}\Delta W\right), \\ s_{xy} &= -\frac{zE}{1-\nu^2}(1-\nu)\frac{\partial^2 W}{\partial x \partial y}, \end{aligned} \tag{4.117}$$

where Δ is the two-dimensional Laplace operator

$$\Delta = \frac{\partial^2}{\partial x^2} + \frac{\partial^2}{\partial y^2}. \tag{4.118}$$

Evaluating the amplitude of the shear stress intensity due to (3.79) gives

$$T_0^2 = (2Gz)^2\left[\frac{1-\nu+\nu^2}{3(1-\nu)^2}(\Delta W)^2 - \frac{\partial^2 W}{\partial x^2}\frac{\partial^2 W}{\partial y^2} + \left(\frac{\partial^2 W}{\partial x \partial y}\right)^2\right]. \tag{4.119}$$

Inserting (4.119) and (4.116) into the expression for the complex rigidity c_c, eq. (4.24), and integrating over z leads to the result

$$c_c = c\,(1 + l_* a^\alpha)\,, \tag{4.120}$$

where the following notation is used

$$c = \int\limits_\Omega D\left\{(\Delta W)^2 - 2\,(1-\nu)\left[\frac{\partial^2 W}{\partial x^2}\frac{\partial^2 W}{\partial y^2} - \left(\frac{\partial^2 W}{\partial x \partial y}\right)^2\right]\right\} d\Omega, \tag{4.121}$$

$$\begin{aligned} l_* \;=\; & \int\limits_\Omega \frac{8Gl}{3+\alpha}\left(\frac{h}{2}\right)^{3+\alpha}\left[\frac{1-\nu+\nu^2}{3\,(1-\nu)^2}\,(\Delta W)^2 - \right. \\ & \left. -\frac{\partial^2 W}{\partial x^2}\frac{\partial^2 W}{\partial y^2} + \left(\frac{\partial^2 W}{\partial x \partial y}\right)^2\right]^{1+\alpha/2} d\Omega \;. \end{aligned} \tag{4.122}$$

Here D stands for the bending rigidity

$$D = \frac{Eh^3}{12\,(1-\nu^2)}\;,$$

and h is the plate thickness. The integration domain in eqs (4.121) and (4.122) is the area of the mid-plane of the plate Ω.

Equation (4.121) yields twice the potential energy of the plate due to a particular vibration mode $W = W\,(x,y)$.

Let us determine the other parameters of eq. (4.17) for this case. Neglecting the displacements in the plane of the plate we obtain, due to (4.19), the following expression for m

$$m = \int\limits_\Omega \rho h W^2 d\Omega, \tag{4.123}$$

which is twice the kinetic energy of the plate caused by the velocity field $W = W\,(x,y)$.

To evaluate the integrals we make the following assumptions which are customary for the theory of bending in thin plates:

i) the lateral surface is assumed to be cylindrical with a generator parallel to the z axis,

ii) one of the surface planes is load-free whereas the second surface plane is loaded by a normal load p.

Considering the above and substituting the displacements due to (4.114) into (4.18) gives the following result

$$p = \int\limits_\Omega qW d\Omega + \oint\limits_\lambda QW d\lambda - \oint\limits_\lambda \left(G_x\frac{\partial W}{\partial x} + G_y\frac{\partial W}{\partial y}\right) d\lambda, \tag{4.124}$$

where

$$Q = \int_{-h/2}^{h/2} p_z dz, \tag{4.125}$$

$$G_x = \int_{-h/2}^{h/2} p_x z dz, \; G_y = \int_{-h/2}^{h/2} p_y z dz. \tag{4.126}$$

Expression (4.124) represents the work performed by external loads on displacements due to the chosen vibration mode. The quantity Q denotes the shear force on the plate contour, while G_x and G_y stand for the external edge moments acting in planes parallel to zx and zy, respectively.

In principle, the expressions derived solve the problem of plate vibrations. Formulae (4.121), (4.123) and (4.124) are relatively simple. Application of the variational principle in the linear theory of elastic vibration leads exactly to these equations. Evaluation of the integrals for l_* may cause some difficulties as the integrand contains powers of rather complicated expressions, the orders not being whole numbers. However, the main difficulty arises when one tries to apply the derived formulae because the closed form expressions for the vibration modes are known only in a few cases. For this reason, one has to use their approximated expressions. This is allowed since the basic equation of dynamics (4.17) is derived from the variational equation. With respect to the accuracy of this approach the following can be said. Numerous problems of linear vibration theory indicate that a rather accurate value for the natural frequency (which is an integral characteristic of the vibration field) is obtained provided that a vibration mode is well-approximated. Therefore, there exist many reasons to suggest that some acceptable results will be obtained for the damping parameter l_*, which is another integral characteristic of the vibration field.

4.7 Vibration of rectangular plates with some boundary conditions

4.7.1 *Simply supported plate*

Let a and b denote the side lengths and let the origin of the coordinate system be placed in one of the plate corners. In this case the vibration modes are given by, cf. [176]

$$W = \sin \lambda x \sin \mu y, \tag{4.127}$$

where λ and μ take the following discrete values

$$\lambda = \frac{\pi i}{a},\ \mu = \frac{\pi k}{b},\quad i,\ k = 1,\ 2,\ 3,\ \dots \tag{4.128}$$

Substituting (4.127) into (4.123) and (4.121) and evaluating the integrals yields

$$m = \rho h \frac{ab}{4},\ c = D\left(\lambda^2 + \mu^2\right)^2 \frac{ab}{4}. \tag{4.129}$$

Substitution of (4.127) in (4.122) gives the following integral

$$l_* = \frac{8Gl}{3+\alpha}\left(\frac{h}{2}\right)^{3+\alpha} \int_0^a \int_0^b \left[2e\sin^2\lambda x \sin^2\mu y + 2d\cos^2\lambda x\cos^2\mu y\right]^{1+\frac{\alpha}{2}} dxdy \tag{4.130}$$

where

$$e = \frac{1}{2}\left[\frac{1-\nu+\nu^2}{3(1-\nu)^2}\left(\lambda^2+\mu^2\right)^2 - \lambda^2\mu^2\right],\ d = \frac{1}{2}\lambda^2\mu^2. \tag{4.131}$$

It can be easily checked that e and d are positive.

Integration over one of the variables in (4.130), say x, may be performed by transformation of the integrand to a form required for the first Laplace integral (4.94). Indeed, the following identity

$$2e\sin^2\lambda x\sin^2\mu y + 2d\cos^2\lambda x\cos^2\mu y = A - B\cos 2\lambda x,$$

holds, in which

$$\begin{aligned} A &= e\sin^2\mu y + d\cos^2\mu y, \\ B &= e\sin^2\mu y - d\cos^2\mu y. \end{aligned}$$

It is evident that

$$|B| < A.$$

The integral over x in (4.130) takes the following form

$$R = \int_0^a (A - B\cos 2\lambda x)^{1+\alpha/2}\, dx.$$

By means of a new variable

$$\xi = 2\lambda x$$

and due to periodicity and evenness of the integrand one obtains

$$R = \frac{a}{\pi}\int_0^\pi (A - B\cos\xi)^{1+\alpha/2}\, d\xi.$$

Replacing ξ by $\pi - \xi$ one can easily see that the integral is insensitive to the sign of B. Thus it can be rewritten as follows

$$R = \frac{a}{\pi} \int_0^\pi (A + |B| \cos \xi)^{1+\alpha/2} \, d\xi.$$

By introducing

$$z = \frac{e \sin^2 \mu y + d \cos^2 \mu y}{2\sqrt{ed} \, |\sin \mu y \cos \mu y|} \tag{4.132}$$

we can rewrite the integral in the following form

$$R = a \left| 2\sqrt{ed} \sin \mu y \cos \mu y \right|^{1+\alpha/2} \frac{1}{\pi} \int_0^\pi \left(z + \sqrt{z^2 - 1} \cos \xi \right)^{1+\alpha/2} d\xi.$$

The first Laplace integral (4.94) is recognized in this expression, hence

$$R = a \left(e \sin^2 \mu y + d \cos^2 \mu y \right)^{1+\alpha/2} \frac{P_{1+\alpha/2}(z)}{z^{1+\alpha/2}}. \tag{4.133}$$

Recalling now that R is an integral over x in (4.130), inserting (4.133) into it and transforming to a new variable

$$\eta = 2\mu y,$$

we obtain, due to the periodicity and evenness of the integrand,

$$l_* = \frac{8Gl}{c(3+\alpha)} \left(\frac{h}{2} \right)^{3+\alpha} ab \left(\frac{e+d}{2} \right)^{1+\alpha/2} m, \tag{4.134}$$

where

$$m = \frac{1}{\pi} \int_0^\pi (1 - \gamma \cos \eta) \frac{P_{1+\alpha/2}(z)}{z^{1+\alpha/2}} d\eta, \tag{4.135}$$

$$\gamma = \frac{e-d}{e+d}. \tag{4.136}$$

Finally, inserting c, eq. (4.129), and $(e+d)/2$ into (4.134) yields

$$l_* = \frac{24l(1-\nu)}{3+\alpha} \left[\frac{1-\nu+\nu^2}{6(1-\nu)^2} \right]^{1+\alpha/2} \left[\frac{h\left(\lambda^2+\mu^2\right)}{2} \right]^{\alpha} m. \tag{4.137}$$

Regretfully, the integral cannot be evaluated in closed form, i.e. a numerical evaluation is required.

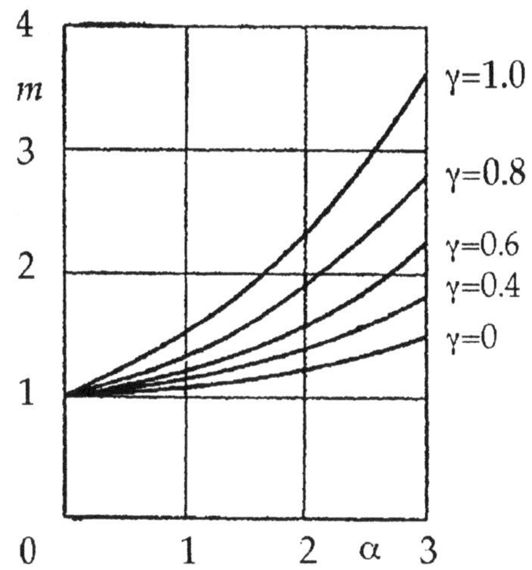

FIGURE 4.6. Parameter m versus α for a simply supported rectangular plate.

Let us now show that coefficient γ depends only on two parameters α and γ. To this end, z must be transformed due to (4.132) and (4.136), to give

$$z = \frac{1 - \gamma \cos \eta}{\sqrt{1 - \gamma^2} \sin \eta}. \tag{4.138}$$

An absolute value sign for $\sin \eta$ is not needed because the integration in (4.135) is over $[0, \pi]$, where $\sin \eta$ is positive.

When using this transformation the range of γ must be considered. Firstly, in order to ensure that expression (4.138) exists the following inequality

$$|\gamma| < 1 \tag{4.139}$$

must hold. The same inequality can be obtained by considering the expression for γ due to (4.136) taking account of the positiveness of e and d. Secondly, by replacing η by $\pi - \eta$ it can be easily shown that m is insensitive to the sign of γ and depends only on its absolute value. Thus, it is necessary to evaluate m only for γ from the interval $[0, 1]$. The results of the numerical calculation are shown in Figs 4.6 and 4.7.

An explicit expression for γ which is obtained by substituting (4.131) in (4.136) is given by

$$\gamma = 1 - \frac{3\left(1-\nu\right)^2}{1 - \nu + \nu^2} \frac{2\lambda^2 \mu^2}{\left(\lambda^2 + \mu^2\right)^2}. \tag{4.140}$$

For example, for the first mode of vibration of a square plate, $\lambda = \mu$ which gives $\gamma = -0.04$ for $\nu = 0.25$.

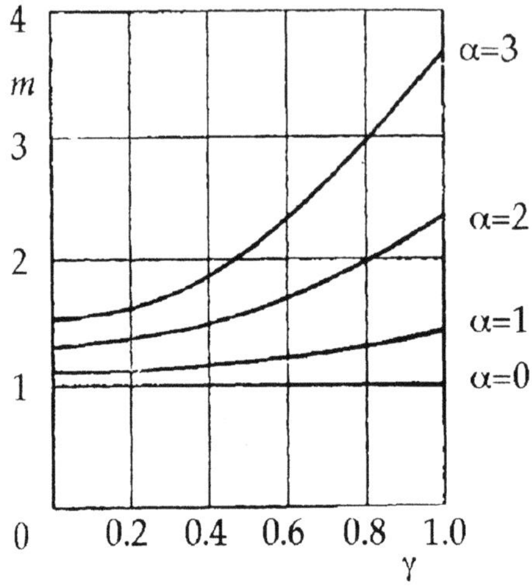

FIGURE 4.7. Parameter m versus γ for a simply supported rectangular plate.

The problem of a simply supported rectangular elastoplastic plate has been studied in [145]. However, the solution reported turns out to be more complicated than the above solution.

4.7.2 A free square plate

Our analysis will be restricted to the first mode of vibration only. Rayleigh [174] has shown that if the origin of the coordinate system is placed in the centre of the plate and the axes are directed parallel to the plate sides, then the first mode may be approximated by

$$W = xy. \tag{4.141}$$

By means of (4.141) we obtain

$$m = \rho h \frac{a^6}{144}, \; c = 2D(1-\nu)\, a^2, \; l_* = \frac{l}{1+\alpha/3} \left(\frac{h}{2}\right)^{\alpha}. \tag{4.142}$$

Though the analysis presented here is restricted to the above examples, other types of support condition may be studied by analogy.

5 Random deformation of elastoplastic materials

5.1 The basic results of the theory of stationary random functions

This and the following Chapters assume that the reader is familiar with the theory of random functions as described in [190].

The strain deviator $\mathbf{e}$ is said to be a stationary random tensor function of time if all its components e_{mn} are: (i) stationary functions of time; and (ii) stationarily related functions of time. In what follows all these components are assumed to have zero mean values.

The correlation moment of components e_{mn} and e_{pq} at the time instant t is denoted by K_{mnpq}, i.e.

$$\mathrm{M}\left[e_{mn}e_{pq}\right] = K_{mnpq} \tag{5.1}$$

where M denotes a mean value or the mathematical expectation. According to the definition of stationary random functions the value of K_{mnpq} does not depend upon time.

The square of the shear strain intensity is given by

$$\gamma^2 = \frac{1}{2}\mathbf{e} : \mathbf{e} = \frac{1}{2}e_{mn}e_{mn}. \tag{5.2}$$

The result of taking the average of this is termed the *mean square of the shear strain intensity* and is denoted by Γ^2. According to its definition

$$\Gamma^2 = \mathrm{M}\left[\gamma^2\right]. \tag{5.3}$$

By virtue of (5.1) and (5.2)

$$\Gamma^2 = \frac{1}{2} K_{mnmn} \,. \tag{5.4}$$

In what follows, for analysis of linear dependencies the spectral theory of stationary random functions is used in the general complex exponential form. For example, a random tensor function $\mathbf{e}$ with a zero mean value will be given by its spectral representation [190]

$$\mathbf{e} = \int_{-\infty}^{\infty} \mathbf{E}(\omega)\, e^{\imath\omega t} d\omega, \tag{5.5}$$

where $\mathbf{E}(\omega)$ is a symmetric tensor, its components being correlated white noises of the argument ω. These white noises possess the following correlational properties

$$\mathrm{M}\left[E_{mn}(\omega)\, E^*_{pq}(\omega_1)\right] = S_{mnpq}(\omega)\, \delta(\omega - \omega_1)\,, \tag{5.6}$$

where δ is the Dirac delta-function and the asterisk denotes a conjugate. Deterministic functions S_{mnpq} for $m = p$ and $n = q$ are called *spectral densities* of components e_{mn} and, in the general case, *cross-spectral densities* of components e_{mn} and e_{pq}.

The *root-mean-square of the shear strain intensity* Γ is obtained by means of the equation

$$\Gamma^2 = \mathrm{M}\left[\frac{1}{2}\mathbf{e} : \mathbf{e}^*\right], \tag{5.7}$$

generalising (5.2) and (5.3) to a complex representation of random processes. Inserting (5.5) and using the formula

$$\mathrm{M}\left[\mathbf{E}(\omega) : \mathbf{E}^*(\omega_1)\right] = S(\omega)\, \delta(\omega - \omega_1)\,, \tag{5.8}$$

which is a sequence of (5.6), we obtain

$$\Gamma^2 = \int_{-\infty}^{\infty} S(\omega)\, d\omega, \tag{5.9}$$

where

$$S(\omega) = \frac{1}{2} S_{mnmn}(\omega)\,. \tag{5.10}$$

The function $S(\omega)$ is seen to characterise to some extent the spectral properties of the strain deviator. For this reason, $S(\omega)$ is conditionally called *the spectral density of strains*.

Characteristics and decompositions of other tensors may be introduced by analogy.

5.2 The method of statistical linearisation for analysing deformations

Consider an elastoplastic material whose behaviour is governed by the set of constitutive equations (3.2) and (3.3)

$$\mathbf{s} = 2G\left[\mathbf{e} - \int_0^\infty \mathbf{e}_h p(h)\, dh\right], \tag{5.11}$$

$$\left[\begin{array}{ll} \dot{\mathbf{e}}_h \neq 0, & h\dot{\mathbf{e}}_h / v_h = \mathbf{e} - \mathbf{e}_h, \\ \dot{\mathbf{e}}_h = 0, & \sqrt{\frac{1}{2}(\mathbf{e} - \mathbf{e}_h):(\mathbf{e} - \mathbf{e}_h)} \leq h. \end{array}\right. \tag{5.12}$$

Assume that the strain deviator $\mathbf{e}$ is a stationary random function of time. For the sake of simplicity, we assume that it has a zero mean value. The material behaviour described by this deformation law is required. Direct application of eqs. (5.11) and (5.12) meets considerable difficulties caused by the nonlinearity of the equations. Under these circumstances it is reasonable to make use of some methods of approximation. One of the most simple and effective methods of analysis of nonlinear control systems under random perturbations is *the method of statistical linearisation* [141], [158] and [190]. This method is applied below to the analysis of material behaviour due to a random deformation law.

The method of statistical linearisation occupies a special place among another methods. A general proof of the method does not exist. General estimates of its accuracy are absent, too. However, comparisons of results obtained by this method with some exact solutions and with solutions obtained by other methods of approximation shows that the accuracy of the method of statistical linearisation, in many problems, is acceptable, cf. [158]. This justifies to some extent an application of the method of statistical linearisation to elastoplastic vibrations.

Letting $\dot{\mathbf{e}}_h \neq 0$, the only nonlinear function in the constitutive equations (5.12) is the tensor function

$$\frac{h}{v_h}\dot{\mathbf{e}}_h, \tag{5.13}$$

where v_h is an intensity of the shear strain rate

$$v_h = \sqrt{\frac{1}{2}\dot{\mathbf{e}}_h : \dot{\mathbf{e}}_h}. \tag{5.14}$$

In accordance with the method of statistical linearisation the nonlinear function (5.13) must be approximated by the following linear one

$$\frac{h}{v_h}\dot{\mathbf{e}}_h \approx k_h \dot{\mathbf{e}}_h, \tag{5.15}$$

where k_h is a constant which is called *the linearisation factor*. The latter is found from the condition of minimising the expectation of the square of the error inherent from approximation (5.15)

$$\mathrm{M}\left[\frac{1}{2}\left(\frac{h}{v_h}\dot{\mathbf{e}}_h - k_h\dot{\mathbf{e}}_h\right):\left(\frac{h}{v_h}\dot{\mathbf{e}}_h - k\dot{\mathbf{e}}_h\right)\right] = \min_{k_h}. \tag{5.16}$$

By using (5.14) we rewrite (5.16) in a simpler form

$$\mathrm{M}\left[\left(\frac{h}{v_h} - k_h\right)^2 v_h^2\right] = \min_{k_h}. \tag{5.17}$$

The requirement of a minimum with respect to k_h leads to the following equation for k_h

$$\mathrm{M}\left[\left(\frac{h}{v_h} - k_h\right) v_h^2\right] = 0. \tag{5.18}$$

It is easy to show that

$$k_h = \frac{h\mathrm{M}\left[v_h\right]}{\mathrm{M}\left[v_h^2\right]}. \tag{5.19}$$

The mean values of v_h and v_h^2 are observed in this equation. To compute them, one needs the probability distribution law of v_h. However, this law remains unknown until the problem is solved. The problem is solved in the method of statistical linearisation by assuming that the probability distribution is already known. As v_h is a non-negative random function an acceptable approximation is the Rayleigh distribution

$$f\left(v_h\right) = \frac{2v_h}{D_h^2}\exp\left(-\frac{v_h^2}{D_h^2}\right), \quad v_h \geq 0, \tag{5.20}$$

where D_h^2 is the mean square of v_h. For this case, evaluation of the linearisation factor due to (5.19) yields the following result

$$k_h = \frac{nh}{D_h}, \tag{5.21}$$

where

$$n = \sqrt{\pi/4} = 0.88.$$

Of course, another distribution law will give a different linearisation factor. However, the value of the linearisation factor is known to be nearly insensitive to a particular distribution law, cf. [141]. It is easy to prove that the structure of eq. (5.21) remains for other distribution laws and only the factor n varies. Its values for some particular distribution laws are collected in the Table 5.1. The normal or the Gaussian distribution law for a random non-negative variable is recognised on the first line. The above Rayleigh law

The distribution law	Its plot	n
$f = \sqrt{\frac{2}{\pi}} \frac{1}{D_h} e^{-\frac{v_h^2}{2D_h^2}}, \ v_h \geq 0$	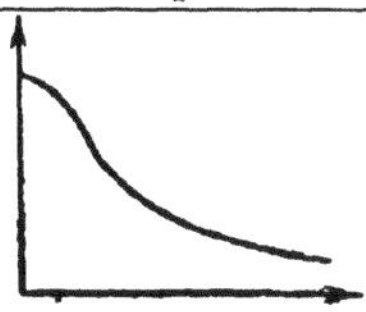	$\sqrt{\frac{2}{\pi}} \approx 0.80$
$f = \frac{2v_h}{D_h^2} e^{-\frac{v_h^2}{D_h^2}}, \ v_h \geq 0$	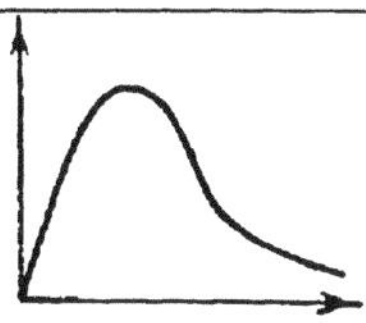	$\sqrt{\frac{\pi}{4}} \approx 0.88$
$f = \frac{1}{\sqrt{3}D_h}, \ 0 \leq v_h \leq \sqrt{3}D_h$	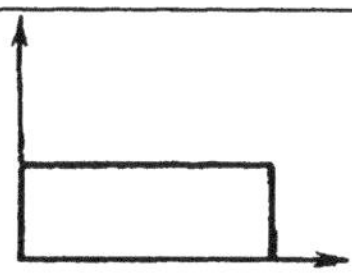	$\frac{\sqrt{3}}{2} \approx 0.85$
$f = \frac{v_h}{D_h^2}, \ 0 \leq v_h \leq \sqrt{2}D_h$	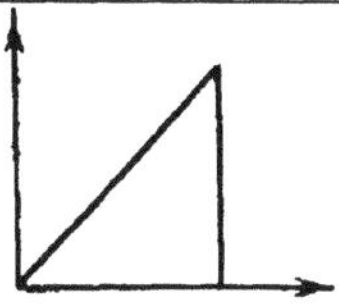	$\frac{2\sqrt{2}}{3} \approx 0.94$
$f = \delta(v_h - D_h), \ v_h \geq 0$	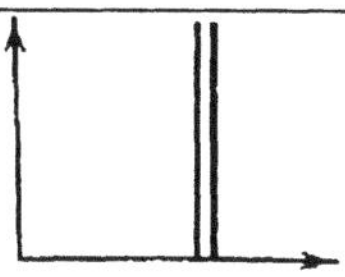	1
$f = \frac{2}{\pi} \frac{1}{\sqrt{2D_h^2 - v_h^2}}, \ 0 \leq v_h \leq \sqrt{2}D_h$	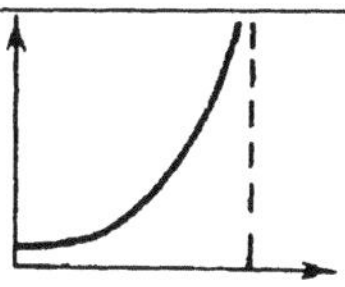	$\frac{2\sqrt{2}}{\pi} \approx 0.90$

TABLE 5.1.

is on the second line. The uniform, linear and delta-function distributions are on the next three. The distribution on the last line is of special interest. It is the law of the distribution of absolute values of harmonics of a fixed amplitude and an uniformly distributed phase. It is apparent that the linearisation factor in this case must coincide with that of the method of harmonic linearisation. Indeed, accounting for the fact that the harmonic amplitude $\omega\Gamma_h$ is related to the root-mean-square D_h by the equation

$$\omega\Gamma_h = \sqrt{2}D_h,$$

and taking n from the last line of Table 5.1 we obtain from (5.21) that

$$k_h = \frac{2\sqrt{2}}{\pi}\frac{1}{D_h} = \frac{4}{\pi}\frac{h}{\omega\Gamma_h}.$$

This coincides with expressions (3.75) and (3.76). Hence, the last distribution allows the relation between the methods of harmonic and statistical linearisation to be established. Analysis of Table 5.1 leads to the conclusion that the value of n depends weakly on the distribution law and is close to the following average value

$$n = 0.90.$$

This value of the factor n should be recommended for use in practical problems.

The first equation after the linearisation has been performed takes the form

$$k_h\dot{\mathbf{e}}_h + \mathbf{e}_h = \mathbf{e}. \tag{5.22}$$

If this equation has no solution, the second line of eq. (5.12) must be taken (see Section 3.1), i.e.

$$\dot{\mathbf{e}}_h = 0. \tag{5.23}$$

Since equations (5.22), (5.23) and (5.11) are linear the technique of spectral theory of random functions is applicable. By using eq. (5.22) and spectral decomposition of $\mathbf{e}$, eq. (5.5), we obtain the following spectral representation

$$\mathbf{e}_h = \int_{-\infty}^{\infty} \frac{e^{i\omega t}}{1 + i\omega k_h}\mathbf{E}(\omega)\, d\omega. \tag{5.24}$$

The square of the absolute value of velocity strain intensity $\dot{\mathbf{e}}_h$ is given by

$$|v_h|^2 = \iint_{-\infty}^{\infty} \frac{1}{2}\mathbf{E}(\omega) : \mathbf{E}(\omega_1)\frac{e^{i(\omega-\omega_1)t}\omega\omega_1 d\omega d\omega_1}{(1 + i\omega k_h)(1 - i\omega_1 k_h)}. \tag{5.25}$$

Taking the expected values in (5.25) yields the mean square

$$D_h^2 = M\left[|v_h|^2\right] = \int_{-\infty}^{\infty} \frac{\omega^2 S(\omega)\, d\omega}{1 + (\omega k_h)^2}. \tag{5.26}$$

Quantity D_h appears twice in the latter equation, explicitly in the left hand side and implicitly in the right hand side in terms of k_h, see eq. (5.21). Hence, expression (5.26) is an equation for D_h. If it has no solution, it means that the original equation (5.22) has no solution for $\dot{\mathbf{e}}_h \neq 0$. In this case one must proceed to eq. (5.23). For zero initial conditions we have

$$\mathbf{e}_h = 0, \tag{5.27}$$

which leads to the following result

$$D_h = 0. \tag{5.28}$$

It is easy to see that solution (5.27) is contained in (5.24) if one formally lets $D_h \to 0$ in (5.24). Indeed, by virtue of (5.21) $k_h \to \infty$ and thus $\mathbf{e}_h \to 0$, as required by (5.27). Hence, the solutions of eqs. (5.22) and (5.23) can be represented by a single formula (5.24).

Substituting expressions (5.24) and (5.5) into (5.11) we obtain a spectral representation for the stress deviator

$$\mathbf{s} = \int_{-\infty}^{\infty} 2G_c \mathbf{E}(\omega) e^{i\omega t} d\omega, \tag{5.29}$$

where G_c denotes the complex shear modulus given by

$$G_c = G \left[1 - \int_0^{\infty} \frac{p(h)\, dh}{1 + i\omega k_h} \right]. \tag{5.30}$$

Thus, the problem is again reduced to the concept of a complex shear modulus. It is worth mentioning that now the complex shear modulus is frequency-dependent, cf. its definition (5.30).

Relation (5.29) may be rewritten in the following conditional form

$$\mathbf{s} = 2G_c \mathbf{e}, \tag{5.31}$$

provided that it is understood as a separate equation for each harmonic component in $\mathbf{e}$ and $\mathbf{s}$. The similarity of eqs. (5.31) and (3.90) indicates that the problem of random vibrations permits a correspondence principle similar to that of Section 3.6.

5.3 A general study of the equation for complex shear modulus

When analysing eq. (5.30) the first and the most important task is to determine the actual integration interval. To solve this problem, we derive one important inequality. To this aim, we insert an explicit expression for the linearisation factor (5.21) into (5.26). The result is

$$D_h^2 = \int_{-\infty}^{\infty} \frac{\omega^2 S(\omega)\, d\omega}{1 + (\omega n h / D_h)^2}. \tag{5.32}$$

Neglecting the one in the denominator of the integrand we arrive at the inequality

$$D_h^2 \leq \frac{D_h^2}{(nh)^2} \int_{-\infty}^{\infty} S(\omega)\, d\omega. \tag{5.33}$$

Simplifying it by means of (5.9) gives

$$D_h^2 \leq D_h^2 \left(\frac{\Gamma}{nh}\right)^2. \tag{5.34}$$

One sees that if $D_h \neq 0$, then

$$\Gamma \geq nh. \tag{5.35}$$

Hence, non-trivial solutions of eq. (5.32) are possible only if inequality (5.35) holds. If it is violated, then due to Section 5.2 we must take $\dot{\mathbf{e}}_h = 0$ or $D_h = 0$. Therefore, the integration domain in (5.30) is as follows

$$0 \leq h \leq \Gamma / n. \tag{5.36}$$

As a result, we find that

$$G_c = G \left[1 - \int_0^{\Gamma/n} \frac{p(h)\, dh}{1 + i\omega k_h}\right]. \tag{5.37}$$

In order to find the structure of the solution of eq. (5.26), we rewrite it in the following form

$$D_h^2 = \frac{1}{k_h^2} \int_{-\infty}^{\infty} \frac{\left[1 + (\omega k_h)^2 - 1\right] S(\omega)\, d\omega}{1 + (\omega k_h)^2}. \tag{5.38}$$

Transforming the integral due to (5.9) and using k_h due to (5.21) we obtain from (5.38)

$$D_h^2 = \frac{D_h^2}{(nh)^2} \left[\Gamma^2 - \int_{-\infty}^{\infty} \frac{S(\omega)\, d\omega}{1 + (\omega k_h)^2}\right]. \tag{5.39}$$

Multiplying both sides of this equation by a finite multiplier $(nh/D_h)^2$ and combining the terms in the obtained equation we obtain

$$\Gamma^2 - (nh)^2 = \int_{-\infty}^{\infty} \frac{S(\omega)\, d\omega}{1 + (\omega k_h)^2}. \tag{5.40}$$

Dividing both sides by Γ^2 yields

$$1 - \left(\frac{nh}{\Gamma}\right)^2 = \int_{-\infty}^{\infty} \frac{S(\omega)}{\Gamma^2} \frac{d\omega}{1 + (\omega k_h)^2}. \tag{5.41}$$

The linearisation factor must be found from this equation. Note that the right hand side of this equation contains a normalised spectral density of strain. Hence, the right hand side can depend on the strain level Γ only by means of k_h. The left hand side depends upon the strain level Γ only by means of a non-dimensional combination (nh/Γ). This prompts the general structure of the solution of eq. (5.41), which is

$$k_h = \varphi\left(\frac{nh}{\Gamma}\right). \tag{5.42}$$

The particular form of function $\varphi(\eta)$ depends on features of the normalised spectral density of strain, but neither h nor Γ.

Substituting the linearisation factor (5.42) into the expression for the complex shear modulus (5.37) gives

$$G_c = G\left[1 - \int_0^{\Gamma/n} \frac{p(h)\, dh}{1 + i\omega\varphi(nh/\Gamma)}\right].$$

Introducing a new integration variable

$$\eta = nh/\Gamma, \tag{5.43}$$

we arrive at the following equation for the complex shear modulus, cf. [128] and [130]

$$G_c = G\left[1 - \int_0^{1} \frac{p\left(\frac{\Gamma}{n}\eta\right) \frac{\Gamma}{n} d\eta}{1 + i\omega\varphi(\eta)}\right]. \tag{5.44}$$

This expression allows the dependence on the strain level to be found for any distribution function. In particular, for a power law of yield stress distribution (2.28)

$$p(h) = \alpha H h^{\alpha-1},$$

we obtain

$$G_c = G\left(1 + l\Gamma^{\alpha}\right), \tag{5.45}$$

where

$$l = -\alpha H n^{-\alpha} \int_0^1 \frac{\eta^{\alpha-1} d\eta}{1 + i\omega\varphi(\eta)}. \tag{5.46}$$

Equation (5.45) for the complex shear modulus predicts a power dependence of the damping properties of the material on the root-mean-square of strain Γ. This conclusion reproduces to some extent the result obtained by analysis of harmonic deformation. But in this case damping proves to vary with frequency, as is seen from (5.46). It is unfortunately impossible to draw some general conclusions on the character of the frequency dependence of energy dissipation by means of (5.46).

For the forthcoming analysis of the frequency-dependent damping, it is convenient to write the expression for l as follows

$$l = H n^{-\alpha}\left(iI - R\right), \tag{5.47}$$

where

$$R - iI = \alpha \int_0^1 \frac{\eta^{\alpha-1} d\eta}{1 + i\omega\varphi(\eta)}. \tag{5.48}$$

We write down the governing equation for $\varphi(\eta)$

$$1 - \eta^2 = \int_{-\infty}^{\infty} \frac{S(\omega)}{\Gamma^2} \frac{d\omega}{1 + (\omega\varphi)^2}. \tag{5.49}$$

The latter has been deduced from (5.41) by using (5.42) and (5.43).

Consider equation (5.49) in some detail. Variable η varies in the interval $[0, 1]$. The unknown function $\varphi(\eta)$ is positive due to (5.21) and (5.42). Its values at the interval bounds are obtained from (5.49) and are given by

$$\begin{aligned} \eta &= 0, \quad \varphi = 0, \\ \eta &= 1, \quad \varphi = \infty. \end{aligned} \tag{5.50}$$

In order to answer the question of how $\varphi(\eta)$ varies within the interval, we differentiate the both sides of (5.49) with respect to η, to obtain

$$-2\eta = -2\varphi\varphi' \int_{-\infty}^{\infty} \frac{S(\omega)}{\Gamma^2} \frac{\omega^2 d\omega}{\left[1 + (\omega\varphi)^2\right]^2}. \tag{5.51}$$

Hence, $\varphi' > 0$ and, as η increases in the interval $[0, 1]$, φ increases from zero to infinity.

In some cases it seems more convenient to express the shear modulus not in terms of the root-mean-square of shear strain intensity Γ but in terms of the root-mean-square of shear stress intensity T given by the formula

$$T = \sqrt{\mathrm{M}\left[\frac{1}{2}\mathbf{s}:\mathbf{s}^*\right]}. \tag{5.52}$$

Inserting here the spectral decomposition of the stress deviator (5.29) and taking the expected values gives

$$T^2 = \int_{-\infty}^{\infty} |2G_c|^2 \, S(\omega)\, d\omega. \tag{5.53}$$

Returning to the spectral representation (5.29) it is easy to realize that the integrand in (5.53) is the spectral density of stresses

$$S_T(\omega) = |2G_c|^2 \, S(\omega), \tag{5.54}$$

and therefore

$$T^2 = \int_{-\infty}^{\infty} S_T(\omega)\, d\omega. \tag{5.55}$$

The term "spectral density of stresses" is conditional and has a meaning identical to that of the "spectral density of strains". Assuming that the effect of plastic strains is very small we obtain with great accuracy that

$$|2G_c| \approx 2G. \tag{5.56}$$

This allows us to simplify expressions (5.53) and (5.54) which finally take the following form

$$S_T = (2G)^2 \, S(\omega), \tag{5.57}$$

$$T = 2G\Gamma. \tag{5.58}$$

The latter equation allows the complex shear modulus (5.45) to be rewritten as follows

$$G_c = G\left[1 + l\left(\frac{T}{2G}\right)^{\alpha}\right]. \tag{5.59}$$

However, the elimination of the deformation parameters from the expression for G_c is not complete since G_c depends on l. The latter depends on $\varphi(\eta)$, cf. (5.46), which is governed by eq. (5.49) containing a normalised spectral density of strains. Removing the latter by means of (5.57) and (5.58) yields

$$1 - \eta^2 = \int_{-\infty}^{\infty} \frac{S_T(\omega)}{T^2} \frac{d\omega}{1 + (\omega\varphi)^2}. \tag{5.60}$$

For a practical realisation of the correspondence principle formulated in Section 5.2 analogies with Young's modulus and Poisson's ratio might be required. By analogy with (3.98) and (3.99) we have

$$E_c = E\left[1 + \frac{2}{3}(1+\nu)\,l\left(\frac{T}{2G}\right)^{\alpha}\right], \tag{5.61}$$

$$\nu_c = \nu - \frac{1-2\nu}{3}(1+\nu)\,l\left(\frac{T}{2G}\right)^{\alpha}. \tag{5.62}$$

These may be used for analysis of dynamic problems with a uniform state of stresses as it has been done above.

5.4 Polyharmonic deformation

A possible application of the method of statistical linearisation for the analysis of polyharmonic processes in nonlinear systems has been pointed out in [86], [87] and [88]. For this reason, when analysing polyharmonic processes we will use some general equations of the previous Sections obtained by the method of statistical linearisation.

For a polyharmonic deformation with harmonic frequencies ω_l and statistically independent phases, the latter being uniformly distributed in the interval $[0, 2\pi]$, equation (5.49) takes the form

$$1 - \eta^2 = \sum_l b_l^2 \left[1 + (\omega_l \varphi)^2\right]^{-1}; \quad \sum_l b_l^2 = 1. \tag{5.63}$$

Here the quantities b_l characterize relations between the intensities of the harmonic components in such a way that $(b_l \Gamma)^2$ is equal to the mean square of intensity of the shear strain with frequency ω_l.

For a single-frequency deformation $l = 1$, $b_1 = 1$ and eq. (5.63) becomes

$$1 - \eta^2 = \left[1 + (\omega_1 \varphi)^2\right]^{-1}. \tag{5.64}$$

The solution to this equation

$$\omega_1 \varphi = \frac{\eta}{\sqrt{1-\eta^2}} \tag{5.65}$$

leads to the following expression for the complex shear modulus

$$G_c = G\left[1 - \int_0^1 \frac{p\left(\frac{\Gamma}{n}\eta\right)\frac{\Gamma}{n}d\eta}{1 + ik\eta/\sqrt{1-\eta^2}}\right], \tag{5.66}$$

where

$$k = \omega/\omega_1. \tag{5.67}$$

As the deformation process is monoharmonic it sounds reasonable to speak about the shear modulus at the process frequency, i.e. put $k = 1$ in eq. (5.66). This gives

$$G_c = G\left[1 - \int_0^1 \frac{p\left(\frac{\Gamma}{n}\eta\right)\frac{\Gamma}{n}d\eta}{1 + i\eta/\sqrt{1-\eta^2}}\right]. \tag{5.68}$$

By means of some elementary transformations one can prove that the latter expression coincides with (3.87) up to the denotations for Γ and n, as was expected.

Let us proceed to the analysis of a biharmonic deformation with frequencies ω_1 and ω_2. Equation (5.63) takes the form

$$1 - \eta^2 = \frac{b_1^2}{1 + (\omega_1\varphi)^2} + \frac{b_2^2}{1 + (\omega_2\varphi)^2}, \tag{5.69}$$

$$b_1^2 + b_2^2 = 1. \tag{5.70}$$

Equation (5.69) can be reduced to a biquadratic equation for φ. Its solution is

$$(\omega_1\varphi)^2 = \frac{B + \sqrt{B^2 + 4k^2\eta^2\left(1-\eta^2\right)}}{2\left(1-\eta^2\right)} \tag{5.71}$$

where

$$k = \omega_1/\omega_2, \; B = b_1^2 + b_2^2k^2 - \left(1-\eta^2\right)\left(1+k^2\right). \tag{5.72}$$

It seems reasonable to speak about the values of the complex shear modulus only for the harmonic components with the frequencies ω_r $(r = 1, 2)$, i.e.

$$G_{cr} = G\left[1 - \int_0^1 \frac{p\left(\frac{\Gamma}{n}\eta\right)\frac{\Gamma}{n}d\eta}{1 + i\omega_r\varphi(\eta)}\right], \tag{5.73}$$

because there are no other components.

Note that $\omega_1\varphi$ is given by (5.71) and $\omega_2\varphi$ is easily obtained by means of the following equation

$$\omega_2\varphi = \omega_1\varphi/k. \tag{5.74}$$

Some numerical computations due to (5.73) will be performed later on. Before we proceed to these, we consider some limiting cases.

1. Let $k \to 1$. Equations (5.71) and (5.74) then yield

$$\omega_r\varphi = \frac{\eta}{\sqrt{1-\eta^2}}.$$

Substituting this into (5.73) leads to the already known result (5.68) for harmonic deformation.

2. Let $k \to 0$. This is the case in which the frequencies of harmonic components differ drastically, ω_2 being the higher frequency. Let us find asymptotic expressions for $\omega_r\varphi$ when $k \to 0$. Equation (5.71) yields the following approximate expression

$$(\omega_1\varphi)^2 = \frac{1}{2(1-\eta^2)}\left\{B + |B|\left[1 + \frac{2k^2\eta^2(1-\eta^2)}{B^2}\right]\right\}.$$

One can see from the latter equation that the asymptotic expression for $\omega_1\varphi$ depends strongly on the sign of $B = b_1^2 + b_2^2k^2 - (1-\eta^2)(1+k^2)$ when $k \to 0$. If $\eta > b_2$, we have $\omega_1\varphi = \sqrt{(\eta^2 - b_2^2)/(1-\eta^2)}$. If $\eta < b_2$, then $\omega_1\varphi = k\eta/\sqrt{(b_2^2 - \eta^2)}$. By using these asymptotic expressions, the following limiting values of $\omega_1\varphi$ and $\omega_2\varphi$ are obtained

$$\omega_1\varphi = \left[\begin{array}{ll} 0, & \eta < b_2, \\ \sqrt{(\eta^2 - b_2^2)/(1-\eta^2)}, & \eta > b_2, \end{array}\right. \tag{5.75}$$

$$\omega_2\varphi = \left[\begin{array}{ll} \eta/\sqrt{b_2^2 - \eta^2}, & \eta < b_2, \\ \infty, & \eta > b_2. \end{array}\right. \tag{5.76}$$

Substituting these into (5.73) leads to the required expressions for the complex shear moduli

$$G_{c1} = G\left[1 - \int_0^{b_2} p\left(\frac{\Gamma}{n}\eta\right)\frac{\Gamma}{n}d\eta - \int_{b_2}^{1} \frac{p\left(\frac{\Gamma}{n}\eta\right)\frac{\Gamma}{n}d\eta}{1 + i\sqrt{(\eta^2 - b_2^2)/(1-\eta^2)}}\right], \tag{5.77}$$

$$G_{c2} = G\left[1 - \int_0^{b_2} \frac{p\left(\frac{\Gamma}{n}\eta\right)\frac{\Gamma}{n}d\eta}{1 + i\eta/\sqrt{b_2^2 - \eta^2}}\right]. \tag{5.78}$$

The first expression gives a complex shear modulus for the component with a very low frequency, whereas the second one is for the component with a high frequency. Replacing in the latter integral the integration variable η by ηb_2 gives

$$G_{c2} = G\left[1 - \int_0^{1} \frac{p(\Gamma b_2\eta/n)\,\Gamma b_2 d\eta}{n\left(1 + i\eta/\sqrt{1-\eta^2}\right)}\right], \tag{5.79}$$

which coincides exactly with the expression for the shear modulus (5.68) for a monoharmonic deformation with intensity Γb_2, the latter being equal to the intensity of high frequency strain component of the biharmonic process.

Thus we can conclude that the presence of a very low frequency component in the deformation law does not influence the damping properties of the material at high frequency.

Another fact that is worth considering is that the imaginary part of the complex shear modulus (5.77) is not zero. Hence, the presence of a high frequency component does not totally suppress the damping ability of the material at low frequency. This fact is especially remarkable since it confirms a hysteretic character of the internal friction. Indeed, if the internal friction were not of a hysteretic character and were dependent on the strain rate, the above affect of suppression would take place.

3. An analysis of situations in which the component of the second frequency considerably prevails over the component of the first frequency is of interest. In order to analyse this situation we let $b_2 \to 1$ in (5.71). The result is

$$\omega_1 \varphi = \frac{k\eta}{\sqrt{1-\eta^2}}, \quad \omega_2 \varphi = \frac{\eta}{\sqrt{1-\eta^2}}.$$

This gives

$$G_{c1} = G\left[1 - \int_0^1 \frac{p\left(\frac{\Gamma}{n}\eta\right)\frac{\Gamma}{n}d\eta}{1 + ik\eta/\sqrt{1-\eta^2}}\right], \tag{5.80}$$

$$G_{c2} = G\left[1 - \int_0^1 \frac{p\left(\frac{\Gamma}{n}\eta\right)\frac{\Gamma}{n}d\eta}{1 + i\eta/\sqrt{1-\eta^2}}\right]. \tag{5.81}$$

As expected, the complex rigidity of the more powerful component G_{c2} does not depend upon the weak component. But the complex modulus of the weak component G_{c1} is totally dependent on the intensity of the strong component and the ratio of both frequencies. Coincidence of expressions (5.80) and (5.66) is somewhat unexpected.

In conclusion we consider some examples of numerical work from equations of the present Section. The complex shear modulus was taken in the form of (5.45), (5.47) and (5.48). Figures 5.1-5.5 show R_1, R_2, I_1 and I_2 versus k for various values of b_2 and $\alpha = 1$. The results were obtained by numerical integration.

Analysis of these figures leads to the following qualitative conclusions.

1. The damping of each frequency component depends on the global intensity Γ, intensities of the harmonic components and the ratio of their frequencies, but it does not depend upon the values of the frequencies.

2. The curve $I_2(k)$ lies above the level $I_2(0)$. By means of analysis of the limiting case it has been shown that $I_2(0)$ characterises the damping of the high frequency component when a low frequency component is absent. Hence, damping I_2 of the high frequency component is always higher with a low frequency component than without it.

3. The dashed lines in Figs 5.1-5.5 show the value of I_1 which corresponds to the damping of the low frequency component when a high frequency

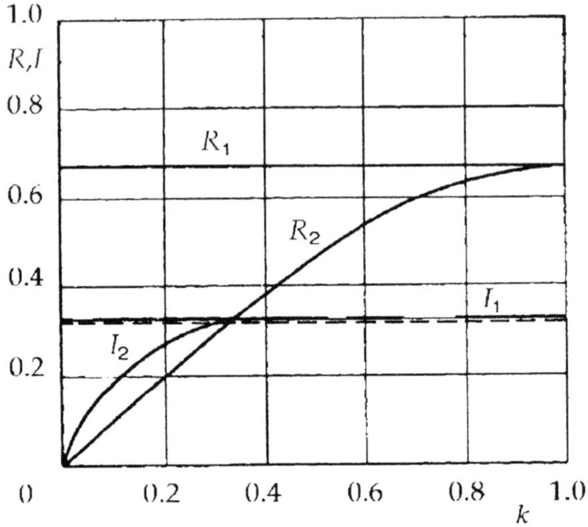

FIGURE 5.1. Parameters R_r and I_r for the first $(r = 1)$ and the second $(r = 2)$ harmonics versus k for $b_2^2 = 0$ and $\alpha = 1$.

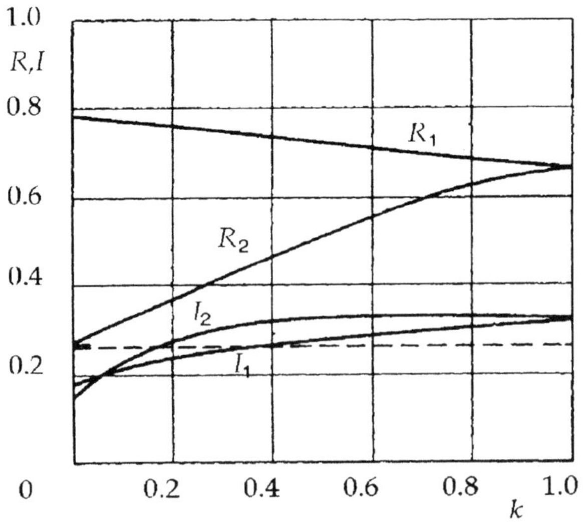

FIGURE 5.2. Parameters R_r and I_r for the first $(r = 1)$ and the second $(r = 2)$ harmonics versus k for $b_2^2 = 0.25$ and $\alpha = 1$.

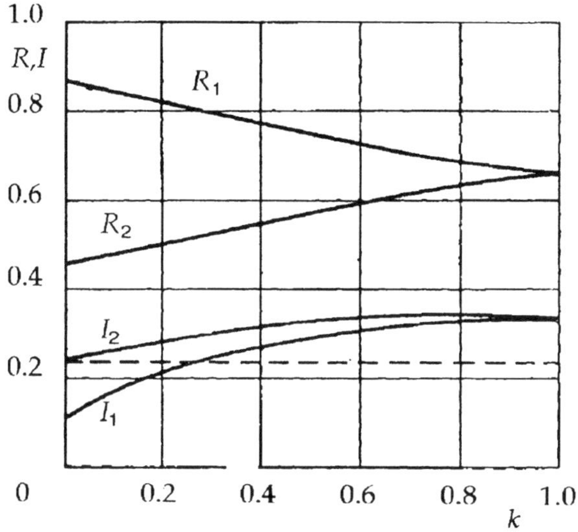

FIGURE 5.3. Parameters R_r and I_r for the first ($r = 1$) and the second ($r = 2$) harmonics versus k for $b_2^2 = 0.5$ and $\alpha = 1$.

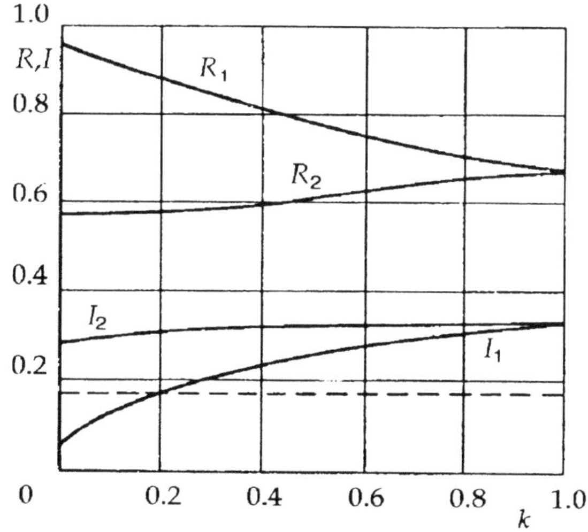

FIGURE 5.4. Parameters R_r and I_r for the first ($r = 1$) and the second ($r = 2$) harmonics versus k for $b_2^2 = 0.75$ and $\alpha = 1$.

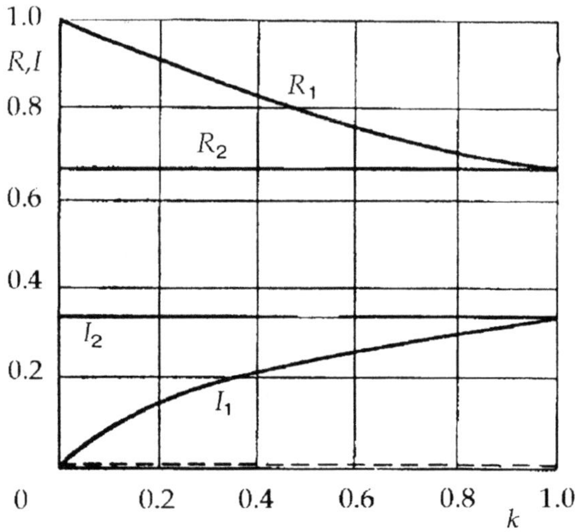

FIGURE 5.5. Parameters R_r and I_r for the first $(r=1)$ and the second $(r=2)$ harmonics versus k for $b_2^2=1$ and $\alpha=1$.

component is absent. This line crosses the curve $I_1(k)$ for any value of b_2 from the interval $[0,1]$. Hence, damping of the low frequency component is always higher with a high frequency component than without it for not very distinct frequencies $(k\approx 1)$ and lower for considerably distinct frequencies $(k\approx 0)$.

4. A total suppression of the damping ability of the lower frequency component $(k\approx 0)$ does not occur, since $I_1(0)\neq 0$ for $0<b_2<1$.

5.5 The simple case of broad band deformation

The simplest broad band random process is white noise. However this process is not suitable for analysis. The reason for this is that it has an unbounded root-mean-square Γ. The latter appears in the main equations of Section 5.3 which are not feasible for unbounded values of Γ. With this in view we consider an analysis of material behaviour for a random deformation with the following spectral density

$$S(\omega)=\frac{\Gamma^2\beta}{\pi\left(1+\beta^2\omega^2\right)}, \tag{5.82}$$

where β is positive. The root-mean-square of this process is bounded and equal to Γ^2. Substitution of (5.82) into (5.49) and evaluation of the integral yields the following equation for φ

$$1-\eta^2=\beta/(\beta+\varphi). \tag{5.83}$$

Its solution is

$$\varphi = \beta\eta^2 / \left(1 - \eta^2\right). \tag{5.84}$$

Inserting this into the expression for the complex shear modulus (5.44) gives

$$G_c = G\left[1 - \int_0^1 \frac{p\left(\frac{\Gamma}{n}\eta\right)\frac{\Gamma}{n}d\eta}{1 + i\Omega\eta^2/\left(1-\eta^2\right)}\right], \tag{5.85}$$

with Ω being a nondimensional frequency $\Omega = \omega\beta$.

For a power function of the yield stress distribution (2.28)

$$p\left(h\right) = \alpha H h^{\alpha-1}, \alpha > 0, \tag{5.86}$$

we arrive at equations for the complex shear modulus (5.45) and (5.47) in which

$$R - iI = \alpha \int_0^1 \frac{\eta^{\alpha-1}d\eta}{1 + i\Omega\eta^2/\left(1-\eta^2\right)}. \tag{5.87}$$

This integral may be expressed in terms of the hypergeometric function F as follows

$$R - iI = F\left(1, \frac{\alpha}{2}, 1 + \frac{\alpha}{2}, 1 - i\Omega\right) - \frac{\alpha}{2+\alpha}F\left(1, 1 + \frac{\alpha}{2}, 2 + \frac{\alpha}{2}, 1 - i\Omega\right). \tag{5.88}$$

As a matter of fact, the existence of this formula is useless because of the absence of suitable tables. For this reason, it seems reasonable to perform a numerical evaluation of the integral in (5.87). For α being a whole number integral (5.87) can be expressed in terms of elementary functions. For $\alpha \longrightarrow 0$ we have

$$R - iI = 1,$$

and the imaginary part is equal to zero. It indicates that a "linear internal damping" does not exist. The previous analysis of "internal damping" under harmonic deformation has already led to this conclusion.

For $\alpha \longrightarrow 1$ we obtain easily that

$$R - iI = \frac{1}{1 - i\Omega}\left[1 + \frac{i\Omega}{2\sqrt{1 - i\Omega}}\ln\frac{i\Omega}{\left(1 + \sqrt{1 - i\Omega}\right)^2}\right]. \tag{5.89}$$

The dependencies of R and I versus Ω are shown in Fig. 5.6.

For $\alpha \longrightarrow 2$ the integration in (5.87) yields

$$R - iI = \frac{1}{1 - i\Omega}\left[1 + \frac{i\Omega\ln i\Omega}{1 - i\Omega}\right]. \tag{5.90}$$

The results of computation due to the latter equation are shown in Fig. 5.7. The dashed lines in Figs 5.6 and 5.7 visualize the values of R_h and I_h

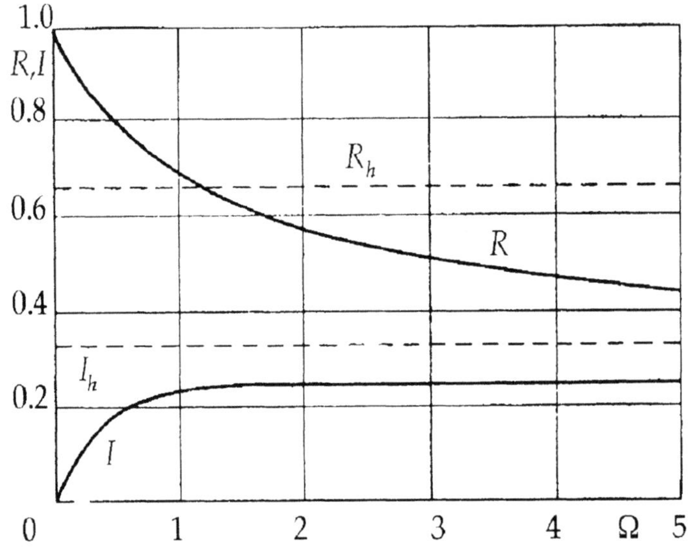

FIGURE 5.6. Parameters R, I and R_h, I_h versus Ω for $\alpha = 1$.

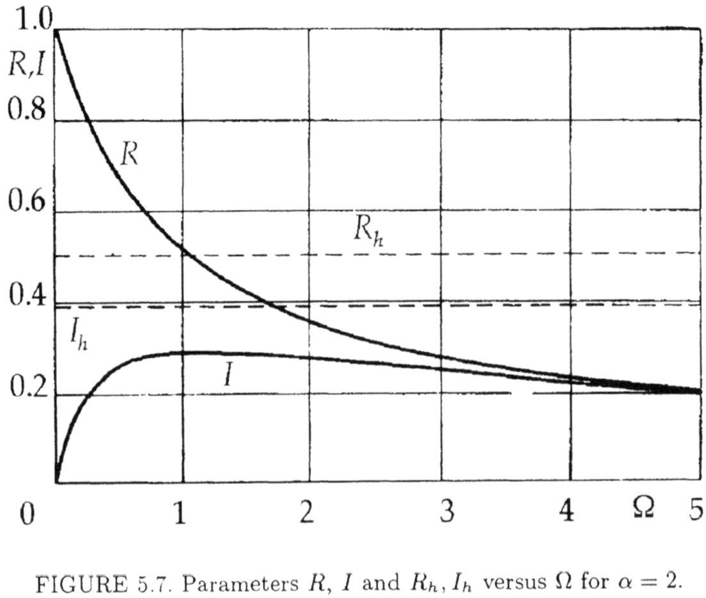

FIGURE 5.7. Parameters R, I and R_h, I_h versus Ω for $\alpha = 2$.

for harmonic deformation with the same value of Γ^2 as for the broad-band process.

Dependencies $I = I(\Omega)$ are of crucial interest, as they determine the damping ability of a material. Their consideration leads to the following qualitative conclusions.

1. The damping properties of a material under broad-band deformation do not strongly depend on frequency, at least in that part of the spectrum which is shown in the above figures. The main portion of energy of the random process is contained in this part of the spectrum.

2. The curves $I = I(\Omega)$ and $I_h = I_h(\Omega)$ are located rather close to each other in the main part of the spectrum. This allows us to draw the important practical conclusion that the error is expected to be relatively small if one takes the damping factor computed for a monoharmonic deformation with the same mean square as for the analysis of a broad-band deformation. Figures 5.6 and 5.7 indicate that the maximum error due to this approximation may be estimated as 50% of I_h.

5.6 On determination of the distribution density of defects $p(h)$

A density of defects distribution is needed for effective use of the equations of this Chapter. In the fourth Chapter wherein the problem of single-frequency vibrations was studied in detail, this question was not considered since for the analysis of vibration in a body of arbitrary form it is sufficient to know only the universal material characteristic, namely a function $\beta(\Gamma)$, which is an integral transform of $p(h)$. The function $\beta(\Gamma)$ allows immediate identification from a test with a uniform state of stress, say, pure shear. However, in random vibration problems the complex shear modulus depends essentially on the spectral density of strains, so one fails to introduce a universal integral characteristic and function $p(h)$ itself is required.

One of the possible ways to determine $p(h)$ is to use experimental data on damping properties of materials for a simple deformation process. Assuming that the latter is a harmonic process, we assume that the dependence of the *energy dissipation factor*

$$v\left(\frac{\Gamma}{n}\right) = \operatorname{Im}\frac{G_c}{G} \tag{5.91}$$

on parameter Γ/n is known, Im denoting an imaginary part.

Introducing the notation that

$$x = \Gamma/n,\ \eta = \sin\varphi,\ u(h) = hp(h) \tag{5.92}$$

and making use of (5.91) and (5.68) yields the following integral equation

$$v(x) = \int_0^{\pi/2} u(x \sin\varphi) \cos^2 \varphi d\varphi. \tag{5.93}$$

The solution to this equation gives the required answer on the posed question. Let us construct the following combination of parameters

$$w(x) = 2v(x) + xv'(x), \tag{5.94}$$

where the prime denotes derivation with respect to x. Then

$$w(x) = 2\int_0^{\pi/2} u(x \sin\varphi) \cos^2 \varphi d\varphi + \int_0^{\pi/2} xu'(x \sin\varphi) \cos^2 \varphi \sin\varphi d\varphi. \tag{5.95}$$

Simplifying the latter integral by means of integration by parts we arrive at the classical Schlömilch equation

$$w(x) = 2\int_0^{\pi/2} u(x \sin\varphi) d\varphi. \tag{5.96}$$

The only smooth solution to this equation, cf. [188], is given by

$$u(x) = \frac{2}{\pi}\left[w(0) + x\int_0^{\pi/2} w'(x \sin\varphi) d\varphi\right]. \tag{5.97}$$

One must put $v(0) = 0$ and thus $w(0) = 0$, otherwise the expression for $p(h)$ has a singularity

$$\frac{2}{\pi}\frac{w(0)}{h}$$

which is not integrable at $h = 0$. This contradicts the physical meaning of $p(h)$ which is a probability density, and, thus must be integrable. Putting $w(0) = 0$ in eq. (5.97) we obtain due to (5.92)

$$p(x) = \frac{2}{\pi}\int_0^{\pi/2} w'(x \sin\varphi) d\varphi. \tag{5.98}$$

The final result in the original notation is as follows

$$p(x) = \frac{2}{\pi}\left[\int_0^{\pi/2} 3v'(x \sin\varphi) + v''(x \sin\varphi)\, x \sin\varphi\right] d\varphi. \tag{5.99}$$

As follows from (5.99) the model of the elastoplastic material under consideration does not admit an amplitude-independent dissipation of energy. If v were constant then (5.99) would render $p = 0$ and thus $v = 0$ due to (5.93). The latter means that energy dissipation is absent. The same conclusion has been drawn in Section 2.4 based on an approximation of the distribution density in the form of a power function.

6

Random vibrations of elastoplastic bodies

6.1 The Galerkin method in the problem of random vibrations

Proceeding to the boundary-value problems of the theory of elastoplastic vibration, we make use of the variational equation of continuum mechanics, eq. (4.9)

$$\int\limits_V [\rho\,(\ddot{\mathbf{u}} - \mathbf{K}) \cdot \delta\mathbf{u} + \boldsymbol{\tau} : \delta\boldsymbol{\varepsilon}]\, dV - \int\limits_O \mathbf{p} \cdot \delta\mathbf{u} dO = 0. \tag{6.1}$$

The vector of displacement is sought in the form of a series expansion in terms of the free elastic normal modes $\mathbf{u}_l$ of the body

$$\mathbf{u} = \sum_{l=1}^{N} \mathbf{u}_l q_l\,(t)\,, \tag{6.2}$$

where q_l are functions of time only and referred to as the *generalised coordinates.*

The normal modes of the elastic vibrations of the body satisfy the known conditions of orthogonality

$$\begin{aligned} &\int\limits_V \rho\mathbf{u}_n \cdot \mathbf{u}_l dV = m_l \delta_{nl}, \\ &\int\limits_V (k\vartheta_n \vartheta_l + 4G\mathbf{e}_n : \mathbf{e}_l)\, dV = c_l \delta_{nl}, \end{aligned} \tag{6.3}$$

where ϑ_l and $\mathbf{e}_l$ denote the dilatation and the strain deviator for l-th normal mode, respectively, and δ_{nl} stands for the Kronecker delta.

The coefficients m_l and c_l are given as follows

$$m_l = \int_V \rho \mathbf{u}_l^2 dV, \; c_l = \int_V \left(k\vartheta_l^2 + 4G\Gamma_l^2\right) dV, \tag{6.4}$$

with Γ_l being the intensity of the shear strain due to l-th normal mode. Coefficients m_l and c_l are called the generalised mass and the generalised rigidity, respectively.

Considering only those variations in (6.1) which belong to class (6.2) and taking into account that the variations δq_l are arbitrary, we obtain by means of the orthogonality conditions (6.3) the following set of equations for the generalised coordinates

$$m_l \ddot{q}_l + B_l = Q_l, \; l = 1, 2, \dots N, \tag{6.5}$$

where

$$Q_l(t) = \int_V \rho \mathbf{K} \cdot \mathbf{u}_l dV + \int_O \mathbf{p} \cdot \mathbf{u}_l dO, \tag{6.6}$$

$$B_l = \int_V \boldsymbol{\tau} : \boldsymbol{\varepsilon}_l dV. \tag{6.7}$$

The system (6.5) is nonlinear since the relation between $\boldsymbol{\tau}$ and $\boldsymbol{\varepsilon}$ is given by the nonlinear equations of Section 3.1. Assuming that the external loads $\mathbf{K}$ and $\mathbf{p}$ are stationary random functions of time, the generalised forces $Q_l(t)$ turn out to be stationary and stationarily related random functions. A steady-state solution of system (6.5) is then governed by stationary and stationarily related random functions q_l. In order to facilitate the problem of their determination we perform a statistical linearization of the nonlinearity in eq. (6.5) before evaluating the volume integral (6.7). Using the linearization results of Section 5.2 yields

$$m_l \ddot{q}_l + c_l q_l + \sum_{n=1}^{N} D_{nl} q_n(t) = Q_l, \tag{6.8}$$

where

$$D_{nl} = \int_V 2(G_c - G)\, \mathbf{e}_n : \mathbf{e}_l dV. \tag{6.9}$$

While deriving eq. (6.8) we converted to the complex exponential form of the variables. The coefficients D_{nl} are complex frequency-dependent values, so that the form of eqs. (6.8) is as conventional as that of eq. (5.31).

In system (6.8) the effect of plastic deformation is taken into account by terms Q_{nl}. Provided that the latter are absent, i.e. $D_{nl} = 0$, system (6.8) takes the following form

$$m_l \ddot{q}_l + c_l q_l = Q_l \,. \tag{6.10}$$

This equation describes the driven vibrations of a set of non-interacting oscillators with the natural frequencies

$$\omega_l = \sqrt{\frac{c_l}{m_l}}. \tag{6.11}$$

It what follows, a typical case in which all natural frequencies are different is studied.

In applied problems of the theory of energy dissipation, the effect of plastic deformations is usually very small, that is $|D_{nl}| \ll c_l$. The asymptotically leading part of the solution is governed by the following system of equations

$$m_l \ddot{q}_l + (c_l + D_{ll})\, q_l = Q_l \,, \tag{6.12}$$

with a diagonal matrix of coefficients, cf. [38] and [189].

By means of a spectral representation of random functions

$$Q_l\,(t) = \int\limits_{-\infty}^{\infty} e^{i\omega t} V_l\,(\omega)\, d\omega, \tag{6.13}$$

with $V_l\,(\omega)$ being correlated white noises, it is easy to find the stationary solution of system (6.12)

$$q_l\,(t) = \int\limits_{-\infty}^{\infty} \frac{e^{i\omega t} V_l\,(\omega)\, d\omega}{c_l + D_{ll} - m_l \omega^2}. \tag{6.14}$$

Expression (6.14) is a spectral representation of the generalised coordinates. The probabilistic properties of the variables of interest can be easily found from the latter equation by means of the standard technique of the spectral theory of random functions. For example, a spectral representation of the strain deviator is as follows

$$\mathbf{e} = \int\limits_{-\infty}^{\infty} \sum_l \frac{\mathbf{e}_l V_l\,(\omega)\, e^{i\omega t} d\omega}{c_l + D_{ll} - m_l \omega^2}. \tag{6.15}$$

Inserting this into an expression for Γ^2 and taking the expected value yields

$$\Gamma^2 = \frac{1}{2} \sum_{n,\,l} \int\limits_{-\infty}^{\infty} \frac{\mathbf{e}_l : \mathbf{e}_n S_{ln}\,(\omega)\, d\omega}{(c_l + D_{ll} - m_l \omega^2)\,(c_n + D^*_{nn} - m_n \omega^2)}, \tag{6.16}$$

where $S_{ln}(\omega)$ is a cross-spectral density of the generalised forces Q_l and Q_n.

It should, however, be kept in mind that the problem of defining Γ is not completed since the expressions for G_c and thus D_{ll} depend on some root-mean-square values. In order to determine these, let us consider eq. (5.49). By virtue of (6.16) the spectral density of strains $S(\omega)$ which appears in this equation is as follows

$$S(\omega) = \frac{1}{2} \sum_{n,l} \frac{\mathbf{e}_l : \mathbf{e}_n S_{ln}(\omega)}{(c_l + D_{ll} - m_l\omega^2)(c_n + D_{nn}^* - m_n\omega^2)}. \tag{6.17}$$

Equation (5.49) then takes the form

$$1 - \eta^2 = \frac{1}{2\Gamma^2} \int_{-\infty}^{\infty} \sum_{n,l} \frac{\left[1 + (\omega\varphi)^2\right]^{-1} \mathbf{e}_l : \mathbf{e}_n S_{ln}(\omega)\, d\omega}{(c_l + D_{ll} - m_l\omega^2)(c_n + D_{nn}^* - m_n\omega^2)}. \tag{6.18}$$

An expression for D_{ll} that appears in the above equations is given by

$$D_{ll} = \int_V 4G\Gamma_l^2 \left\{ \int_0^1 \frac{1}{1 + i\omega\varphi(\eta)} p\left(\frac{\Gamma}{n}\eta\right) \frac{\Gamma}{n} d\eta \right\} dV, \tag{6.19}$$

with Γ_l being the intensity of the shear strain due to l-th normal mode.

Equations (6.16), (6.18) and (6.19) form a closed system of equations for the unknown variables Γ, D_{ll} and $\varphi(\eta)$. It should be noted that Γ is not a constant, but is dependent on the spatial coordinates, cf. (6.16). Function $\varphi(\eta)$ also depends upon the spatial coordinates, while coefficients D_{ll} are frequency-dependent. For this reason, the problem of solving the present system of equations is very difficult in general.

By formally substituting eqs. (6.16) and (6.19) into (6.18) the whole problem is reduced to a single equation for φ. Function φ is present in this equation not only explicitly, but also implicitly by means of the integral in D_{ll}. Therefore, this equation is an implicit integral equation.

Let us consider some particular external loads.

A practical situation of considerable interest is the case in which the external loads (6.13) are wide-band, i.e. they are assumed to have smooth spectral and cross-spectral densities. Consider first expression (6.16). For $D_{ll} \to 0$ such an integral

$$\int_{-\infty}^{\infty} \frac{S_{ln}(\omega)\, d\omega}{(c_l + D_{ll} - m_l\omega^2)(c_n + D_{nn}^* - m_n\omega^2)}$$

is bounded if $l \neq n$ whereas the following integral

$$I_l = \int_{-\infty}^{\infty} \frac{S_{ll}(\omega)\, d\omega}{|c_l + D_{ll} - m_l\omega^2|^2} \tag{6.20}$$

increases without bound for a smooth spectral density $S_{ll}(\omega)$. Therefore, for small D_{ll} the terms $n = l$ form the major contribution to the sum in (6.16), so that we have asymptotically

$$\Gamma^2 = \sum_l \Gamma_l^2 I_l. \tag{6.21}$$

Let us find an asymptotic representation of integral (6.20) for small D_{ll}. The main contribution (in accordance to the terminology of [36]) is due to the points $\omega = \pm\omega_l$, i.e. the following approximate equation is obtained

$$I_l = 2S_l \int_0^\infty \frac{d\omega}{\left|c_l + D_{ll}(\omega_l) - m_l\omega^2\right|^2}, \tag{6.22}$$

where

$$S_l = S_{ll}(\omega_l)$$

and $D_{ll}(\omega_l)$ stands for D_{ll} at $\omega = \omega_l$. As the denominator of the integrand in (6.22) is a quadratic function of ω^2 use is made of the following integral, cf. [44]

$$\int_0^\infty \frac{dz}{a + bz^2 + cz^4} = \frac{\pi \cos(\alpha/2)}{2cq^3 \sin\alpha}, \tag{6.23}$$

where

$$q = \sqrt[4]{\frac{a}{c}}, \quad \cos\alpha = -\frac{b}{2\sqrt{ac}}. \tag{6.24}$$

This leads to the following asymptotic value of the integral

$$I_l = \frac{\pi S_l}{m_l\omega_l d_l}, \tag{6.25}$$

where

$$d_l = \int_V 4G\Gamma_l^2 \left\{ \int_0^1 \frac{\omega_l\varphi(\eta)}{1 + [\omega_l\varphi(\eta)]^2} p\left(\frac{\Gamma}{n}\eta\right)\frac{\Gamma}{n}d\eta \right\} dV. \tag{6.26}$$

Equation (6.21) then takes the form

$$\Gamma^2 = \sum_l \frac{\pi\Gamma_l^2 S_l}{m_l\omega_l d_l}. \tag{6.27}$$

Similar evaluations allow eq. (6.18) to be simplified and represented in the form

$$1 - \eta^2 = \sum_l \frac{\pi\Gamma_l^2 S_l}{\Gamma^2 m_l\omega_l d_l} \frac{1}{1 + [\omega_l\varphi]^2}. \tag{6.28}$$

Equations (6.26)-(6.28) generate a system of equations for Γ, φ and d_l in the case of a wide-band loading.

It is worth noting that eq. (6.28) for φ has the same structure as eq. (5.63) in the case of a polyharmonic deformation provided that

$$b_l^2 = \frac{\pi \Gamma_l^2 S_l}{\Gamma^2 m_l \omega_l d_l}. \tag{6.29}$$

This is certainly not a coincidence, but reflects the fact that the vibrations of a lightly damped body under broad-band load are a set of narrow-band vibrations, their mean frequencies being close to the body eigenfrequencies.

The following case is of some interest to experimentalists. Several harmonics are tuned in resonances with some natural frequencies, but have random and independent phases. Assuming that the each phase is uniformly distributed in $[0, 2\pi]$ and retaining only asymptotically leading terms we obtain from (6.16) and (6.17)

$$\Gamma^2 = \sum_l \frac{\Gamma_l^2 a_l^2}{2}, \ 1 - \eta^2 = \sum_l \frac{\Gamma_l^2 a_l^2}{2\Gamma^2} \frac{1}{1 + [\omega_l \varphi]^2}, \tag{6.30}$$

where

$$a_l = A_l / d_l. \tag{6.31}$$

In these equations a_l and A_l imply amplitudes of harmonics with frequency ω_l in generalised coordinate q_l and generalised force Q_l, respectively. Equations (6.30) together with eq. (6.26) generate a system for Γ, φ and d_l in this particular case of loading.

Equations (6.12) can also be used for analysis of a slowly-decaying free vibration in an elastoplastic body. Putting $Q_l = 0$ in them and looking for a solution in the form

$$q_l(t) = \operatorname{Re}\left[a_l(t) e^{i\varphi_l(t)}\right], \tag{6.32}$$

where a_l is an amplitude and φ_l is a phase, we obtain by the method of slowly varying coefficients, see Section 2.6, the following equations for a_l and φ_l

$$\dot{\varphi}_l \approx \omega_l, \ \dot{a}_l \dot{\varphi}_l = -\frac{a_l d_l}{2m_l}. \tag{6.33}$$

Only asymptotically leading terms are retained in these equations, d_l being a small parameter. These equations enable us to find a logarithmic decrement of slowly-decaying vibration due to each mode of vibration

$$\Delta = \pi d_l / c_l. \tag{6.34}$$

We evaluate the integrals in eqs. (6.18) and (6.16) under the assumption that the initial values of the phases $\varphi_l(0)$ are statistically independent and uniformly distributed in the interval $[0, 2\pi]$. As a result we arrive at eq. (6.30) which together with eqs. (6.26) indicate a very complicated dependence of decrements (6.34) on the amplitude of the harmonics a_l.

Sometimes it is convenient to use the relations in which the stresses appear. The following formulae, cf. eqs. (5.58) and (3.1)

$$\begin{aligned} T &= 2G\Gamma, \\ T_l &= 2G\Gamma_l, \\ \sigma_l &= k\vartheta_l \end{aligned} \tag{6.35}$$

solve this problem. In these equations T stands for the root-mean-square of the shear stress intensity, whilst T_l and σ_l denote intensities of shear stress and mean normal stress for the l-th mode of elastic vibration, respectively.

As mentioned above, the performed analysis is valid only for sufficiently distinct natural frequencies of the body. The case of equal or a rather close natural frequencies gives rise to additional difficulties. The main complication is that the system (6.8) cannot be replaced by (6.12) because one has to take into account the non-diagonal elements D_{nl}. This question will not be studied here. It will only be pointed out that the equation derived may be applied to the case of equal natural frequencies $\omega_n = \omega_l$ provided that the non-diagonal elements D_{nl} vanish and the corresponding generalised forces Q_n and Q_l are statistically independent, i.e. $S_{ln} = 0$.

6.2 Random vibrations of an oscillator

Consider a simple example of an elastoplastic body under a broad band random loading. The system under consideration is depicted in Fig. 2.5, $p(t)$ implying now a broad-band random load. The system body consists of a lumped mass m and a cylindrical elastoplastic element of length L and cross-sectional area F. By analogy with Section 2.5 we assume that the mass of the elastoplastic element is much smaller than the lumped mass. In this case, the fundamental frequency of longitudinal vibration is essentially lower than the other eigenfrequencies of longitudinal vibration. Assuming that the loading spectrum is broad but does not reach the frequency domain of higher eigenfrequencies, we conclude that only the fundamental frequency may be included in the analysis.

Let us denote the vertical displacement of the lumped mass by q_1. The fundamental mode of the deformable element w_1 is then as follows

$$w_1 = z/L, \tag{6.36}$$

where z is the vertical coordinate, $z = 0$ corresponding to the fixed end. The only non-zero stress is the normal stress σ_z which is constant over the volume of the deformable element when the system vibrates due to its fundamental mode. Hence

$$T_1 = \sigma_z/\sqrt{3}, \tag{6.37}$$

and σ_z is expressed in terms of w_1 as follows

$$\sigma_z = E\frac{dw_1}{dz} = \frac{E}{L}, \tag{6.38}$$

where E is the Young's modulus of the material of the deformable element.

Neglecting the mass of the deformable elements and using the equations of Section 6.1, we obtain

$$m_1 = m, \tag{6.39}$$

$$c_1 = EF/L, \tag{6.40}$$

$$Q_1(t) = p(t). \tag{6.41}$$

Equations (6.26)-(6.28) together with eq. (6.35) lead to the following expressions

$$d_1 = FL\frac{T_1^2}{G}\int_0^1 \frac{\omega_1\varphi(\eta)}{1+[\omega_1\varphi(\eta)]^2} p\left(\frac{T\eta}{2Gn}\right)\frac{T}{2Gn}d\eta, \tag{6.42}$$

$$T^2 = \frac{\pi T_1^2 S_1}{m_1\omega_1 d_1}, \tag{6.43}$$

$$1-\eta^2 = \left\{1+[\omega_1\varphi(\eta)]^2\right\}^{-1}. \tag{6.44}$$

The latter equation coincides exactly with the equation for φ under monoharmonic deformation at frequency ω_1. The only value of frequency that appears in eq. (6.42) is ω_1. Hence, analysing the vibration of an oscillator under a broad-band excitation we can use the results of the energy dissipation theory derived for a harmonic deformation. This reflects the fact that a lightly damped oscillator driven by a wide-band random loading exhibits narrow-band random vibrations with a central frequency equal to the oscillator's eigenfrequency.

As a matter of fact, the example studied is concerned with the random vibration of a hysteretic oscillator. This problem has been reported in the literature, cf. [23], [43], [101] etc., and an approximate solution has been found. The solution relied very much on the assumption of narrow-band vibration of the oscillator which allows the dependencies for harmonic vibrations of the hysteretic element to be applied. The present analysis actually gives a formal substantiation of this technique. Moreover, the general equations of Section 6.1 allow us to omit the assumption of small hysteretic effects or narrow-band vibration. By means of eqs. (3.114) and (3.94), expression (6.42) can be transformed to the familiar form

$$d_1 = FL\frac{T_1^2}{G}D(T).$$

Function $D(T)$ has already appeared in the analysis of single-frequency vibrations of an elastoplastic body. It should, however, be kept in mind that in this case one must put $n=0.90$ in the equation for D, and not $n = 1.27$ as in Chapters 3 and 4. It has already been pointed out in Section 5.2. There also exists a difference in the arguments of the function D. Here T denotes a root-mean-square of intensity of shear stress whereas in the above Chapters it denoted amplitude of the shear stress intensity.

The above-said is completely applicable to vibrations in an arbitrary body provided that it is known *a priori* that one vibration mode is sufficient for the modelling. The resulting equations are

$$d_1 = \int\limits_V \frac{T_1^2}{G}D(T)\,dV, \tag{6.45}$$

$$T^2 = \frac{\pi T_1^2 S_1}{m_1\omega_1 d_1}. \tag{6.46}$$

Equations (6.45) and (6.46) provide us with a system of equations for unknown T and d_1. Substitution of the expression for T due to (6.46) into (6.45) yields an algebraic equation for d_1. After its solution, T can be found by means of (6.46). For example, for a power function (2.28) and a homogeneous material we obtain the following equation for d_1

$$d_1^{1+\alpha/2} = 4Gg\left(\frac{\pi S_1}{m_1\omega_1}\right)^{\alpha/2}\int\limits_V\left(\frac{T_1}{2G}\right)^{2+\alpha}dV. \tag{6.47}$$

Here g is given by (3.89) taking account of the above difference in values of n. Some examples of the evaluation of the integrals for certain bodies can be found in Chapter 4.

By means of the solution of eq. (6.47), the value of T is given by

$$T = AT_1S_1^{1/(2+\alpha)}, \tag{6.48}$$

where A is a constant. As seen from (6.48), an increase in the load intensity does not lead to a proportional increase in stresses. The latter grows more slowly than the load.

Let us proceed to considering of the resonant vibration of a body due to its arbitrary single mode. The first equation in (6.30) and equations (6.35) yield

$$T^2 = \frac{T_1^2A_1^2}{2d_1^2}. \tag{6.49}$$

Inserting this into (6.45) we obtain an equation for d_1 for this particular load. For a power dependence $p(h)$ it takes the form

$$d_1^{1+\alpha/2} = 4Gg\left(\frac{A_1}{\sqrt{2}}\right)^{\alpha}\int\limits_V\left(\frac{T_1}{2G}\right)^{2+\alpha} dV. \tag{6.50}$$

The stresses are found by means of (6.49) due to d_1 given by (6.50). The result is

$$T = BT_1A_1^{1/(1+\alpha)}, \tag{6.51}$$

with B being a constant. Comparison of eqs. (6.48) and (6.51) leads to the conclusion that the stresses grow more rapidly in the second case than in the first one for the same increasing load.

An analysis of decaying vibrations in a single-degree-of-freedom-system is performed due to the general equation (6.34). The vibration decrement is

$$\Delta = \left\{\int\limits_V \frac{1}{G}\left[\frac{3(1-2\nu)}{2(1+\nu)}\Sigma^2 + T^2\right] dV\right\}^{-1}\int\limits_V \frac{T^2}{G} D(T)\, dV, \tag{6.52}$$

where ν is the Poisson ratio and Σ denotes the root-mean-square of the mean normal stress. As might be expected, expressions (6.52) and (4.30) formally coincide. One should however be reminded of the different denotations in these expressions.

6.3 Analysis of experiments of biharmonic vibrations

The following experiment on a cantilever thin-walled tube-like specimen with a massive disc on its end has been reported in [143]. Assuming that the disc mass is much greater than the specimen mass, a practically uniform state of stresses is observed in the specimen when the system vibrates due to its first torsional and first longitudinal modes.

The torsional vibration causes a state of pure stress with an amplitude of shear stress A_t in a thin walled tube, hence

$$T = T_t,\ \Sigma = 0,\ T_t = A_t/\sqrt{2}. \tag{6.53}$$

Substituting these values into the general formula (6.52) we obtain the following expression for the decrement of torsional vibration

$$\Delta_t = D(T_t)\,. \tag{6.54}$$

Vibration due to the first mode of longitudinal vibration causes an uniaxial stress state, i.e.

$$T = T_l = \frac{A_l}{\sqrt{6}}, \; \Sigma = \frac{T}{\sqrt{3}}, \tag{6.55}$$

where A_l stands for the amplitude of the normal stress. By virtue of eqs. (6.55) and (6.52) we obtain the decrement of longitudinal vibration

$$\Delta_l = \frac{2(1+\nu)}{3} D(T_l). \tag{6.56}$$

In the case of joint torsional-longitudinal vibration due to the above modes we arrive at equations (6.30) and (6.35) which are as follows

$$T^2 = T_t^2 + T_l^2, \tag{6.57}$$

$$1 - \eta^2 = \frac{b_t^2}{1 + (\omega_t \varphi)^2} + \frac{b_l^2}{1 + (\omega_l \varphi)^2}, \; b_t = \frac{T_t}{T}, \; b_l = \frac{T_l}{T}, \tag{6.58}$$

with ω_t and ω_l denoting the frequency of torsional and longitudinal vibrations, respectively.

Direct evaluation shows that the following identity

$$\mathbf{e}_l : \mathbf{e}_t = 0 \tag{6.59}$$

holds, where $\mathbf{e}_l$ and $\mathbf{e}_t$ stand for the strain deviator of torsional and longitudinal vibration mode, respectively. Therefore, due to (6.9) $D_{nl} = 0$ for $n \neq l$. Taking into account the remark at the end of Section 6.1, we come to the conclusion that the equations of Section 6.1 are applicable in this case even if the frequencies of the torsional and longitudinal vibrations coincide. Equations for the decrements for each mode in the case of joint decaying vibrations are obtained by means of eqs. (6.34), (6.26) and (6.35). By virtue of eqs. (6.53) and (6.55) we find

$$\begin{aligned} \Delta_t &= \pi \int_0^1 \frac{\omega_t \varphi(\eta)}{1 + [\omega_t \varphi(\eta)]^2} p\left(\frac{T\eta}{2Gn}\right) \frac{T}{2Gn} d\eta, \\ \Delta_l &= \frac{2(1+\nu)}{3} \pi \int_0^1 \frac{\omega_l \varphi(\eta)}{1 + [\omega_l \varphi(\eta)]^2} p\left(\frac{T\eta}{2Gn}\right) \frac{T}{2Gn} d\eta. \end{aligned} \tag{6.60}$$

Consider some limiting cases. Let the frequencies of torsional and longitudinal vibrations coincide. Equation (6.58) is then converted into (6.44) which by virtue of general formulae (6.60), and leads to the following vibration decrements

$$\Delta_t = D\left(\sqrt{T_t^2 + T_l^2}\right), \ \Delta_l = \frac{2(1+\nu)}{3} D\left(\sqrt{T_t^2 + T_l^2}\right). \qquad (6.61)$$

Thus, they can be estimated by using dependence (6.54) for pure torsional vibrations. For the sake of simplicity, we restrict ourselves to the case of a linear dependence for $D(T)$, the latter being rather popular in practical applications and a good approximation to available experimental data. This enables expressions (6.61) to be represented in the form

$$\Delta_t = \Delta_{t0} B_t (T_l/T_t), \ \Delta_{t0} = D(T_t), \qquad (6.62)$$

$$\Delta_l = \Delta_{l0} B_l (T_t/T_l), \ \Delta_{l0} = \frac{2(1+\nu)}{3} D(T_l), \qquad (6.63)$$

where

$$B_t\left(\frac{T_l}{T_t}\right) = \sqrt{1 + \left(\frac{T_l}{T_t}\right)^2}, \ B_l\left(\frac{T_t}{T_l}\right) = \sqrt{1 + \left(\frac{T_t}{T_l}\right)^2}. \qquad (6.64)$$

The decrement of torsional vibration without a longitudinal one is denoted by Δ_{t0}, whilst B_t characterises the change caused by the presence of the longitudinal vibration. By analogy, Δ_{l0} means the decrement of longitudinal vibration without a torsional one, while B_l describes the change caused by joint torsional-longitudinal vibration.

Dependencies B_t and B_l versus their arguments are represented by the upper curves in Figs 6.1 and 6.2. These indicate an increase in the decrement of torsional vibration after superposition of a longitudinal vibration, and vice versa.

Let us consider the limiting case in which the longitudinal frequency is much higher than the torsional one. Considering the ratio ω_t/ω_l to be small and finding an asymptotic solution of eq. (6.58), we obtain as in Section 5.4

$$\omega_t \varphi = \left[\begin{array}{ll} \eta(\omega_t/\omega_l)/\sqrt{b_l^2 - \eta^2} & \eta < b_l, \\ \sqrt{(\eta^2 - b_l^2)/(1-\eta^2)}, & \eta > b_l. \end{array} \right. \qquad (6.65)$$

Substituting (6.65) into (6.60), we come to the following asymptotic expressions for the vibration decrements

$$\Delta_t = \frac{\pi}{b_t^2} \int_{b_l}^{1} \sqrt{(1-\eta^2)(\eta^2 - b_l^2)} p\left(\frac{T\eta}{2Gn}\right) \frac{T}{2Gn} d\eta, \qquad (6.66)$$

$$\Delta_l = \frac{2\pi(1+\nu)}{3b_l^2} \int_0^{b_l} \eta\sqrt{b_l^2 - \eta^2} p\left(\frac{T\eta}{2Gn}\right) \frac{T}{2Gn} d\eta. \qquad (6.67)$$

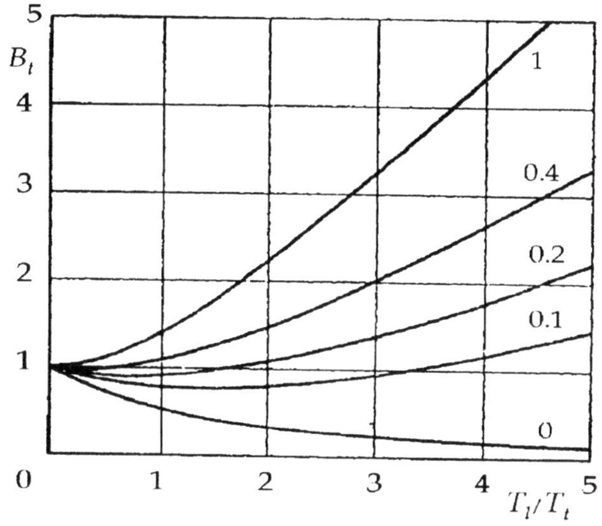

FIGURE 6.1. Dependence B_t versus ratio of shear stress intensities for some frequency ratios ω_t/ω_l.

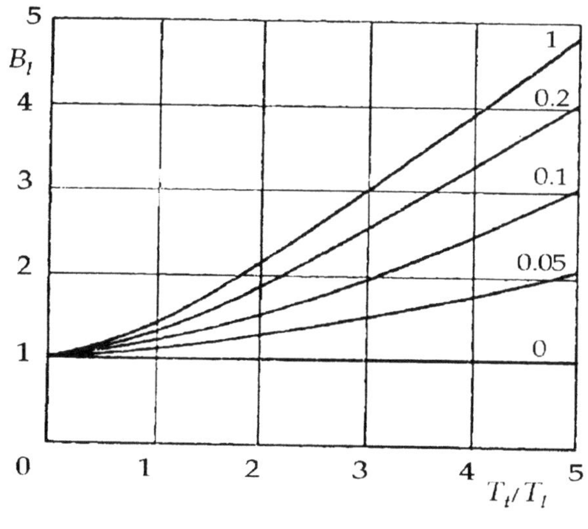

FIGURE 6.2. Dependence B_l versus ratio of shear stress intensities for some frequency ratios ω_t/ω_l.

Replacing in the latter integral its integration variable η by ηb_l leads to the following result

$$\Delta_l = \frac{2(1+\nu)}{3} D(T_l). \tag{6.68}$$

Expression (6.68) coincides exactly with the decrement of longitudinal vibration (6.56). Hence, the presence of a low frequency torsional vibration does not influence the decay of high frequency longitudinal vibration. The following fact is also worth noting. The decrement of torsional vibration (6.66) is never zero, i.e. the presence of high-frequency longitudinal vibration does not completely suppress the damping ability of a material at low frequency torsional vibration. However, it is rather difficult to give an answer on the question as to whether this damping ability increases or decreases due to superposition of the longitudinal vibration. The reason for this is the complexity of expression (6.66). Therefore, we study a particular case p=const in which the dependence $D = D(T)$ is linear due to eqs. (3.94) and (3.114). The integral in (6.66) is evaluated, yielding equations (6.62) and (6.63) in which

$$B_l = 1, \; B_t = \frac{1}{b_t}\left[2\mathrm{B}(b_t) - \mathrm{E}(b_t)\right], \tag{6.69}$$

where B and E are complete elliptic integrals. The dependencies (6.69) are represented by the lower curves in Figs. 6.1 and 6.2. The Figures indicate that superposition of a high frequency longitudinal vibration leads to a decrease in decrement of low frequency torsional vibration.

A theoretical analysis of the specimen behaviour for another frequency ratio and p= const has been performed by means of numerical integration in the general formulae (6.58) and (6.60). The result is given by eqs. (6.62) and (6.63), and the dependencies B_t and B_l versus their arguments are shown in Figs. 6.1 and 6.2. The numbers near the curves indicate the frequency ratio ω_t/ω_l.

The case in which the torsional frequency is equal to 20% of the longitudinal frequency has been observed in a test by Novikov [143] and [112]. Based on the test results the following qualitative conclusions were drawn:

a) Superposition of torsional vibration causes no change in the decay of a longitudinal one.

b) Superposition of longitudinal vibration causes a change in the decay of torsional vibration, this decay increasing with increase of amplitude of normal stress.

Generally speaking, Figures 6.1 and 6.2 lead to the opposite conclusions. The conclusions of the above test and the present theory coincide if $T_l/T_t > 2$, which was not provided by the test. This discrepancy between the theoretical and experimental conclusions does not give sufficient reason to reject the model of elastoplastic material in the theory of internal damping. Only one material has been tested in the above experiment. Therefore, one can only assert that this particular material exhibits more complicated behaviour than the present theory predicts.

Let us proceed now to other experimental investigations. Bending-torsional vibration of a clamped-clamped solid round rod with a massive disc in the middle has been studied in [147], [83] and [82]. The disc was attached to the rod without eccentricity. Two vibration modes were excited, namely, the first mode of torsion and the first mode of bending. The decrements of each mode were measured separately.

Theoretical analysis in this case must be done by means of the general formulae (6.34), (6.26) and (6.4). In cylindrical coordinates r, θ, z, the z axis being the axial coordinate of the rod, the stress distribution due to the first torsional mode is given by

$$T = T_t \rho, \;\; \Sigma = 0. \tag{6.70}$$

Here ρ is a nondimensional radius, $0 < \rho < 1$, and T_t is the root-mean-square of intensity of shear stress on the rod surface. When the rod vibrates due to its first bending mode, the stress distribution is as follows

$$T = T_b \rho \left|(1 - |\zeta|) \cos\theta\right|, \;\; \Sigma = \frac{T_b \rho}{\sqrt{3}} \left|(1 - |\zeta|) \cos\theta\right|, \tag{6.71}$$

a non-dimensional axial coordinate ζ changing within the interval $[-2, 2]$ and T_b denoting the maximum root-mean-square of shear stress intensity (at clamping).

Direct evaluation shows that in this case the following equation holds, cf. (6.59)

$$\mathbf{e}_b : \mathbf{e}_t = 0, \tag{6.72}$$

where $\mathbf{e}_b$ and $\mathbf{e}_t$ stand for the strain deviator of the bending and longitudinal vibration modes, respectively. Therefore, the equations of Section 6.1 are applicable for any frequencies of the torsional and longitudinal vibrations even if they coincide.

For joint bending-torsional vibration due to the above modes, we obtain by means of eqs. (6.30) and (6.35)

$$T = \rho \sqrt{T_t^2 + T_b^2 (1 - |\zeta|)^2 \cos^2\theta}, \tag{6.73}$$

$$1 - \eta^2 = \frac{b_t^2}{1 + (\omega_t \varphi)^2} + \frac{b_b^2}{1 + (\omega_b \varphi)^2}, \tag{6.74}$$

where

$$b_t = \frac{T_t \rho}{T}, \;\; b_l = \rho \frac{T_b \left|(1 - |\zeta|) \cos\theta\right|}{T}, \tag{6.75}$$

and ω_t and ω_b are the eigenfrequencies of torsional and bending vibration, respectively. The decrements of torsional and bending vibration are obtained by eqs. (6.34), (6.26) and (6.4). By virtue of the stress distributions

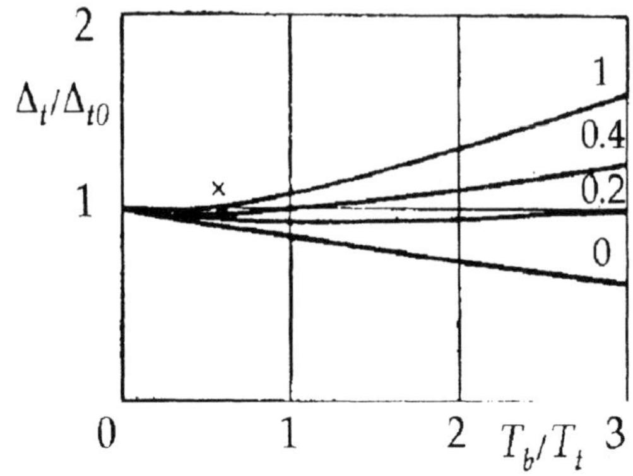

FIGURE 6.3. Ratio of torsional decrements versus ratio of shear stress intensities for some values of k.

(6.70) and (6.71), we obtain

$$\Delta_t = \int_0^1\!\!\int\!\!\int\int_0^{\pi/2} \frac{8\omega_t\varphi(\eta)}{1+[\omega_t\varphi(\eta)]^2} p\left(\frac{T\eta}{2Gn}\right)\frac{T\rho^3}{2Gn} d\eta d\rho d\zeta d\theta, \tag{6.76}$$

$$\Delta_b = \int_0^1\!\!\int\!\!\int\int_0^{\pi/2} \frac{32(1+\nu)\omega_b\varphi(\eta)}{1+[\omega_b\varphi(\eta)]^2} p\left(\frac{T\eta}{2Gn}\right)\frac{T\rho^3}{2Gn}(1-\zeta)^2\cos^2\theta d\eta d\rho d\zeta d\theta \tag{6.77}$$

where T is given by (6.73).

The computations due to (6.76), (6.77) and (6.74) can be done only by means of numerical integration. The results are depicted in Figs. 6.3 and 6.4, Δ_{t0} denoting the decrement of the torsional vibration without the bending one, and Δ_{b0} denoting the decrement of the bending vibration without the torsional one. The numbers near the curves correspond to the following ratio

$$k = \omega_t/\omega_b.$$

The computations were performed under the assumption that function $p(h)$ was independent of its argument h and the plastic properties of the material may depend on the radius ρ. The latter two diagrams are similar to the curves of Figs. 6.1 and 6.2. One difference is however striking, namely the influence of the bending vibration on the torsional one becomes weaker, whereas the influence of the torsional vibration on the bending becomes stronger.

The tests reported in the papers [147], [83] and [185] dealt with the case where the frequency ratio k for all materials under consideration was close to 0.5, i.e. $k \approx 0.5$. The experimental results are shown in Figs. 6.3 and

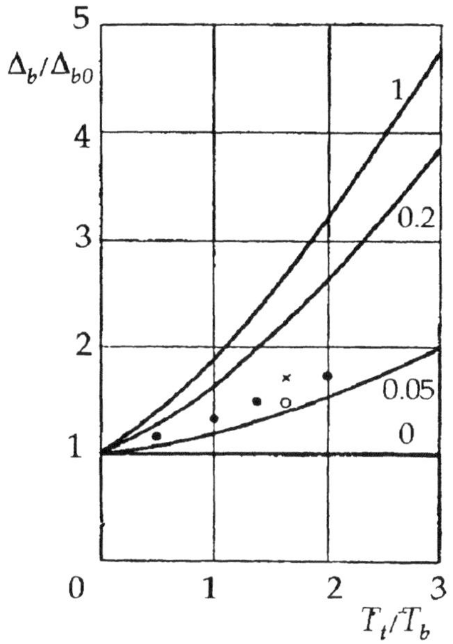

FIGURE 6.4. Ratio of bending decrements versus ratio of shear stress intensities for some values of k.

6.4 by points [83], crosses [147] and circles [185]. A rather good qualitative agreement of tests and theory is seen, however a considerable quantitative discrepancy for Δ_b is observed.

Taking into account what was said in Section 4.2, we may draw the general conclusion that the plasticity theory enables numerous fine effects of the energy dissipation in vibrating materials to be correctly predicted.

6.4 Differential equations of vibration in thin plates

Analysis of vibration in thin plates and thin shells driven by a broad-band distributed load produces considerable difficulties even in the framework of linear theory, e.g. for a "linear damping". The main complication is the broad band spectrum of external load and high modal density, i.e. a very large numbers of modes of vibration must be taken into account when the solution is sought in the form of an expansion in terms of normal modes, cf. [17]. On this account, such values of interest as deflections, accelerations, bending moments etc. require a very large number of terms in the expansion, making it difficult to draw any general conclusions about the influence of various physical parameters.

An idea for the approximate computation of sums for typical variables can be found in [18], [19] and [31]. For example, some rather simple expressions for the spectral density of a plate driven by excitation which is a spatial white noise and a stationary function of time with an arbitrary spectral density, are derived in paper [18] by means of the concept of dynamic fringe effects [18], [21] and [22]. An approximate evaluation of the series for a cylindrical shell is given in [31].

As shown in [118], [119], [138], [139] and [120] by means of linear damping theory, the same simple equations are obtained by consideration of an infinite and a semi-infinite plate, or an "infinite" and "semi-infinite" shell. Essential simplifications have been gained by using the spectral theory of spatial-temporal uniform random fields. The same equations can be successfully applied to plates and shells of finite but considerably large sizes. Indeed, the perturbations caused by boundary conditions decay with the distance from the boundary due to damping. Hence, for considerably extended plates and shells these perturbation are essential near the boundaries and not essential far from the boundaries. Therefore, the boundary conditions can be completely ignored when dealing with the probabilistic properties of deflection, acceleration and bending moment at points remote from the boundary, i.e. it allows us to consider an "infinite" plate or an "infinite" shell. Further, if one needs the probabilistic properties near a boundary it is necessary to take into account the appropriate boundary condition. Altogether this leads to the concept of a "semi-infinite" plate or shell.

The above-said is relevant to "linear friction". Nonlinearity of the internal friction in materials causes some additional complications, as will be shown below. For this reason, we start with the simplest case of an infinite model.

Consider a thin plate. Let the plane xy of the orthogonal coordinate system lie in the mid-plane of the plate. Let w denote the transverse deflection of the plate. It is shown in the theory of thin plate bending that the principal stresses in a thin plate are given by

$$\begin{aligned} \sigma_x &= -\frac{zE}{1-\nu^2}\left(\frac{\partial^2 w}{\partial x^2}+\nu\frac{\partial^2 w}{\partial y^2}\right), \\ \sigma_y &= -\frac{zE}{1-\nu^2}\left(\frac{\partial^2 w}{\partial y^2}+\nu\frac{\partial^2 w}{\partial x^2}\right), \\ \tau_{xy} &= -\frac{zE}{1-\nu^2}\left(1-\nu\right)\frac{\partial^2 w}{\partial x \partial y}. \end{aligned} \tag{6.78}$$

Using the correspondence principle, cf. Sections 5.2 and 5.3 we state that in order to take into account the effects of plastic deformations, the latter equations must be replaced by the following ones

$$\begin{aligned}
\sigma_x &= -\frac{zE_c}{1-\nu_c^2}\left(\frac{\partial^2 w}{\partial x^2}+\nu_c\frac{\partial^2 w}{\partial y^2}\right), \\
\sigma_y &= -\frac{zE_c}{1-\nu_c^2}\left(\frac{\partial^2 w}{\partial y^2}+\nu_c\frac{\partial^2 w}{\partial x^2}\right), \\
\tau_{xy} &= -\frac{zE_c}{1-\nu_c^2}\left(1-\nu_c\right)\frac{\partial^2 w}{\partial x \partial y},
\end{aligned} \tag{6.79}$$

the complex moduli E_c and ν_c being given by eqs. (5.61) and (5.62). These contain the mean square of the shear stress intensity T^2. The latter can be found from its definition (5.52) by virtue of (6.79) with terms of the order of the plastic strains being neglected. The simplest, most convenient, and practical expression for T^2 is given by

$$T^2 = (2GzN)^2, \tag{6.80}$$

where

$$N^2 = \frac{1}{2}\mathrm{M}\left[\left|\frac{\partial^2 w}{\partial x^2}\right|^2+\left|\frac{\partial^2 w}{\partial y^2}\right|^2+2\left|\frac{\partial^2 w}{\partial x \partial y}\right|^2-\frac{1-4\nu+\nu^2}{3\left(1-\nu\right)^2}\left|\Delta w\right|^2\right]. \tag{6.81}$$

Here M denotes the operation of mathematical expectation. Equation (6.81) is suitable for using spectral representations of random processes both in real and complex form.

With the help of eqs. (6.80) and (6.81) it is easy to understand that the spectral density of stresses has the following expression

$$S_T^2 = (2Gz)^2 S_N, \tag{6.82}$$

where

$$S_N = \frac{1}{2}\left[S_{xx}+S_{yy}+2S_{xy}-\frac{1-4\nu+\nu^2}{3\left(1-\nu\right)^2}S_\Delta\right]. \tag{6.83}$$

The spectral densities of the second derivatives of deflection with respect to the spatial coordinates and the Laplace operator appear in the latter equation. It is worth noting that eqs. (6.80) and (6.82) are explicitly dependent on z. Variables N and S_N do not depend upon z.

The coefficients in eq. (6.79) have the following asymptotic expressions

$$\frac{E_c}{1-\nu_c^2} = \frac{E}{1-\nu^2}\left[1+\frac{2}{3}\frac{1-\nu+\nu^2}{(1-\nu)}l\left(\omega\right)\left(\frac{T}{2G}\right)^\alpha\right], \tag{6.84}$$

$$\frac{E_c\nu_c}{1-\nu_c^2} = \frac{E\nu}{1-\nu^2}\left[1-\frac{1-4\nu+\nu^2}{3\nu\left(1-\nu\right)^2}l\left(\omega\right)\left(\frac{T}{2G}\right)^\alpha\right]. \tag{6.85}$$

Inserting these into (6.79) and taking into account the representation for T due to (6.80), we obtain an equation for the normal stress σ_x

$$\begin{aligned}\sigma_x &= -\frac{Ez}{1-\nu^2}\left[1+\frac{2}{3}\frac{1-\nu+\nu^2}{(1-\nu)}lN^\alpha |z|^\alpha\right]\frac{\partial^2 w}{\partial x^2} - \\ &\quad -\frac{E\nu z}{1-\nu^2}\left[1-\frac{1-4\nu+\nu^2}{3\nu(1-\nu)^2}lN^\alpha |z|^\alpha\right]\frac{\partial^2 w}{\partial y^2}.\end{aligned} \tag{6.86}$$

Similar equations for σ_y and τ_{xy} can also be determined.

In accordance with the traditional scheme for the theory of thin plates we introduce two bending moments M_x and M_y and a torsional moment H such that

$$M_x = \int_{-h/2}^{h/2} \sigma_x z dz, \quad M_y = \int_{-h/2}^{h/2} \sigma_y z dz, \; H = \int_{-h/2}^{h/2} \tau_{xy} z dz. \tag{6.87}$$

Substituting equations for σ_x, σ_y and τ_{xy} (see for example eq. (6.86)) into the latter equations and evaluating the integrals over z, we obtain

$$\begin{aligned} M_x &= -D_c\left(\frac{\partial^2 w}{\partial x^2}+\sigma_c\frac{\partial^2 w}{\partial y^2}\right), \\ M_y &= -D_c\left(\frac{\partial^2 w}{\partial y^2}+\sigma_c\frac{\partial^2 w}{\partial x^2}\right), \\ H &= -D_c(1-\sigma_c)\frac{\partial^2 w}{\partial x \partial y}. \end{aligned} \tag{6.88}$$

The factors D_c and σ_c are introduced in such a way that equations (6.88) have the structure conventional in plate theory. The factors D_c and σ_c are given by

$$\begin{aligned} D_c &= D[1+R(\omega)], \\ D_c\sigma_c &= D\nu[1+P(\omega)], \end{aligned} \tag{6.89}$$

where

$$R(\omega) = \frac{1-\nu+\nu^2}{(1-\nu)}\frac{2}{\alpha+3}\left(\frac{hN}{2}\right)^\alpha l(\omega), \tag{6.90}$$

$$P(\omega) = -\frac{1-4\nu+\nu^2}{\nu(1-\nu)(\alpha+3)}\left(\frac{hN}{2}\right)^\alpha l(\omega). \tag{6.91}$$

In what follows, D_c is referred to as the *complex bending rigidity*.

The introduced moments satisfy the governing equation of the plate dynamics

$$m\ddot{w} - \frac{\partial^2 M_x}{\partial x^2} - 2\frac{\partial^2 H}{\partial x \partial y} - \frac{\partial^2 M_y}{\partial y^2} = p, \tag{6.92}$$

where m is the mass per unit area and $p(x, y, t)$ denotes an external transverse distributed load.

Equations (6.88) and (6.92) formally coincide with the equations of elastic plate. The factors D_c and σ_c depend upon parameter N which is, in general, a function of the x and y coordinates. For this reason, equations (6.88) and (6.92) correspond to the equations of a heterogeneous plate. The law of heterogeneity is not known in advance as it depends upon an unknown parameter N.

Note, finally, that by virtue of (6.80) and (6.82), eq. (5.60) takes the form

$$1 - \eta^2 = \int_{-\infty}^{\infty} \frac{S_N}{N^2} \frac{d\omega}{1 + (\omega\varphi)^2}. \tag{6.93}$$

Coordinate z does not appear in this equation.

6.5 Vibrations in an infinite thin plate under broad band random loading

Consider an infinite plate in both directions under a distributed transverse load $p(x, y, t)$. The plate is assumed to be clamped at infinity which is equivalent to the requirement of a bounded deflection at infinity. Let the load $p(x, y, t)$ represent a statistically homogeneous spatial-temporal random field with a zero mean value.

It is evident that under these conditions the variables are homogeneous random fields, i.e. they have constant root-mean-squares and mean values. Therefore, the value of N and also the parameters D_c and σ_c which depend upon N are independent on x and y. In this case, by substitution of eq. (6.88) into (6.92), we reduce the problem to the analysis of a single differential equation of vibration for a homogeneous infinite plate

$$m\ddot{w} + D_c \Delta\Delta w = p(x,\, y,\, t)\,. \tag{6.94}$$

This equation coincides with that for vibration of a plate with linear damping.

Basically, the forthcoming analysis reproduces the results obtained by the author in [118] where a linear problem was considered. We represent the load by its spectral decomposition

$$p = \iiint_{-\infty}^{\infty} e^{i(\omega t + \lambda x + \mu y)} V(\omega,\, \lambda,\, \mu)\, d\omega d\lambda d\mu, \tag{6.95}$$

where $V(\omega, \lambda, \mu)$ denotes a random function of its arguments and is assumed to be a sort of three-dimensional white noise with the correlation function

$$\mathrm{M}\left[V\left(\omega,\lambda,\mu\right)V^{*}\left(\omega_1,\lambda_1,\mu_1\right)\right]=S_p\left(\omega,\lambda,\mu\right)\delta\left(\omega-\omega_1\right)\delta\left(\lambda-\lambda_1\right)\delta\left(\mu-\mu_1\right). \tag{6.96}$$

Function $S_p(\omega,\ \lambda,\ \mu)$ is termed the spectral density of the pressure field. It is known, cf. [190], that the correlation function K_p is related directly to the spectral density S_p by means of the Fourier transform

$$K_p\left(\tau,\ \xi,\eta\right)=\iiint\limits_{-\infty}^{\infty} e^{i(\omega\tau+\lambda\xi+\mu\eta)}S_p\left(\omega,\ \lambda,\ \mu\right)d\omega d\lambda d\mu, \tag{6.97}$$

$$S_p\left(\omega,\ \lambda,\ \mu\right)=\frac{1}{(2\pi)^3}\iiint\limits_{-\infty}^{\infty} e^{i(\omega\tau+\lambda\xi+\mu\eta)}K_p\left(\tau,\ \xi,\eta\right)d\omega d\lambda d\mu, \tag{6.98}$$

where τ is the time difference, while ξ and η denote the differences of the spatial coordinates.

Spectral decomposition for the deflection is sought in the form

$$w=\iiint\limits_{-\infty}^{\infty} e^{i(\omega t+\lambda x+\mu y)}F\left(\omega,\ \lambda,\ \mu\right)d\omega d\lambda d\mu. \tag{6.99}$$

Inserting (6.99) and (6.95) into (6.94) we obtain F, and then using (6.99) the following spectral representation for the deflection is obtained

$$w=\iiint\limits_{-\infty}^{\infty}\frac{e^{i(\omega t+\lambda x+\mu y)}}{D_c\left(\lambda^2+\mu^2\right)^2-m\omega^2}V\left(\omega,\ \lambda,\ \mu\right)d\omega d\lambda d\mu. \tag{6.100}$$

Given the spectral decomposition for the deflection (6.100), other probabilistic properties of the deflection and other variables of interest can be derived with the help of the standard spectral theory of random fields. For example, the correlation function of the acceleration is given by

$$K_{\ddot{w}}\left(\tau,\ \xi,\eta\right)=\iiint\limits_{-\infty}^{\infty}\frac{\omega^4 e^{i(\omega t+\lambda\xi+\mu\eta)}}{\left|D_c\left(\lambda^2+\mu^2\right)^2-m\omega^2\right|^2}S_p\left(\omega,\ \lambda,\ \mu\right)d\omega d\lambda d\mu. \tag{6.101}$$

Formula (6.101) is too complicated for a direct analysis. However, it can be used to produce other, less general probabilistic properties. Let us restrict our consideration to the estimation of the correlation function at $\xi=\eta=0$, i.e.

$$K_{\ddot{w}}\left(\tau\right)=K_{\ddot{w}}\left(\tau,0,0\right). \tag{6.102}$$

Thus $K_{\ddot{w}}(\tau)$ is the correlation function of acceleration at a single, arbitrary point of the plate. Comparing eqs. (6.102) and (6.101) it is easy to derive the following equation for the spectral density of acceleration

$$\Phi_{\ddot{w}}(\omega) = \iint\limits_{-\infty}^{\infty} \frac{\omega^4 S_p(\omega, \lambda, \mu)}{\left|D_c\left(\lambda^2+\mu^2\right)^2 - m\omega^2\right|^2} d\lambda d\mu. \tag{6.103}$$

Of considerable interest are the bending moments in the plate. Their spectral decompositions are obtained by inserting (6.100) in (6.88), and the result is

$$M_x = \iiint\limits_{-\infty}^{\infty} \frac{D_c\left(\lambda^2+\sigma_c\mu^2\right)V}{D_c\left(\lambda^2+\mu^2\right)^2 - m\omega^2} e^{i(\omega t+\lambda x+\mu y)} d\omega d\lambda d\mu, \tag{6.104}$$

$$M_y = \iiint\limits_{-\infty}^{\infty} \frac{D_c\left(\mu^2+\sigma_c\lambda^2\right)V}{D_c\left(\lambda^2+\mu^2\right)^2 - m\omega^2} e^{i(\omega t+\lambda x+\mu y)} d\omega d\lambda d\mu, \tag{6.105}$$

$$H = \iiint\limits_{-\infty}^{\infty} \frac{D_c(1-\sigma_c)\lambda\mu V}{D_c\left(\lambda^2+\mu^2\right)^2 - m\omega^2} e^{i(\omega t+\lambda x+\mu y)} d\omega d\lambda d\mu. \tag{6.106}$$

We limit our analysis now to considering the moments as functions of time. The spectral densities of the moments at an arbitrary point are easily found by means of the spectral representation (6.104) etc., to give

$$\Phi_{M_x} = \iint\limits_{-\infty}^{\infty} \left|\frac{D_c\left(\lambda^2+\sigma_c\mu^2\right)}{D_c\left(\lambda^2+\mu^2\right)^2 - m\omega^2}\right|^2 S_p(\omega, \lambda, \mu)\, d\lambda d\mu,$$

$$\Phi_{M_y} = \iint\limits_{-\infty}^{\infty} \left|\frac{D_c\left(\mu^2+\sigma_c\lambda^2\right)}{D_c\left(\lambda^2+\mu^2\right)^2 - m\omega^2}\right|^2 S_p(\omega, \lambda, \mu)\, d\lambda d\mu, \tag{6.107}$$

$$\Phi_H = \iint\limits_{-\infty}^{\infty} \left|\frac{D_c(1-\sigma_c)\lambda\mu}{D_c\left(\lambda^2+\mu^2\right)^2 - m\omega^2}\right|^2 S_p(\omega, \lambda, \mu)\, d\lambda d\mu.$$

The cross-spectral densities are given as follows

$$\Phi_{M_xM_y} = \iint\limits_{-\infty}^{\infty} \frac{|D_c|^2\left(\lambda^2+\sigma_c\mu^2\right)\left(\mu^2+\sigma_c^*\lambda^2\right)}{\left|D_c\left(\lambda^2+\mu^2\right)^2 - m\omega^2\right|^2} S_p d\lambda d\mu,$$

$$\Phi_{M_x H} = \iint\limits_{-\infty}^{\infty} \frac{|D_c|^2 \left(\lambda^2 + \sigma_c \mu^2\right) \left(1 - \sigma_c^*\right) \lambda \mu}{\left|D_c \left(\lambda^2 + \mu^2\right)^2 - m\omega^2\right|^2} S_p d\lambda d\mu, \tag{6.108}$$

$$\Phi_{M_y H} = \iint\limits_{-\infty}^{\infty} \frac{|D_c|^2 \left(\mu^2 + \sigma_c \lambda^2\right) \left(1 - \sigma_c^*\right) \lambda \mu}{\left|D_c \left(\lambda^2 + \mu^2\right)^2 - m\omega^2\right|^2} S_p d\lambda d\mu.$$

Then, by substituting the spectral decomposition of deflection (6.100) into (6.81) we obtain the following expression

$$N^2 = \frac{1 - \nu + \nu^2}{3\left(1 - \nu\right)^2} \iiint\limits_{-\infty}^{\infty} \frac{\left(\lambda^2 + \mu^2\right)^2 S_p \left(\omega, \lambda, \mu\right)}{\left|D_c \left(\lambda^2 + \mu^2\right)^2 - m\omega^2\right|^2} d\omega d\lambda d\mu. \tag{6.109}$$

The mean square of the shear stress intensity is known to depend upon N^2. In particular, the maximum mean square of the shear stress intensity is obtained by putting $z = h/2$ in eq. (6.80), which gives

$$T_m^2 = \left(GhN\right)^2. \tag{6.110}$$

Finally, by means of (6.83), it is easy to find the spectral density

$$S_N\left(\omega\right) = \frac{1 - \nu + \nu^2}{3\left(1 - \nu\right)^2} \iint\limits_{-\infty}^{\infty} \frac{\left(\lambda^2 + \mu^2\right)^2 S_p \left(\omega, \lambda, \mu\right)}{\left|D_c \left(\lambda^2 + \mu^2\right)^2 - m\omega^2\right|^2} d\lambda d\mu. \tag{6.111}$$

This spectral density appears to be necessary to evaluate the spectral density of stress S_T by means of eq. (6.82).

The equations obtained are of general character, and are sufficient for further evaluation for particular pressure fields.

6.6 Vibrations in an infinite plate subjected to spatial white noise loading

Let the load be spatially uncorrelated, i.e. its correlation function and its spectral density are given by

$$K_p\left(\tau, \xi, \eta\right) = K_p\left(\tau\right) \delta\left(\xi\right) \delta\left(\eta\right), \tag{6.112}$$

$$S_p\left(\omega, \lambda, \mu\right) = \Psi\left(\omega\right), \tag{6.113}$$

with $\Psi\left(\omega\right)$ being a frequency-dependent function ($\Psi\left(\omega\right) \geq 0$).

Let us evaluate the integrals of the previous Section. For instance, we substitute (6.113) into (6.103), to obtain

$$\Phi_{\ddot{w}}(\omega) = w^4 \Psi(\omega) \iint\limits_{-\infty}^{\infty} \frac{d\lambda d\mu}{\left| D_c \left(\lambda^2 + \mu^2\right)^2 - m\omega^2 \right|^2}. \tag{6.114}$$

In order to evaluate the latter integral we transform to the polar coordinates

$$\lambda = r\cos\theta,\ \mu = r\sin\theta, \tag{6.115}$$

and then introduce a new integration variable $z = r^2$. The result is

$$A = \iint\limits_{-\infty}^{\infty} \frac{d\lambda d\mu}{\left| D_c \left(\lambda^2 + \mu^2\right)^2 - m\omega^2 \right|^2} = \frac{1}{2} \int\limits_0^{2\pi} d\theta \int\limits_0^{\infty} \frac{dz}{\left| D_c z^2 - m\omega^2 \right|^2}. \tag{6.116}$$

To evaluate integral (6.116), use is made of the following integral [44]

$$\int\limits_0^{\infty} \frac{dz}{a + bz^2 + cz^4} = \frac{\pi\cos(\alpha/2)}{2cq^3 \sin\alpha},$$

where

$$q = \sqrt[4]{\frac{a}{c}},\ \cos\alpha = -\frac{b}{2\sqrt{ac}}.$$

Evaluation yields the following asymptotic expression for integral (6.116) for a small value of damping ($R(\omega) \ll 1$)

$$A \approx \frac{\pi^2}{2m\omega^3 \psi \sqrt{Dm}}, \tag{6.117}$$

where

$$\psi = \psi(\omega) = \operatorname{Im} R(\omega), \tag{6.118}$$

function $R(\omega)$ being defined by (6.90). Inserting (6.117) in (6.114), we obtain the result

$$\Phi_{\ddot{w}}(\omega) = \frac{\pi^2 \omega \Psi(\omega)}{2m\sqrt{Dm}\psi}. \tag{6.119}$$

An analogous manipulation leads to the following equations for spectral densities of moments and their cross-spectral densities

$$\Phi_{M_x} = \Phi_{M_y} = \frac{\pi^2}{16}\left(3 + 2\nu + 3\nu^2\right)\sqrt{\frac{D}{m}} \frac{\Psi(\omega)}{\omega\psi},$$

$$\Phi_H = \frac{\pi^2}{16}(1-\nu)^2 \sqrt{\frac{D}{m}} \frac{\Psi(\omega)}{\omega\psi}, \tag{6.120}$$

$$\Phi_{M_xM_y}=\frac{1+6\nu+\nu^2}{3+2\nu+3\nu^2}\Phi_{M_x},\ \Phi_{M_xH}=\Phi_{M_yH}=0.$$

It is known that the main stresses on the plate surfaces are expressed in terms of the bending moments as follows

$$\sigma_x=\frac{6M_x}{h^2},\sigma_y=\frac{6M_y}{h^2},\tau_{xy}=\frac{6H}{h^2},\tag{6.121}$$

where h is the plate thickness. Using (6.120) and (6.121) it is easy to derive the spectral densities of stresses on the plate outer surfaces

$$\Phi_\sigma=\frac{9\pi^2}{4h^4}\left(3+2\nu+3\nu^2\right)\sqrt{\frac{D}{m}}\frac{\Psi(\omega)}{\omega\psi(\omega)},\tag{6.122}$$

$$\Phi_\tau=\frac{9\pi^2}{4h^4}\left(1-\nu\right)^2\sqrt{\frac{D}{m}}\frac{\Psi(\omega)}{\omega\psi(\omega)}.\tag{6.123}$$

The first of these formulae coincides with that obtained by another method by Bolotin [18].

Evaluation of the integrals in (6.109) and (6.111) gives the following results

$$N^2=\frac{1-\nu+\nu^2}{6\left(1-\nu\right)^2}\frac{\pi^2}{D\sqrt{Dm}}\int\limits_{-\infty}^{\infty}\frac{\Psi(\omega)}{\omega\psi(\omega)}d\omega,\tag{6.124}$$

$$S_N(\omega)=\frac{1-\nu+\nu^2}{6\left(1-\nu\right)^2}\frac{\pi^2}{D\sqrt{Dm}}\frac{\Psi(\omega)}{\omega\psi(\omega)}.\tag{6.125}$$

Up to now our evaluations are valid regardless of the sort of damping, "linear" or nonlinear. This result is remarkable. It has been shown that for the studied broad-band load all probabilistic properties of both acceleration field and stress field depend crucially on the material damping. The latter is determined by function $\psi(\omega)$ in the denominator of the latter equations. The mentioned fact once again highlights the importance of studies on energy dissipation. Indeed, if there exists no way to compute function $\psi(\omega)$, the value of the analysis of the forced vibration is completely lost. This observation is equally applicable to other dynamical problems. If damping is linear, then a dependence $\psi=\psi(\omega)$ is given and the analysis reduces to the equations obtained.

Let us proceed to the problem of estimating ψ in the case of a nonlinear internal damping. Recall that ψ is found from the chain of equations (6.118), (6.90) and (5.47). The estimation yields

$$\psi=\frac{\left(1-\nu+\nu^2\right)2\alpha H}{(1-\nu)(\alpha+3)}\left(\frac{hN}{2n}\right)^\alpha\int\limits_0^1\frac{\eta^{\alpha-1}\omega\varphi(\eta)\,d\eta}{1+\left[\omega\varphi(\eta)\right]^2}.\tag{6.126}$$

Function $\varphi(\eta)$ which appears in this equation is the solution of eq. (5.60). Due to (6.80), (6.82), (6.124) and (6.125), equation (6.126) takes the form

$$1-\eta^2=\frac{\int\limits_{-\infty}^{\infty}\frac{\Psi(\omega)}{\omega\psi}\frac{1}{1+(\omega\varphi)^2}d\omega}{\int\limits_{-\infty}^{\infty}\frac{\Psi(\omega)}{\omega\psi}d\omega}. \tag{6.127}$$

The difficulty is that ψ depends on N which is not given in advance. In order to determine it, equation (6.124) must be used.

Summarising we can say that the problem is reduced to the determination of the unknown parameters N, ψ and φ from the system of equations (6.126), (6.127) and (6.124). Let us rewrite this system in a simpler form. To this end, we let

$$I(\omega)=\alpha\int\limits_0^1\frac{\eta^{\alpha-1}\omega\varphi(\eta)\,d\eta}{1+[\omega\varphi(\eta)]^2}, \tag{6.128}$$

to obtain

$$\psi=\frac{(1-\nu+\nu^2)\,2H}{(1-\nu)(\alpha+3)}\left(\frac{hN}{2n}\right)^{\alpha}I(\omega). \tag{6.129}$$

Substituting this expression into eqs. (6.124) and (6.127) gives

$$\left(\frac{hN}{2n}\right)^{2+\alpha}=\frac{\pi^2(\alpha+3)}{12H(1-\nu)D\sqrt{Dm}}\left(\frac{h}{2n}\right)^2\int\limits_{-\infty}^{\infty}\int\limits_{-\infty}^{\infty}\frac{\Psi(\omega)}{\omega I(\omega)}d\omega, \tag{6.130}$$

$$1-\eta^2=\frac{\int\limits_{-\infty}^{\infty}\frac{\Psi(\omega)}{\omega I(\omega)}\frac{1}{1+(\omega\varphi)^2}d\omega}{\int\limits_{-\infty}^{\infty}\frac{\Psi(\omega)}{\omega I(\omega)}d\omega}. \tag{6.131}$$

It is easy to see that the system of equations is split into two uncoupled systems. Functions $\eta(\omega)$ and $I(\omega)$ are governed by eqs. (6.128) and (6.131). After having solved this system, we obtain the value of N from the algebraic equation (6.130).

Let us proceed to the system of equations (6.128) and (6.131). By substituting (6.128) into (6.131) this system reduces to a single equation for $\varphi(\eta)$. The latter should be qualified as an implicit nonlinear integral equation. Given a spectrum of external load, its solution can be obtained by numerical integration.

One of the ways of analytically constructing a solution to the above system is by using the inverse method. Using this method, we prescribe an expression

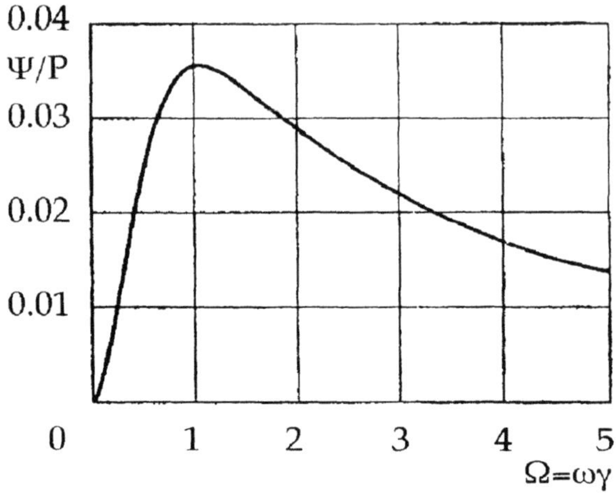

FIGURE 6.5. Spectral density Ψ due to eq. (6.136).

$$C(\omega) = \frac{\Psi(\omega)}{\omega I(\omega)} \tag{6.132}$$

in a convenient form which allows the integrals in (6.131) to be easily evaluated.

For example, putting

$$\frac{\Psi(\omega)}{\omega I(\omega)} = \frac{P\gamma}{\pi\left(1+\gamma^2\omega^2\right)}, \tag{6.133}$$

where P and γ are positive constants, we obtain from (6.131)

$$1-\eta^2 = \gamma(\gamma+\varphi)^{-1}. \tag{6.134}$$

Substituting this solution into (6.128) when $\alpha = 1$ leads to an integral which can be evaluated in closed form, to give

$$I(\omega) = \operatorname{Im}\left\{\frac{-1}{1-i\Omega}\left[1+\frac{i\Omega}{2\sqrt{1-i\Omega}}\ln\frac{i\Omega}{\left(1+\sqrt{1-i\Omega}\right)^2}\right]\right\}, \tag{6.135}$$

where $\Omega = \omega\gamma$ is a non-dimensional frequency. The spectral density of the load which fits assumption (6.133) is obtained from this equation, to give

$$\Psi(\omega) = \frac{P\Omega I}{\pi\left(1+\Omega^2\right)}, \quad \Omega = \omega\gamma. \tag{6.136}$$

The latter equation is seen in Fig. 6.5, and analysis of it shows that $\Psi(\omega)$ is the spectral density of a typical broad-band random process.

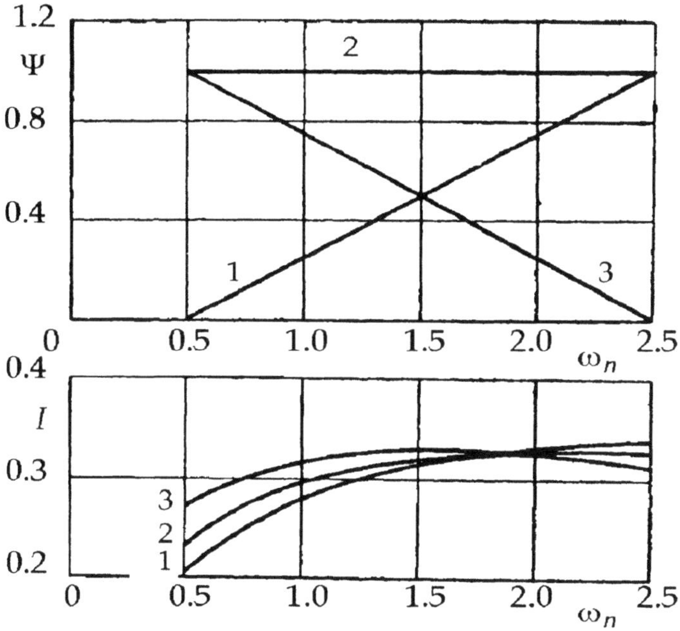

FIGURE 6.6. Dependences Ψ and I versus ω_n.

Inserting (6.132) in (6.130) leads to the following equation for N

$$\left(\frac{hN}{2n}\right)^{2+\alpha} = \frac{\pi^2(\alpha+3)}{12H(1-\nu)D\sqrt{Dm}}\left(\frac{h}{2n}\right)^2 P. \tag{6.137}$$

A closed form solution of this equation indicates that there is no direct proportionality between the load P and the stress intensity parameter N. Further, because of (6.110) there exists no direct proportionality between the load P and the maximum of the mean square of the stress intensity.

Given a spectrum $\Psi(\omega)$, the solution of the system of equations (6.128) and (6.131) must be obtained numerically. The iteration method is one of a number of possible ways, and is described next. We prescribe a constant value of I for $\omega > 0$. Equation (6.131) is solved to yield a dependence $\varphi = \varphi(\eta)$ and then due to (6.128) $I = I(\omega)$ is refined. Some specific computations have shown that this process converges very quickly, namely the fourth approximation gives a near-exact result. The results of computations for some particular spectral densities are displayed in Fig. 6.6. The density $\Psi(\omega)$ is assumed to be nontrivial only in the interval $0.5 < |\omega_n| < 2.5$ where ω_n is a properly normalised non-dimensional frequency. The dependencies $\Psi(\omega)$ selected for analysis are depicted in the upper diagram whereas the lower diagram shows the corresponding curves $I = I(\omega)$. Analysis of these curves indicates that in all cases I varies insignificantly in the interval ω_n of interest, and is close to the value $I_h = 0.33$ for the harmonic deformation.

For this reason, we expect no considerable error if our analysis is limited to the first approximation, i.e. if we take $I = I_h$.

6.7 Action of a wave load on a plate

An oscillating load with the following correlation function

$$K_p(\tau, \xi, \eta) = K_p\left(\tau - \frac{\xi\cos\gamma + \eta\sin\gamma}{c}\right), \tag{6.138}$$

is termed a wave load. Here c is the velocity of propagation of a pressure wave and γ is an angle between the wave vector and the axis x.

The load spectral density, defined as follows

$$K_p(\tau, \xi, \eta) = \iiint\limits_{-\infty}^{\infty} e^{i(\omega\tau+\lambda\xi+\mu\eta)} S_p(\omega, \lambda, \mu)\, d\omega d\lambda d\mu, \tag{6.139}$$

for the particular load with the correlation function (6.138) is given by

$$S_p(\omega, \lambda, \mu) = \Phi_p(\omega)\,\delta\left(\lambda + \frac{\omega}{c}\cos\gamma\right)\delta\left(\mu + \frac{\omega}{c}\sin\gamma\right). \tag{6.140}$$

Here $\Phi_p(\omega)$ is the spectral density of pressure acting on an arbitrary point of the plane.

Substituting (6.140) into (6.103) and evaluating the integral yields

$$\Phi_{\ddot{w}}(\omega) = \frac{\omega^4\Phi_p(\omega)}{\left|D_c(\omega/c)^4 - m\omega^2\right|^2}. \tag{6.141}$$

This result in well-known in acoustics. But in this particular case, the complex rigidity of the plate depends upon the unknown quantities of the problem. In order to determine these quantities eq. (6.93) should be addressed. All vibrational parameters in this equation are easily computed by means of eqs. (6.109), (6.111) and (6.140), to give

$$S_N(\omega) = \frac{1-\nu+\nu^2}{3(1-\nu)^2}\frac{(\omega/c)^4\,\Phi_p(\omega)}{\left|D_c(\omega/c)^4 - m\omega^2\right|^2}, \tag{6.142}$$

$$N^2 = \frac{1-\nu+\nu^2}{3(1-\nu)^2}\int\limits_{-\infty}^{\infty}\frac{(\omega/c)^4\,\Phi_p(\omega)\,d\omega}{\left|D_c(\omega/c)^4 - m\omega^2\right|^2}. \tag{6.143}$$

It leads to the following form of eq. (6.93)

$$1-\eta^2 = \frac{\int\limits_{-\infty}^{\infty} \frac{c^4 \Phi_p(\omega)\, d\omega}{|D_c\omega^2 - mc^4|^2} \frac{d\omega}{1+(\omega\varphi)^2}}{\int\limits_{-\infty}^{\infty} \frac{c^4 \Phi_p(\omega)}{|D_c\omega^2 - mc^4|^2} d\omega}. \tag{6.144}$$

The integrals in the latter equation have a structure similar to that in (6.20) which has been derived in the framework of the analysis of an arbitrary body. Hence, we can use the approach of Section 6.1 in this case as well. Assuming that the load is broad-band and has a smooth spectral density $\Phi_p(\omega)$ and evaluating asymptotically the integrals in (6.144) gives

$$1-\eta^2 = \frac{1}{1+(\omega_*\varphi)^2}. \tag{6.145}$$

The frequency

$$\omega_* = \sqrt{\frac{mc^4}{D}} \tag{6.146}$$

is referred to as the *coincidence frequency* in acoustics. The velocity of propagation of a stationary harmonic bending wave "coincides" with the velocity of the loading wave at this frequency. Equation (6.145) is the same as the equation for φ for a harmonic deformation law with the frequency ω_*. Further on, a sharp resonant character of dependence (6.141) etc. indicates that the account of damping turns out to be essential near the coincidence frequency only. Hence, in this case we can use the equations of internal friction derived for a harmonic deformation.

6.8 Differential equations for elastoplastic vibrations in shallow shells

Consider a shallow shell. Let the Cartesian coordinates x and y locate a generic point in the middle surface of the shell. The normal distance from the mid-surface is given by z. The transverse displacement at coordinates x and y is denoted by w which is assumed to be a random function of time.

By virtue of the correspondence principle we find that the distribution of the principal stresses in an elastoplastic shallow shell is as follows (by analogy with [41])

$$\sigma_x = \frac{E_c}{1-\nu_c^2}\left[-z\left(\frac{\partial^2 w}{\partial x^2} + \nu_c \frac{\partial^2 w}{\partial y^2}\right) + \varepsilon_x + \nu_c \varepsilon_y\right],$$

$$\sigma_y = \frac{E_c}{1-\nu_c^2}\left[-z\left(\frac{\partial^2 w}{\partial y^2} + \nu_c \frac{\partial^2 w}{\partial x^2}\right) + \varepsilon_y + \nu_c \varepsilon_x\right], \tag{6.147}$$

$$\tau_{xy} = \frac{E_c}{1-\nu_c^2}\left[-z\left(1-\nu_c\right)\frac{\partial^2 w}{\partial x \partial y} + \left(1-\nu_c\right)\varepsilon_{xy}\right].$$

Here ε_x, ε_y and ε_{xy} are components of the strain tensor in the shell mid-surface. As shown in the theory of shallow shells these components must satisfy the following compatibility equation

$$\frac{\partial^2 \varepsilon_x}{\partial y^2} + \frac{\partial^2 \varepsilon_y}{\partial x^2} - 2\frac{\partial^2 \varepsilon_{xy}}{\partial x \partial y} - \left(k_1 \frac{\partial^2 w}{\partial y^2} + k_2 \frac{\partial^2 w}{\partial x^2}\right) = 0, \tag{6.148}$$

where k_1 and k_2 are the curvatures of the coordinate curves x and y of the shell. In what follows, k_1 and k_2 are assumed to be the principal curvatures of the shell.

Inserting (6.147) into (5.52) and retaining only asymptotically leading terms renders the mean square of the shear stress intensity

$$T^2 = \frac{(2G)^2}{2}\mathrm{M}\left[\left|z\frac{\partial^2 w}{\partial x^2} - \varepsilon_x\right|^2 + \left|z\frac{\partial^2 w}{\partial y^2} - \varepsilon_y\right|^2 + \right.$$

$$\left. 2\left|z\frac{\partial^2 w}{\partial x \partial y} - \varepsilon_{xy}\right|^2 - \frac{1-4\nu+\nu^2}{3\left(1-\nu\right)^2}\left|z\Delta w - \varepsilon_x - \varepsilon_y\right|^2\right]. \tag{6.149}$$

Here, as usual, M is the symbol of mathematical expectation.

Next, we introduce the moments and the forces in the mid-surface

$$M_x = \int\limits_{-h/2}^{h/2} \sigma_x z dz, \quad M_y = \int\limits_{-h/2}^{h/2} \sigma_y z dz, \quad H = \int\limits_{-h/2}^{h/2} \tau_{xy} z dz, \tag{6.150}$$

$$T_x = \int\limits_{-h/2}^{h/2} \sigma_x dz, \quad T_y = \int\limits_{-h/2}^{h/2} \sigma_y dz, \quad S = \int\limits_{-h/2}^{h/2} \tau_{xy} dz. \tag{6.151}$$

Substituting the equations for the stresses (6.147) into the latter equations we obtain due to (6.84) and (6.85)

$$\begin{aligned} M_x &= -D_c\left(\frac{\partial^2 w}{\partial x^2} + \sigma_c \frac{\partial^2 w}{\partial y^2}\right) + C_c\left(\varepsilon_x + \delta\varepsilon_y\right), \\ M_y &= -D_c\left(\frac{\partial^2 w}{\partial y^2} + \sigma_c \frac{\partial^2 w}{\partial x^2}\right) + C_c\left(\varepsilon_y + \delta\varepsilon_x\right), \\ H &= -D_c\left(1-\sigma_c\right)\frac{\partial^2 w}{\partial x \partial y} + C_c\left(1-\delta\right)\varepsilon_{xy}, \end{aligned} \tag{6.152}$$

$$
\begin{aligned}
T_x &= B_c\left(\varepsilon_x + \kappa_c \varepsilon_y\right) - C_c\left(\frac{\partial^2 w}{\partial x^2} + \delta \frac{\partial^2 w}{\partial y^2}\right), \\
T_y &= B_c\left(\varepsilon_y + \kappa_c \varepsilon_x\right) - C_c\left(\frac{\partial^2 w}{\partial y^2} + \delta \frac{\partial^2 w}{\partial x^2}\right), \\
S &= B_c\left(1-\kappa_c\right)\varepsilon_{xy} + C_c\left(1-\delta\right)\frac{\partial^2 w}{\partial x \partial y}.
\end{aligned}
\tag{6.153}
$$

The introduced factors are given by

$$
D_c = D\left[1 + \frac{1-\nu+\nu^2}{1-\nu}\frac{8}{h^3}\int_{-h/2}^{h/2} lz^2\left(\frac{T}{2G}\right)^{\alpha} dz\right],
$$

$$
D_c\sigma_c = D\nu\left[1 + \frac{1-4\nu+\nu^2}{\nu\left(1-\nu\right)}\frac{4}{h^3}\int_{-h/2}^{h/2} lz^2\left(\frac{T}{2G}\right)^{\alpha} dz\right],
$$

$$
B_c = \frac{Eh}{1-\nu^2}\left[1 + \frac{2\left(1-\nu+\nu^2\right)}{3\left(1-\nu\right)}\frac{1}{h}\int_{-h/2}^{h/2} l\left(\frac{T}{2G}\right)^{\alpha} dz\right],
$$

$$
B_c\kappa_c = \frac{Eh\nu}{1-\nu^2}\left[1 - \frac{1-4\nu+\nu^2}{3\nu\left(1-\nu\right)}\frac{1}{h}\int_{-h/2}^{h/2} l\left(\frac{T}{2G}\right)^{\alpha} dz\right],
$$

$$
C_c = \frac{2E}{3\left(1-\nu^2\right)}\frac{1-\nu+\nu^2}{1-\nu}\frac{1}{h}\int_{-h/2}^{h/2} lz\left(\frac{T}{2G}\right)^{\alpha} dz,
$$

$$
D = \frac{Eh^3}{12\left(1-\nu^2\right)}, \ \delta = -\frac{1-4\nu+\nu^2}{2\left(1-\nu+\nu^2\right)}. \tag{6.154}
$$

Here D_c denotes complex bending rigidity, B_c stands for the complex stiffness of the shell in its tangential plane and C_c has no analogy in the theory of elastic shells. Parameters σ_c, δ and κ_c are analogues to the Poisson's ratio.

The introduced forces and moments must satisfy the governing equations of a shallow shell

$$
m\ddot{w} - \frac{\partial^2 M_x}{\partial x^2} - 2\frac{\partial^2 H}{\partial x \partial y} - \frac{\partial^2 M_y}{\partial y^2} + k_1 T_x + k_2 T_y = p, \tag{6.155}
$$

$$\frac{\partial T_x}{\partial x} + \frac{\partial S}{\partial y} = 0, \quad \frac{\partial S}{\partial x} + \frac{\partial T_y}{\partial y} = 0, \tag{6.156}$$

where m is the mass per surface area and p denotes an external random transverse load. Inertial forces in the tangential plane of the shell are neglected in the latter two equations.

This system of equations must be completed by including an additional equation, namely eq. (5.60)

$$1 - \eta^2 = \int_{-\infty}^{\infty} \frac{S_T(\omega)}{T^2} \frac{d\omega}{1 + (\omega\varphi)^2}. \tag{6.157}$$

By means of this equation we find function φ which appears in the expression for l (5.46). As seen from eqs. (6.157) and (5.46), function l depends not only on its main argument ω, but also on the spatial coordinates x, y and z, since S_T and T^2 may depend upon the spatial coordinates.

The forthcoming estimations will be performed for the case in which the vibrational field is a statistically homogeneous function of coordinates x and y and time t. The root-mean-square of the shear stress T is independent of x and y, that is all coefficients in eqs. (6.152) and (6.153) turn out to be constant. As pointed out in Section 6.4 the condition of statistical homogeneity is approximately fulfilled far from the boundaries of a sufficiently extended shallow shell subject to a broad-band high-frequency excitation.

By analogy with the theory of thin elastic shells we express the forces in the mid-surface in terms of the stress function Φ

$$T_x = \frac{\partial^2 \Phi}{\partial y^2}, \quad T_y = \frac{\partial^2 \Phi}{\partial x^2}, \quad S = -\frac{\partial^2 \Phi}{\partial x \partial y}. \tag{6.158}$$

Equations (6.156) are then satisfied identically. Inserting (6.158) in (6.153) and solving the obtained equations for the strains of the mid-surface yields

$$\begin{aligned}
B_c\left(1-\kappa_c^2\right)\varepsilon_x &= \frac{\partial^2 \Phi}{\partial y^2} - \kappa_c \frac{\partial^2 \Phi}{\partial x^2} + \\
&\quad + C_c\left(\frac{\partial^2 w}{\partial x^2} + \delta \frac{\partial^2 w}{\partial y^2}\right) - C_c\kappa_c\left(\frac{\partial^2 w}{\partial y^2} + \delta \frac{\partial^2 w}{\partial x^2}\right), \\
B_c\left(1-\kappa_c^2\right)\varepsilon_y &= \frac{\partial^2 \Phi}{\partial x^2} - \kappa_c \frac{\partial^2 \Phi}{\partial y^2} + \\
&\quad + C_c\left(\frac{\partial^2 w}{\partial y^2} + \delta \frac{\partial^2 w}{\partial x^2}\right) - C_c\kappa_c\left(\frac{\partial^2 w}{\partial x^2} + \delta \frac{\partial^2 w}{\partial y^2}\right), \\
B_c\left(1-\kappa_c^2\right)\varepsilon_{xy} &= -\left(1+\kappa_c\right)\frac{\partial^2 \Phi}{\partial x \partial y} + C_c\left(1-\delta\right)\left(1+\kappa_c\right)\frac{\partial^2 w}{\partial x \partial y}.
\end{aligned} \tag{6.159}$$

Substituting ε_x, ε_y and ε_{xy} into the compatibility equation (6.148) gives the first equation for the unknown variables w and Φ

$$\Delta\Delta\Phi - B_c\left(1-\kappa_c^2\right)\left(k_1\frac{\partial^2 w}{\partial y^2}+k_2\frac{\partial^2 w}{\partial x^2}\right)+C_c\left(\delta-\kappa_c\right)\Delta\Delta w=0. \quad (6.160)$$

The second equation for these variables is derived by substitution of equations for forces (6.158) and moments (6.152) into the dynamics equation (6.155)

$$m\ddot{w}+\left(D_c-\frac{C_c^2}{B_c}\right)\Delta\Delta w+k_1\frac{\partial^2\Phi}{\partial y^2}+k_2\frac{\partial^2\Phi}{\partial x^2}-$$
$$-C_c\left(\delta-\kappa_c\right)\left(k_1\frac{\partial^2 w}{\partial y^2}+k_2\frac{\partial^2 w}{\partial x^2}\right)=p. \quad (6.161)$$

This equation may be simplified by neglecting terms of higher order in the plastic deformation effect. Indeed, we have from (6.154) that C_c is first order in l. Thus, C_c^2 is a value of second order and may be omitted. Next, in order to retain terms up to first order in (6.161) it is sufficient to put $\kappa_c=\nu$ in the expression for $C_c\left(\delta-\kappa_c\right)$. An analogous simplification can also be performed in eq. (6.160), to give

$$m\ddot{w}+D_c\Delta\Delta w+k_1\frac{\partial^2\Phi}{\partial y^2}+k_2\frac{\partial^2\Phi}{\partial x^2}-C_c\left(\delta-\nu\right)\left(k_1\frac{\partial^2 w}{\partial y^2}+k_2\frac{\partial^2 w}{\partial x^2}\right)=p, \quad (6.162)$$

$$\Delta\Delta\Phi - B_c\left(1-\kappa_c^2\right)\left(k_1\frac{\partial^2 w}{\partial y^2}+k_2\frac{\partial^2 w}{\partial x^2}\right)+C_c\left(\delta-\nu\right)\Delta\Delta w=0. \quad (6.163)$$

The coefficients of the latter equations admit the following relatively simple asymptotic expressions to be obtained by virtue of (6.154)

$$B_c\left(1-\kappa_c^2\right)=Eh\left[1+\frac{2\left(1+\nu\right)}{3h}\int\limits_{-h/2}^{h/2} l\left(\frac{T}{2G}\right)^{\alpha}dz\right],$$

$$C_c\left(\delta-\nu\right)=-\frac{E\left(1-2\nu\right)}{3\left(1-\nu\right)}\int\limits_{-h/2}^{h/2} lz\left(\frac{T}{2G}\right)^{\alpha}dz. \quad (6.164)$$

Assuming that the load p is modelled by a homogeneous random field with a zero mean value we write down its spectral decomposition

$$p=\iiint\limits_{-\infty}^{\infty} e^{i(\omega t+\lambda x+\mu y)}V\left(\omega,\,\lambda,\,\mu\right)d\omega d\lambda d\mu. \quad (6.165)$$

Substituting this into the system (6.162) and (6.163) and constraining the solution to be bounded at infinity, we obtain

$$w = \iiint_{-\infty}^{\infty} \frac{V(\omega,\ \lambda,\ \mu)}{\Omega(\omega,\ \lambda,\ \mu)} e^{i(\omega t+\lambda x+\mu y)} d\omega d\lambda d\mu, \tag{6.166}$$

$$\Phi = \iiint_{-\infty}^{\infty} \frac{AV(\omega,\ \lambda,\ \mu)}{\Omega(\omega,\ \lambda,\ \mu)} e^{i(\omega t+\lambda x+\mu y)} d\omega d\lambda d\mu, \tag{6.167}$$

where

$$\begin{aligned} \Omega &= -m\omega^2 + D_c\left(\lambda^2+\mu^2\right)^2 + B_c\left(1-\kappa_c^2\right)\frac{\left(k_1\mu^2+k_2\lambda^2\right)^2}{\left(\lambda^2+\mu^2\right)^2} + \\ &\quad +2\left(k_1\mu^2+k_2\lambda^2\right)C_c\left(\delta-\nu\right), \end{aligned} \tag{6.168}$$

$$A = C_c\left(\delta-\nu\right) + B_c\left(1-\kappa_c^2\right)\frac{k_1\mu^2+k_2\lambda^2}{\left(\lambda^2+\mu^2\right)^2}. \tag{6.169}$$

Let us next consider the mean square of the shear stress intensity. Inserting (6.166) and (6.167) into (6.158) and (6.152), and then into (6.149) and neglecting terms of higher order leads to the result

$$T^2 = G\iiint_{-\infty}^{\infty} \frac{\Lambda(\lambda,\ \mu,\ z)}{\left|\Omega(\omega,\ \lambda,\ \mu)\right|^2} S_p(\omega,\ \lambda,\ \mu)\, d\omega d\lambda d\mu, \tag{6.170}$$

where

$$\begin{aligned} \Lambda(\lambda,\ \mu,\ z) &= \frac{4G}{3}\left[z^2\left(\lambda^2+\mu^2\right)^2\frac{1-\nu+\nu^2}{(1-\nu)^2} - \right. \\ &\left. -z\left(k_1\mu^2+k_2\lambda^2\right)\frac{1-\nu+\nu^2}{(1-\nu)^2} + (1+\nu)^2\frac{\left(k_1\mu^2+k_2\lambda^2\right)^2}{\left(\lambda^2+\mu^2\right)^2}\right]. \end{aligned} \tag{6.171}$$

By virtue of (6.170) we get an expression for the spectral density of stresses

$$S_T = G\iint_{-\infty}^{\infty} \frac{\Lambda(\lambda,\ \mu,\ z)}{\left|\Omega(\omega,\ \lambda,\ \mu)\right|^2} S_p(\omega,\ \lambda,\ \mu)\, d\lambda d\mu. \tag{6.172}$$

Finally, we derive an asymptotic expression for Ω. After a little algebra due to (6.168) and (6.164) we obtain an unexpected result

$$\Omega = -m\omega^2 + D(1+R)\left[\left(\lambda^2+\mu^2\right)^2 + \frac{Eh}{D}\frac{\left(k_1\mu^2+k_2\lambda^2\right)^2}{\left(\lambda^2+\mu^2\right)^2}\right], \tag{6.173}$$

where

$$R = \frac{\int\limits_{-h/2}^{h/2} l\Lambda\left(\lambda,\, \mu,\, z\right)\left(T/2G\right)^{\alpha} dz}{D\left[\left(\lambda^2+\mu^2\right)^2 + \frac{Eh}{D}\left(k_1\mu^2 + k_2\lambda^2\right)^2 / \left(\lambda^2+\mu^2\right)^2\right]}. \tag{6.174}$$

The presence of function $\Lambda\left(\lambda,\, \mu,\, z\right)$, which has already appeared in (6.170) and (6.172), is somewhat unexpected.

Equations (6.170), (6.172), (6.174), (6.157) and (5.46) give us a system of equation for determination of the unknown parameters φ, l, R, S_T and T^2. This system in its general form is extremely complicated.

6.9 Vibrations of a shallow shell under broad band random load

Let the load be spatial white noise. In this case its spectral density is given by

$$S_p\left(\omega,\, \lambda,\, \mu\right) = \Psi\left(\omega\right). \tag{6.175}$$

Insertion of this expression into (6.172) yields

$$S_T = G\Psi\left(\omega\right) \iint\limits_{-\infty}^{\infty} \frac{\Lambda\left(\lambda,\, \mu,\, z\right)}{\left|\Omega\left(\omega,\, \lambda,\, \mu\right)\right|^2} d\lambda d\mu. \tag{6.176}$$

In order to evaluate this integral it is necessary to transform to new integration variables by means of the following transformation

$$\lambda = \sqrt{x}\cos\theta,\ \mu = \sqrt{x}\sin\theta. \tag{6.177}$$

Then, due to (6.173) we obtain

$$S_T = \int\limits_0^{2\pi}\int\limits_0^{\infty} \frac{\left[G\Psi\left(\omega\right)/2\right]\Lambda^* dx d\theta}{\left|-m\omega^2 + D\left(1+R^*\right)\left[x^2 + \frac{Eh}{D}\left(k_2\cos^2\theta + k_1\sin^2\theta\right)^2\right]\right|^2}. \tag{6.178}$$

where

$$\begin{aligned} \Lambda^* &= \Lambda\left(\sqrt{x}\cos\theta, \sqrt{x}\sin\theta, z\right), \\ R^* &= R\left(\omega, \sqrt{x}\cos\theta, \sqrt{x}\sin\theta\right). \end{aligned} \tag{6.179}$$

We compute the integral over x numerically following [119] and using the fact that the plastic strain effect is small. The latter is determined by function R^*. If $R^* = 0$ and $\omega < \sqrt{Eh\left(k_2\cos^2\theta + k_1\sin^2\theta\right)^2/mD}$, the integral

(6.178) converges. But if $R^* = 0$ and $\omega > \sqrt{Eh\left(k_2\cos^2\theta + k_1\sin^2\theta\right)^2/mD}$, the integral over x in (6.178) diverges because of the singularity at the point

$$x = x^* = \sqrt{\left[m\omega^2 - Eh\left(k_2\cos^2\theta + k_1\sin^2\theta\right)^2\right]/D}. \tag{6.180}$$

The integral (6.178) is bounded for a finite R^*. For small R^* the main contribution to its value gives a vicinity of the point $x = x^*$ as the denominator of the integrand achieves its minimum near this point. With this in view, estimating the contribution in the vicinity of the point x^* to the integral we replace Λ^* and R^* by their values at $x = x^*$. Denoting $\Lambda^{**} = \Lambda^*(x^*)$ and $R^{**} = R^*(x^*)$ we obtain asymptotically

$$S_T = \int_0^{2\pi}\int_0^{\infty} \frac{\left[G\Psi(\omega)/2\right]\Lambda^{**}dxd\theta}{\left|-m\omega^2 + D\left(1+R^{**}\right)\left[x^2 + \frac{Eh}{D}\left(k_2\cos^2\theta + k_1\sin^2\theta\right)^2\right]\right|^2}. \tag{6.181}$$

Now we point out that it is the imaginary part that has an essential influence on the asymptotic behaviour of the integral over x in (6.181). Denoting

$$\psi = \operatorname{Im} R^{**} \tag{6.182}$$

and completely ignoring the real part of R^{**} we obtain

$$S_T = \int_0^{2\pi}\int_0^{\infty} \frac{\left[G\Psi(\omega)/2\right]\Lambda^{**}dxd\theta}{\left|-m\omega^2 + D\left(1+i\psi\right)\left[x^2 + \frac{Eh}{D}\left(k_2\cos^2\theta + k_1\sin^2\theta\right)^2\right]\right|^2}. \tag{6.183}$$

Evaluation of this integral over x yields a closed form expression, cf. [119]. To this aim, we make use of the formula, see [44]

$$\int_0^{\infty} \frac{dx}{a + bx^2 + cx^4} = \frac{\pi}{2q^2 c\sin\alpha}\,\frac{\cos(\alpha/2)}{q}, \tag{6.184}$$

where

$$q = \sqrt[4]{\frac{a}{c}}, \quad \cos\alpha = -\frac{b}{2\sqrt{ac}}, \quad 0 \le \alpha \le \pi.$$

Comparing (6.184) and the integral over x in (6.183), we find the following expressions for the coefficients a, b, and c

$$\begin{aligned} &c = \left(1+\psi^2\right)D,\ b = \left[2\left(d^2 - \delta^2\right) + 2d^2\psi^2\right]D,\\ &a = \left(d^2 + \delta^2\right)^2 + d^4\psi^2,\ \delta^2 = m\omega^2,\\ &d^2 = d^2(\theta) = Eh\left(k_2\cos^2\theta + k_1\sin^2\theta\right)^2. \end{aligned} \tag{6.185}$$

Evaluation of the first quotient in the right hand side of eq. (6.184) yields

$$\frac{\pi}{2cq^2 \sin\alpha} = \frac{\pi}{2m\omega^2 D\,|\psi|}. \tag{6.186}$$

The second quotient is too unwieldy. However for $\psi \to 0$ it is bounded for nearly all values of frequency ω. For this reason, we will find its asymptotic representation for $\psi \to 0$. Formulae (6.185) then give

$$\cos\alpha = \sin(\delta - d), \cos\frac{\alpha}{2} = \left[\begin{array}{ll} 1, & \delta > d, \\ 0, & \delta < d, \end{array}\right. \tag{6.187}$$

$$q = \sqrt{\left|d^2 - \delta^2\right|/D}.$$

Inserting (6.187) and (6.186) into (6.184) we obtain an approximate value for the integral I_1 over x in (6.183) such that

$$I_1 = \left[\begin{array}{ll} \pi\left[2\psi m\sqrt{Dm}\omega^2\sqrt{\omega^2 - d^2(\theta)/m}\right]^{-1}, & \omega^2 > d^2/m, \\ 0, & \omega^2 < d^2/m. \end{array}\right. \tag{6.188}$$

Substituting this into (6.183) we obtain finally

$$S_T = \frac{\pi G \Psi(\omega)}{m\sqrt{Dm}\omega^3} \int\limits_0^{\pi/2} \frac{\Lambda^{**} d\theta}{\psi\sqrt{1 - \beta^2\left(\cos^2\theta + \chi\sin^2\theta\right)^2}}, \tag{6.189}$$

where

$$\beta = \left(\frac{Ehk_2^2}{m\omega^2}\right)^{1/2}, \quad \chi = \frac{k_1}{k_2}. \tag{6.190}$$

We integrate in (6.189) over that domain in which the integrand is real. It is easy to see that the integrand has an integrable singularity for any β, except $\beta = 1$, provided that the integration bound lies in the interval $[0, \pi/2]$.

The mean square of the shear stress intensity is found from the equation

$$T^2 = \int\limits_{-\infty}^{\infty} S_T(\omega)\, d\omega. \tag{6.191}$$

An expression for factor ψ, which is determined by means of (5.49), (5.48) and (6.182), is given by

$$\psi = \frac{Hn^{-\alpha}}{m\omega^2} \int\limits_{-h/2}^{h/2} I(\omega)\,\Lambda^{**}\left(\frac{T}{2G}\right)^{\alpha} dz. \tag{6.192}$$

Here $I(\omega)$ is given by (5.48) and has the following form

$$I(\omega) = \alpha \int_0^1 \frac{\eta^{\alpha-1}\omega\varphi(\eta)\,d\omega}{1+[\omega\varphi(\eta)]^2}. \tag{6.193}$$

Function $\varphi(\eta)$ in this equation is determined from eq. (6.157).

Equations (6.189), (6.190), (6.157), (6.192) and (6.193) give us a system of equations for the unknown variables. It is worthwhile noting that the spectral density (6.189) depends on the coordinate z besides the main argument ω. This leads to the conclusion that the solution of eq. (6.157) also depends on z . Consider next one possible means of simplification. For example, let us look for an approximate solution of the system such that the function $\varphi(\eta)$ does not depend upon z. We can find such a function from the equation

$$1-\eta^2 = \int_{-\infty}^{\infty} \frac{\int_{-h/2}^{h/2} S_T\left(\frac{T}{2G}\right)^\alpha dz}{\int_{-h/2}^{h/2} T^2\left(\frac{T}{2G}\right)^\alpha dz} \frac{d\omega}{1+(\omega\varphi)^2}, \tag{6.194}$$

which says that eq. (6.157) is satisfied on average with a weight $(T/2G)^{2+\alpha}$. Substituting eqs. (6.189) and (6.191) into the latter equation leads, due to (6.192), to a surprisingly simple result

$$1-\eta^2 = \frac{\int_{-\infty}^{\infty} \frac{\Psi(\omega)}{\omega I(\omega)} \frac{\Pi(\beta,\chi)\,d\omega}{1+(\omega\varphi)^2}}{\int_{-\infty}^{\infty} \frac{\Psi(\omega)}{\omega I(\omega)}\Pi(\beta,\chi)\,d\omega}, \tag{6.195}$$

where

$$\Pi(\beta,\chi) = \frac{2}{\pi} \int_0^{\pi/2} \frac{d\theta}{\sqrt{1-\beta^2\left(\cos^2\theta+\chi\sin^2\theta\right)^2}}. \tag{6.196}$$

Equations (6.195) and (6.193) give us a system of equations for $\varphi(\eta)$ and $I(\omega)$. This system differs from the corresponding system of equations in the theory of plates, see Section 6.6, only in a factor $\Pi(\beta,\chi)$ in the integrand of eq. (6.196). This function was first introduced by Bolotin in [18] and has a clear physical meaning. $\Pi(\beta,\chi)$ is equal to the ratio of the density of the shell eigenfrequency at frequency ω to that as $\omega \to \infty$. By means of a standard transformation it may be expressed in terms of elliptic integrals

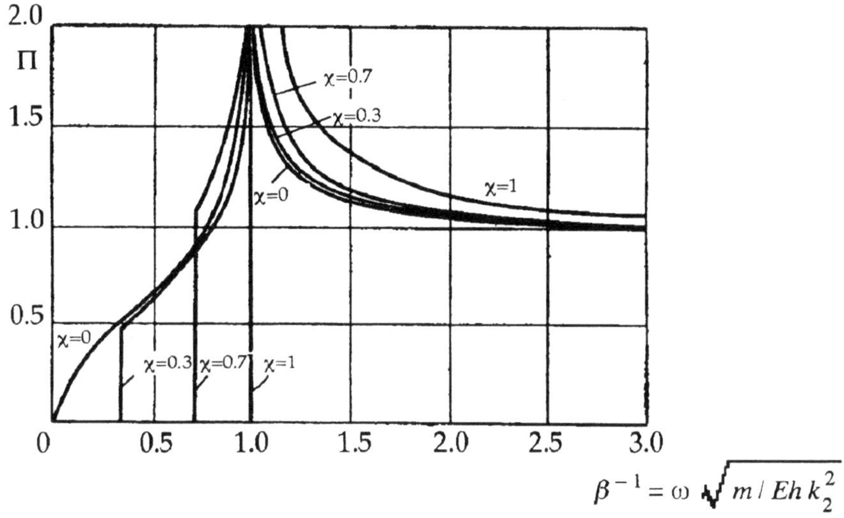

FIGURE 6.7. Modal density$\Pi(\beta, \gamma)$ versus $\beta^{-1} = \omega\sqrt{m/Ehk_2^2}$ for some values of χ.

in normal form. The result of the evaluations for $0 < \chi < 1$ is given below

$$\Pi(\beta, \chi) = \left[\begin{array}{ll} 0, & 0 < \beta^{-1} < \chi, \\ \frac{\sqrt{2}}{\pi\sqrt{\beta(1-\chi)}} \mathrm{K}\left(\sqrt{\frac{(1+\beta)(1-\beta\chi)}{2\beta(1-\chi)}}\right), & \chi < \beta^{-1} < 1, \\ \frac{2}{\pi\sqrt{(1+\beta)(1-\beta\chi)}} \mathrm{K}\left(\sqrt{\frac{2\beta(1-\chi)}{(1+\beta)(1-\beta\chi)}}\right), & \beta^{-1} > 1. \end{array} \right. \quad (6.197)$$

Here K (...) is a complete elliptic integral of the first kind. For a plate $\beta \to 0$, and therefore $\Pi = 1$. For a spherical shell $\chi = 1$ and the relative modal density Π takes on an especially simple expression

$$\Pi(\beta, 1) = \left[\begin{array}{ll} 0, & \beta^{-1} < 1, \\ \left(1 - \beta^2\right)^{-1/2}, & \beta^{-1} > 1. \end{array} \right. \quad (6.198)$$

The results of calculating $\Pi(\beta, \chi)$ are displayed in Fig. 6.7 versus non-dimensional frequency $\beta^{-1} = \omega\sqrt{m/Ehk_2^2}$ for various values of χ.

Analysis of expressions (6.197) shows that for any χ in the interval $0 < \chi < 1$, Π has an integrable singularity at $\beta = 1$. The curves of Fig. 6.7 confirm this.

Solution of the system of equations (6.193) and (6.195) is not the last complication of this problem. In order to determine T^2 we must make use of (6.191) which, by virtue of eqs. (6.189) and (6.171), gives

$$T^2 = \sum_{n=0}^{2} B_n z^n. \quad (6.199)$$

The factors of this decomposition are obtained by means of eqs. (6.189) and (6.171), to give

$$B_l = \frac{\pi n^\alpha}{4G\sqrt{DmH}} \int_{-\infty}^{\infty} \frac{\Psi(\omega)\, C_l(\omega, B_s)\, d\omega}{\omega I(\omega)}, \tag{6.200}$$

where

$$C_l(\omega, B_s) = \int_0^{\pi/2} \frac{A_l \left[1 - \beta^2 \left(\cos^2\theta + \chi \sin^2\theta\right)^2\right]^{-1/2} d\theta}{\int\limits_{-h/2}^{h/2} \sum\limits_{n=0}^{2} A_n z^n \left(\frac{1}{2G} \sum\limits_{m=0}^{2} B_m z^m\right)^\alpha dz}, \tag{6.201}$$

$$A_2 = \frac{4G}{3}(x^*)^2 \frac{1 - \nu + \nu^2}{(1-\nu)^2},$$

$$A_1 = \frac{4G}{3} x^* \frac{1 - \nu - 2\nu^2}{1-\nu}\left(k_2 \cos^2\theta + k_1 \sin^2\theta\right), \tag{6.202}$$

$$A_0 = \frac{4G}{3}(1+\nu)^2 \left(k_2 \cos^2\theta + k_1 \sin^2\theta\right)^2.$$

Regretfully, the integral over z in (6.201) can only be evaluated numerically for arbitrary α which considerably impedes the solution of this system of transcendental equations for B_0, B_1 and B_2.

The spectral density of the acceleration of the transverse deflection is of considerable interest for the application of vibration technology. The corresponding spectral representation of $\ddot{w}$ is obtained by means of eq. (6.166) and is given by

$$\ddot{w} = -\iiint_{-\infty}^{\infty} \frac{\omega^2 V(\omega, \lambda, \mu)}{\Omega(\omega, \lambda, \mu)} e^{i(\omega t + \lambda x + \mu y)} d\omega d\lambda d\mu. \tag{6.203}$$

The correlation function of $\ddot{w}$ is derived, due to eq. (6.97), to give

$$K_{\ddot{w}} = \iiint_{-\infty}^{\infty} \frac{\omega^4 S_p(\omega, \lambda, \mu)}{|\Omega(\omega, \lambda, \mu)|^2} e^{i(\omega t + \lambda x + \mu y)} d\omega d\lambda d\mu. \tag{6.204}$$

For a delta-correlated surface load (6.112), the spectral density S_p is given by eq. (6.113). By analogy with eqs. (6.101) and (6.103) we obtain the spectral density of the deflection acceleration $\ddot{w}$ for a shallow shell

$$\Phi_{\ddot{w}}(\omega) = \iint_{-\infty}^{\infty} \frac{\omega^4 \Psi(\omega)}{|\Omega(\omega, \lambda, \mu)|^2} d\lambda d\mu. \tag{6.205}$$

The integral in (6.205) can be evaluated by analogy with the previous evaluations of this Section. In the case of small values of $\psi(\omega)$, the asymptotic representation for this is such that

$$\Phi_{\ddot{w}}(\omega) = \frac{\pi^2 \Psi(\omega)\,\omega}{2m\sqrt{Dm}\psi(\omega)} \Pi(\beta, \chi), \tag{6.206}$$

where $\Pi(\beta, \chi)$ is the relative modal density of the shell, cf. eq. (6.196).

The spectral density (6.206) differs from the spectral density of a plate only by a factor Π. Expression (6.206) for the case of a "linear" energy dissipation has been first obtained in the author's paper [119]. Due to the ideas of this book, function $\psi(\omega)$ is to be calculated by means of plasticity theory.

7

Propagation of vibration in a nonlinear dissipative medium

A simple problem, namely the problem of longitudinal vibration in a homogeneous rod, will be studied in this chapter. Most attention will be given to the analysis of vibration in a semi-infinite rod loaded at one of its ends. A traditional method of nonlinear mechanics, namely a series expansion in terms of normal modes, which was used in Chapters 4 and 6 is not applicable here. The approach used in this chapter is based on the method of propagating waves which overcomes some difficulties caused mainly by the wave heterogeneity due to the energy dissipation in the rod's material. The problem of the vibration of a very long rod excited by a high frequency load is a related problem. A peculiarity of this problem lies in the fact that distortion of the normal modes due to small dissipative forces cannot be ignored.

7.1 Propagation of vibration in a linear viscoelastic rod

The linear vibration in a dissipative rod, cf. [122], is studied in this Section. Consider a semi-infinite rod $x > 0$. The governing equation for its longitudinal vibration is given by

$$Q' - m\ddot{u} = 0, \tag{7.1}$$

where Q denotes a tensile axial force in the cross-section with coordinate x, u is an axial displacement, and m is the mass per unit length. The dot and the prime indicate differentiation with respect to time and spatial coordinate x, respectively.

The axial strain of the rod is as follows

$$\varepsilon = u' . \tag{7.2}$$

The constitutive equation for the rod's material is assumed to be given by

$$Q = c\left[1 + R\left(\partial/\partial t\right)\right]\varepsilon, \tag{7.3}$$

where c is the static axial rigidity of the rod, and $R\left(\partial/\partial t\right)$ is an operator of viscoelasticity. The axial elastic force in eq. (7.3) is given by the term $c\varepsilon$, while the dissipative force is represented by the term $cR\varepsilon$. In what follows, the dissipative force is assumed to be small compared to the elastic force, i.e. the following inequality

$$\left|R\left(\partial/\partial t\right)\varepsilon\right| \ll \left|\varepsilon\right| \tag{7.4}$$

holds.

Combining eqs. (7.1), (7.2) and (7.3) yields a single equation for u

$$c\left[1 + R\left(\partial/\partial t\right)\right]u'' - m\ddot{u} = 0. \tag{7.5}$$

Let the cross-section $x = 0$ be subject to a harmonic force, that is the boundary condition at $x = 0$ is

$$Q = B_0 e^{i\omega t}. \tag{7.6}$$

We require a decrease in vibration as $x \to \infty$. A steady-state vibration of the rod is sought in the form

$$u = U\left(\omega, x\right) e^{i\omega t}. \tag{7.7}$$

Inserting (7.7) into (7.5) gives the following equation for U

$$U'' + \lambda^2 U = 0, \quad \lambda^2 = \frac{m\omega^2}{c\left[1 + R\left(i\omega\right)\right]}. \tag{7.8}$$

A solution satisfying the above condition of decay at infinity has the form

$$U = De^{-i\lambda x}, \tag{7.9}$$

where λ is that root of the second equation in (7.8) which has a negative imaginary part.

One can easily obtain from the second equation in (7.8) the following representation for λ

$$\lambda = \frac{\omega}{a}\left(\gamma - i\eta\right), \quad a = \sqrt{\frac{c}{m}}. \tag{7.10}$$

Here a denotes the velocity of propagation of small disturbances in the undamped rod $(R=0)$, while the parameters γ and η are obtained from the equation

$$(\gamma - i\eta)^2 = [1 + R(i\omega)]^{-1}, \ \eta > 0. \tag{7.11}$$

Substituting (7.9) into (7.7) and satisfying the boundary condition (7.6) gives

$$u = -\frac{B_0}{i\lambda c [1 + R(i\omega)]} e^{i(\omega t - \lambda x)}. \tag{7.12}$$

The value of acceleration is of interest in many problems of vibrational technology

$$\ddot{u} = -\frac{i\lambda B_0}{m} e^{i(\omega t - \lambda x)}. \tag{7.13}$$

The value of force is required for analysis of the dynamical strength

$$Q = B_0 e^{i(\omega t - \lambda x)}. \tag{7.14}$$

Inserting λ due to (7.10) into the latter equation yields

$$Q = B_0 e^{-\omega \eta x/a} e^{i\omega (t - \gamma x/a)}. \tag{7.15}$$

As follows from this equation the rod motion is a decaying travelling wave.

Note that the wave solution (7.15) cannot be principally obtained by the approximate methods of Chapters 4 and 6 for the simple reason that the concept of a normal mode of free elastic vibration does not exist for a semi-infinite rod.

Consider now a finite rod of length L whose end $x=0$ is free whereas the end $x=L$ is subject to the load (7.6). The solution, as above, is sought in the form (7.7), with eq. (7.8) being the governing equation for U. The boundary conditions for U are obtained by substitution of (7.6) and (7.7) into the boundary conditions of the problem, i.e.

$$\begin{aligned} x &= 0, \quad \frac{dU}{dx} = 0, \\ x &= L, \quad c[1 + R(i\omega)] \frac{dU}{dx} = B_0. \end{aligned} \tag{7.16}$$

The solution of eq. (7.8) satisfying the above boundary conditions is

$$U = -\frac{\lambda B_0}{m\omega^2} \frac{\cos \lambda x}{\sin \lambda L}. \tag{7.17}$$

Using this equation we find the following expression for the acceleration

$$\ddot{u} = \Phi(\omega, x) B_0 e^{i\omega t}, \tag{7.18}$$

where the transfer function Φ is given by

$$\Phi(\omega, x) = \frac{\lambda \cos \lambda x}{m \sin \lambda L}. \tag{7.19}$$

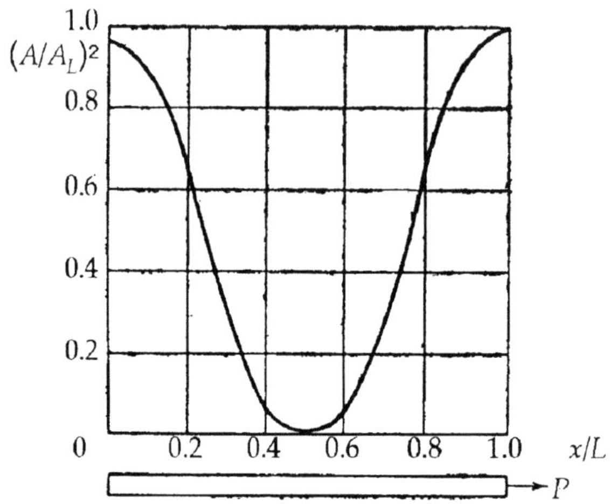

FIGURE 7.1. Square of the nondimensional amplitude $(A/A_L)^2$ along the rod for the first eigenfrequency.

The square of the vibration amplitude takes the form

$$A^2 = B_0^2 \left|\Phi(\omega, x)\right|^2. \tag{7.20}$$

Using (7.19) and expression (7.10) for λ we obtain the following result

$$\left|\Phi(\omega, x)\right|^2 = \frac{|\lambda|^2}{m^2} \frac{\cosh 2\omega\eta x/a + \cos 2\omega\gamma x/a}{\cosh 2\omega\eta L/a - \cos 2\omega\gamma L/a}. \tag{7.21}$$

Expressions (7.20) and (7.21) enable us to see the difference between the mode of free elastic vibration and the modes of forced vibration of a viscoelastic rod. For example, Figs. 7.1, 7.2 and 7.3a show plots of $(A/A_L)^2$ versus the non-dimensional variable x/L at $\gamma = 1$ and $\eta = 0,05$ for the first, fifth and tenth natural frequencies. Examination of Fig. 7.1 indicates that the energy dissipation changes the shape of the low-frequency normal modes by a negligibly small amount. The influence of the energy dissipation on the shape of high frequency modes is considerable, see Figs. 7.2 and 7.3a.

It is instructive to compare Figs. 7.3a and 7.3b. Figure 7.3a plots $(A/A_L)^2$ with damping being taken into account, while Fig. 7.3b shows the same dependence without damping. Both diagrams are plotted for the same frequency equal to the tenth eigenfrequency of the rod. In the first case a considerable spatial decay of vibration is observed whereas in the second case there is no decay.

We conclude from this that even a small amount of damping leads to considerable alteration of high frequency modes of forced vibration. This fact is well-known in the literature, see [165].

The results of these comparisons enables us to assess critically the potential for using the approximate methods described in Chapters 4 and 6

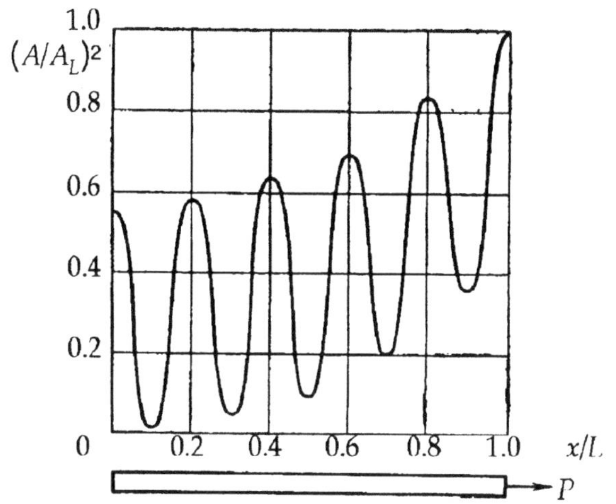

FIGURE 7.2. Square of the non-dimensional amplitude $(A/A_L)^2$ along the rod for the fifth eigenfrequency.

for this particular problem. It is clear that the method of Chapter 4 cannot in principle reflect the difference between the mode of forced vibration in the viscoelastic rod and the free elastic mode of the rod since this method operates with an undistorted mode. Hence, the method of Chapter 4 is applicable, in this problem, to analysis only the first few natural frequencies. The method of Chapter 6 is preferable as the solution is sought in the form of a sum in terms of normal modes of elastic vibration. Hence, this method gives us, in principle, a possible means of obtaining a proper mode of forced vibration. But we have to pay a high price for this success since we have to solve the complete system of equations (6.8). Besides, the effect of the spatial decay of vibration is achieved by summation of a large number of terms in the series expansion in terms of normal modes, each mode possessing no effect of spatial decay. With this in view, this strategy is deemed to be ineffective. Note, that considerable simplification of the solution which is obtained by means of the simplified equations (6.10) is prohibited here. Indeed, using the simplified equations implies an asymptotically small energy dissipation, but under this assumption even the exact solution (7.21) does not give a spatial decay of vibration at the resonant frequencies.

Let us proceed to consideration of a random load. Assume that the load $p(t)$ applied at $x = 0$ is a stationary random function of time with a zero mean value. Prescribing the load by its spectral decomposition we obtain the following boundary condition

$$Q = p(t) = \int_{-\infty}^{\infty} e^{i\omega t} V(\omega)\, d\omega, \tag{7.22}$$

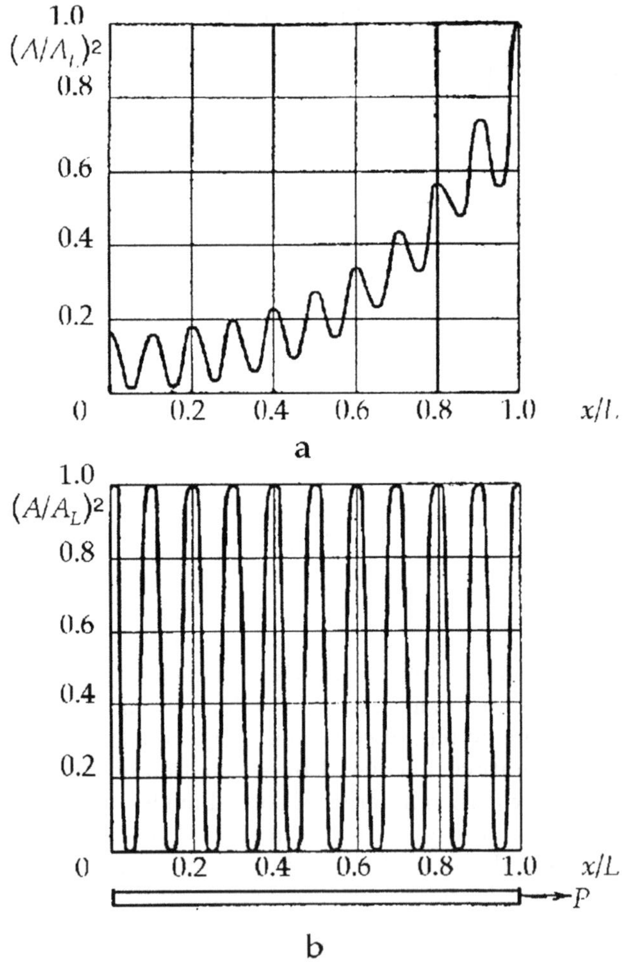

FIGURE 7.3. Dependence $(A/A_L)^2$ versus x/L with (a) and without (b) damping.

with $V(\omega)$ being a white noise random function with intensity $S(\omega)$. A deterministic function $S(\omega)$ is the spectral density of the load $p(t)$. A steady-state vibration of the rod is sought in the form of an integral canonical representation

$$u(x, t) = \int_{-\infty}^{\infty} e^{i\omega t} U(\omega, x)\, d\omega, \tag{7.23}$$

where U satisfies eq. (7.8). As a semi-infinite rod is considered, the solution U must be taken in the form (7.9). Inserting this into (7.23) and into the boundary condition (7.22) yields

$$u = -\int_{-\infty}^{\infty} e^{i(\omega t - \lambda x)} \frac{V(\omega)\, d\omega}{i\lambda c\,[1 + R(i\omega)]}. \tag{7.24}$$

The acceleration is given by

$$\ddot{u} = -\int_{-\infty}^{\infty} \frac{i\lambda}{m} e^{i(\omega t - \lambda x)} V(\omega)\, d\omega. \tag{7.25}$$

The latter equation is an integral canonical representation of the acceleration. With its help we can easily find the correlational function of acceleration

$$K_{\ddot{u}}(x_1, x_2, \tau) = \int_{-\infty}^{\infty} e^{i\omega\tau - i\gamma\omega(x_1 - x_2)/a} e^{-\omega\eta(x_1 + x_2)/a} \frac{|\lambda|^2}{m^2} S(\omega)\, d\omega, \tag{7.26}$$

where $\tau = t_1 - t_2$ is the time difference. In particular, we see that the correlation function of acceleration at $x_1 = x_2 = x$ is given by

$$K(\tau, x) = K_{\ddot{u}}(x, x, \tau) = \int_{-\infty}^{\infty} \frac{|\lambda|^2}{m^2} e^{i\omega\tau} e^{-2\omega\eta x/a} S(\omega)\, d\omega. \tag{7.27}$$

Consider now a rod of length L having a free end $x = 0$, while end $x = L$ is subject to a stationary random force $p(t)$. Prescribing the latter by its spectral representation (7.22), looking for a solution in form (7.23) and satisfying the above boundary conditions we obtain the following integral canonical representation for the acceleration

$$\ddot{u} = \int_{-\infty}^{\infty} e^{i\omega\tau} \Phi(\omega, x) V(\omega)\, d\omega, \tag{7.28}$$

where $\Phi(\omega, x)$ is given by eq. (7.19). By means of the integral canonical representation (7.28) we obtain the correlation function of acceleration

$$K_{\ddot{u}}(x_1, x_2, \tau) = \int_{-\infty}^{\infty} e^{i\omega\tau} \Phi(\omega, x_1) \Phi(\omega, x_2) S(\omega)\, d\omega. \tag{7.29}$$

Consider, for example, the acceleration at a point with coordinate x as a function of time only. By virtue of eq. (7.29) the correlation function of acceleration is as follows

$$K(\tau, x) = \int_{-\infty}^{\infty} e^{i\omega\tau} |\Phi(\omega, x)|^2 S(\omega)\, d\omega, \tag{7.30}$$

with the square of the absolute value of the transfer function Φ being given by formula (7.21).

In principle, equations (7.29) and (7.30) solve the problem of the correlation function of the vibration field. Numerical computations due to these equations are simple provided that the frequency domain of the load contains a few eigenfrequencies of the rod. If the number of eigenfrequencies in this domain is large, the numerical computations become exceedingly complicated since the function $|\Phi(\omega, x)|^2$ has a maximum in the vicinity of each eigenfrequency. However, for this case it is possible to simplify the analysis by incorporating an average over a frequency band.

In mechanical problems, the idea of averaging over a frequency band was first applied by Bolotin to vibrations in plates and shells, cf. [17]. The equations derived in [17] are valid for asymptotically small damping under the assumption that the correlation between the different modes is negligible.

The equations which will be derived below, cf. [122], are valid for both small and intensive damping, the correlation between the different modes being automatically taken into account because the present approach requires no expansion in terms of normal modes at all.

Consider, for instance, the correlation function (7.30). Assume that the load p is a broad-band random process with a smooth spectral density $S(\omega)$. Under this assumption, the integrand in (7.30) is a highly oscillating function of frequency ω with a large numbers of maxima and minima even for small τ. This is explained by the presence of trigonometric functions in expression (7.21). In order to derive some simpler formulae it is reasonable to perform the integration in (7.30) in two stages, namely, first average the integrand in (7.30) over a few eigenfrequencies, say one or two, and then integrate over ω. We proceed now to a realisation of this idea, i.e. $|\Phi(\omega, x)|^2$ is to be averaged. As all functions in expression (7.21) for $|\Phi(\omega, x)|^2$, except the trigonometric functions are smooth, an averaging must be performed over the arguments of these trigonometric functions, the other frequency functions being kept constant. This leads to the following expression

$$|\Phi(\omega, x)|_{av}^2 = \frac{1}{(2\pi)^2}\int_0^{2\pi}\int_0^{2\pi}\frac{|\lambda|^2}{m^2}\frac{\cosh 2\omega\eta x/a + \cos z_1}{\cosh 2\omega\eta L/a - \cos z_2}dz_1 z_2. \tag{7.31}$$

A consequent averaging over each argument is allowed since these arguments are incommensurable in the general case. Evaluation of the integrals in (7.31) gives the following result

$$|\Phi(\omega, x)|_{av}^2 = \frac{|\lambda|^2}{m^2}\frac{\cosh 2\omega\eta x/a}{\sinh 2\omega\eta L/a}\quad . \tag{7.32}$$

It has already been mentioned above that the smallness of τ is required for validity of the integration in two stages. Strictly speaking, the value of τ must be so small that a change of product $\omega\tau$ will be not larger than unity in the frequency band between two adjacent maxima of the integrand $(\Delta\omega \approx \pi a/\gamma L)$. This leads to the following restriction imposed on τ

$$\tau < \frac{1}{\omega}\min\frac{\gamma L}{\pi a}. \tag{7.33}$$

For $\gamma \approx 1$ the right hand side of this equation is about six times smaller than the period of the fundamental frequency of the rod. This estimation shows that the above averaging of the integrand is valid for dispersion in any case.

Inserting the averaged expression (7.32) into (7.30) we obtain a simplified equation for the correlation function of vibration

$$K(\tau, x) = \int_{-\infty}^{\infty}\frac{|\lambda|^2}{m^2}\frac{\cosh 2\omega\eta x/a}{\sinh 2\omega\eta L/a}e^{i\omega\tau}S(\omega)\,d\omega. \tag{7.34}$$

Let us consider some particular cases of this equation.

1. Let the damping be small, i.e. $\eta \ll 1$, $\gamma \approx 1$. In addition, if for any frequency in the loading spectrum, the following inequality holds

$$2\frac{\omega\eta L}{a} \ll 1, \tag{7.35}$$

we obtain from eq. (7.34)

$$K(\tau, x) = \int_{-\infty}^{\infty}\frac{\omega S(\omega)}{2a\eta L m^2}e^{i\omega\tau}d\omega. \tag{7.36}$$

From this equation we conclude that the spectral density of vibration is the same in all cross-sections and is given by

$$S_{\ddot{u}} = \frac{\omega S(\omega)}{2a\eta L m^2}. \tag{7.37}$$

It is worth noting that the asymptotic method by Bolotin [17] leads to the same result.

2. Let the damping be small. In addition, we assume that, for all frequencies of the loading spectrum, the following inequality holds

$$2\frac{\omega\eta L}{a} \gg 1. \tag{7.38}$$

Using asymptotic equations for the hyperbolic functions yields

$$K(\tau, x) = \int_{-\infty}^{\infty} e^{i\omega\tau} \frac{|\lambda|^2}{m^2} e^{-2\omega\eta(L-x)/a} S(\omega)\, d\omega. \tag{7.39}$$

This expression coincides with the expression for the correlation function of vibration in a semi-infinite rod (7.27).

In the general case when a frequency spectrum of the load is so broad that neither (7.35) nor (7.38) holds for the whole frequency domain, equation (7.34) should be used.

The methods of this Section will next be modified and adapted to analyse the longitudinal vibration of a rod with nonlinear properties and, in particular, analyse the vibration in elastoplastic rods.

7.2 Statement of the problem for a nonlinear dissipative medium

The problem of vibration propagation in nonlinear media has great practical interest. It is studied in nonlinear acoustics [191] and nonlinear optics [15]. Such an important effect as internal damping is generally governed by nonlinear equations. Problems of vibrations in such classical nonlinear media as elastoplastic, rigid-plastic, elastoviscoplastic media as well as in media with hardening belong to the same range of problems. Similar problems appear in the theory of gravitational waves on a fluid surface [97].

In what follows, we consider dissipative media, i.e. media with an energy dissipation. The main problem related to these media is to reveal the laws of vibration decay with distance from the energy source.

In this Section we study the simplest problem of this kind, namely the problem of vibration propagation in a semi-infinite rod. We will derive an approximate solution which enables a clear picture of the vibration field to be obtained in some particular cases.

Consider longitudinal vibrations in a semi-infinite rod $x > 0$.

The equation governing the dynamics of the rod is given by eq. (7.1)

$$Q' - m\ddot{u} = 0. \tag{7.40}$$

We keep the same notation as in Section 7.1. Assume that the strain is small and it is given by the equation

$$\varepsilon = u' \tag{7.41}$$

which has no nonlinear terms.

The constitutive equation is assumed to have the following form

$$Q = Q\{\varepsilon, \dot{\varepsilon}\}, \tag{7.42}$$

where $Q\{\varepsilon, \dot{\varepsilon}\}$ is an odd functional of its arguments.

For the sake of simplicity, force Q is modelled as a function of strain and its velocity. The forthcoming analysis will, however, be valid for a broader class of dependences Q, e.g. for non-single-valued ones, looped ones etc. Moreover, the latter dependences are of crucial practical interest.

One of these two conditions

$$Q = B_0 \cos \omega t, \; u = A_0 \cos \omega t \tag{7.43}$$

will be taken as a boundary condition at $x = 0$. The variables Q and u must satisfy the Sommerfeld radiation conditions as $x \to \infty$.

The method of harmonic linearization will be applied to solve this problem. This method is one of the simplest and most effective methods of the theory of nonlinear vibration and is actively used in the theory of automatic control [149], theory of vibration protection [86] etc.

Note that the method of harmonic linearisation possesses unquestionable advantages over the Krylov-Bogolyubov method for multi-degree-of-freedom systems [16] when applied to the problem of determining the forced vibration modes. This fact has been mentioned in [175] where it has also been shown how to modify the equation of Krylov-Bogolyubov's method in order to ensure the coincidence with the Ritz method (actually with the method of harmonic linearization) in the form of the steady-state forced vibration up to values of the first order of smallness.

An essential feature of the method of harmonic linearization is its simplicity and clearness of the physical results obtained. In the present book this method is applied to the analysis of wave processes in a rod whose material is governed by nonlinear equations.

Using the method of harmonic linearization the strain ε is assumed to be a nearly harmonic process in each cross-section

$$\varepsilon = a(x) \cos [\omega t - \varphi(x)], \tag{7.44}$$

where a is the strain amplitude and φ is the phase in cross-section x.

Next, a nonlinear functional $Q\{\varepsilon, \dot{\varepsilon}\}$ is approximated by a linear function

$$Q \approx q\varepsilon + \frac{h}{\omega}\dot{\varepsilon}. \tag{7.45}$$

The linearisation coefficients q and h do not depend on time and are chosen from the condition of equality of amplitude and phase of the harmonics of frequency ω in the left and right hand sides of eq. (7.45) due to harmonical motion (7.44). It can be shown, cf. [149] and [86], that q and h which satisfy this requirement are as follows

$$\begin{aligned} q &= \frac{1}{\pi a}\int_0^{2\pi} Q\left(a\cos\kappa,\ -a\omega\sin\kappa\right)\cos\kappa d\kappa, \\ h &= -\frac{1}{\pi a}\int_0^{2\pi} Q\left(a\cos\kappa,\ -a\omega\sin\kappa\right)\sin\kappa d\kappa. \end{aligned} \tag{7.46}$$

As a result of the linearisation the nonlinear relation (7.42) is replaced by an approximate linear equation (7.45). Inserting it into (7.40) and accounting for (7.41) yields

$$\frac{\partial}{\partial x}\left(q\frac{\partial u}{\partial x}+\frac{h}{\omega}\frac{\partial^2 u}{\partial x\partial t}\right)-m\frac{\partial^2 u}{\partial t^2}=0. \tag{7.47}$$

This equation is similar to an equation for the longitudinal vibration in a linear viscoelastic rod. The difference is that in this case q and h depend upon the strain amplitude a, i.e. they are not known in advance. As q and h are dependent on the strain amplitude it is expedient to differentiate eq. (7.47) with respect to x and consider strain ε as the unknown variable. The result is

$$\frac{\partial^2}{\partial x^2}\left(q\varepsilon+\frac{h}{\omega}\frac{\partial\varepsilon}{\partial t}\right)-m\frac{\partial^2\varepsilon}{\partial t^2}=0. \tag{7.48}$$

We substitute the assumed solution (7.44) into (7.48). Equating the factors in sine and cosine to zero we obtain two equations for the unknown a and φ, cf. [121]

$$\begin{aligned} &(aq)^{''}-aq\left(\varphi^{'}\right)^2+2\,(ah)^{'}\varphi^{'}+ah\varphi^{''}+m\omega^2 a=0, \\ &(ah)^{''}-ah\left(\varphi^{'}\right)^2-2\,(aq)^{'}\varphi^{'}-aq\varphi^{''}=0. \end{aligned} \tag{7.49}$$

It should be mentioned that not each solution of the latter system is a solution of the posed problem. According to the Sommerfeld radiation condition waves are outgoing at infinity and do not come back. This gives the following condition:

$$\varphi^{'}\geq 0 \tag{7.50}$$

as $x\to\infty$. Sometimes it is more convenient to replace this condition by a more rigid one, namely the condition of the non-increasing of the strain amplitude

$$a\geq 0,\ a^{'}\leq 0. \tag{7.51}$$

It is clear that it is not always possible to get a closed form solution of system (7.49) even after the simplifications have been made. However, for some particular dependences $q = q(a)$ and $h = h(a)$, a closed form solution can be obtained.

The equations for strains are derived and displacement is obtained by integration over x in eq. (7.41).

Let us consider the result of a formal application of the method of harmonic linearisation. The starting point is the system of equations (7.40), (7.41) and (7.42) which are equivalent to a single nonlinear partial differential equation. Applying the method of statistical linearisation gives two nonlinear ordinary differential equations (7.49).

The system of equations (7.49) is not always suitable for analysis of the problem. In some cases, another equivalent form turns out to be useful. Some of these are considered below.

Note, first, that after linearisation of the only nonlinear relation (7.42) one can transform to the general complex exponential form for the problem variables

$$\varepsilon = ae^{i(\omega t-\varphi)}, \; Q = Be^{i(\omega t-\psi)}, \tag{7.52}$$

where a and B are amplitudes, while φ and ψ are phases. As always, only the real parts of eq. (7.52) have physical meaning.

Using the linearised formula (7.45) it is easy to establish a relation between the complex strain and the complex force

$$Q = c_c \varepsilon, \tag{7.53}$$

with c_c being the complex rigidity

$$c_c = q + ih. \tag{7.54}$$

Let us find now the equations for amplitude and tensile force Q. To this end, we take derivatives of both sides of eq. (7.40) with respect to x and take into account eqs. (7.41) and (7.53). We will have

$$Q'' - \frac{m}{c_c}\ddot{Q} = 0.$$

Inserting an expression for the force Q due to (7.52) into the latter equation and separating real and imaginary parts we arrive at the required equations, see [126]

$$\begin{aligned} B'' - B\left(\psi'\right)^2 &= -m\omega^2 BR, \\ \tfrac{1}{B}\left(B^2\psi'\right)' &= -m\omega^2 BI, \end{aligned} \tag{7.55}$$

where R and I are the real and imaginary parts of the inverse function for the complex rigidity

$$R - iI = c_c^{-1}. \tag{7.56}$$

Note that R and I depend upon the strain amplitude a which is still unknown. In order to find their expressions in terms of B the following equation

$$B^2 = \left|c_c(a)\right|^2 a^2 \tag{7.57}$$

must be solved for a and its solution must be substituted into R and I.

Equation (7.57) is easily obtained by means of eq. (7.53).

The boundary conditions for the system of equations (7.55) at $x = 0$ are

$$B = B_0, \ \psi = 0, \tag{7.58}$$

and the Sommerfeld radiation condition as $x \to \infty$

$$\psi' \geq 0. \tag{7.59}$$

It is worth noting that q, h, R and I are non-negative for physically meaningful problems. This fact will be used in what follows.

Let us consider an interesting analogy. Equations (7.55) are identical with the equations for dynamics of a point mass in a plane non-central field. The right hand side of the first equation in (7.55) is analogous to the force of attraction to the center, while the right hand side of the second equation is analogous to a transverse perturbing force in the dynamics of a point mass. Therefore the original complex problem of the wave propagation in a nonlinear medium reduces, after a harmonic linearisation, to an equally difficult problem of the dynamics of a point mass. In addition to this, the problem is complicated by the character of the boundary conditions. The Cauchy problem is stated in the dynamics of a point mass whereas in our case it is a boundary problem with a condition at infinity.

The established analogy is useful from two points of view. Firstly, it gives an idea on the character of the difficulties one faces when integrating the system (7.55). Secondly, it allows the exact and approximate methods developed in the dynamics of a point mass, in particular, in the dynamics of satellites and spacecraft, to be applied. One of these approximate methods is applied below.

Provided that a trajectory of a spacecraft with a small transverse thrust is close to a circular path, the radial acceleration in the equations of dynamics can be neglected. This assumption has been used in [181] when studying the motion of a spacecraft with a solar sail. This method may be used in the problem of wave propagation.

The assumption of smallness of the transverse thrust is analogous to the assumption that I is small compared to R. Neglecting the radial acceleration in the dynamics of a point mass is analogous to neglecting the second derivative in the first equation in (7.55). This equation then gives

$$\psi' = \sqrt{m\omega^2 R}, \tag{7.60}$$

where the sign of the square root is chosen to be positive because of the radiation condition (7.59).

Substitution of this expression into the second equation (7.55) leads to an equation for B

$$\left(B^2\sqrt{R}\right)' = -\sqrt{m\omega^2}B^2 I. \tag{7.61}$$

If the product in the parentheses is a monotonically increasing function of B, then due to the non-negativeness of I we deduce that the force amplitude does not increase as x increases. Integrating (7.61) we get B and then ψ from eq. (7.60).

Equations (7.60) and (7.61) may be rewritten in a form containing amplitude and phase of strain. To this aim, we make use of eqs. (7.53) and (7.57) and take into account that condition

$$I \ll R$$

is equivalent to the following condition

$$h \ll q. \tag{7.62}$$

Neglecting small values yields

$$\varphi' = \sqrt{\frac{m\omega^2}{q}}, \tag{7.63}$$

$$a' = -\frac{\sqrt{m\omega^2}ah}{2q^{3/2}\left[1 + \frac{3}{4}\frac{a}{q}\frac{\partial q}{\partial a}\right]}. \tag{7.64}$$

Consider now this problem from another perspective. Using the complex form of the variables, we separate the variables directly in eq. (7.47) assuming that

$$u(x, t) = U(x)\, e^{i\omega t}, \tag{7.65}$$

where U is a complex function of x.

The result is an ordinary differential equation

$$\left(c_c U'\right)' + m\omega^2 U = 0. \tag{7.66}$$

Here c_c has already been introduced as the complex rigidity. Note that the complex rigidity depends on the strain amplitude a by virtue of eqs. (7.54) and (7.46) and also that the strain amplitude depends upon x and is not known *a priori*.

Equation (7.66) may be considered a linear equation with varying coefficient $c_c(x)$. *Prima facie*, this interpretation seems to be useless as it is impossible to find a general closed form solution of eq. (7.66) for arbitrary $c_c(x)$. There exist, however, two cases in which this approach leads to such solutions.

1. Assume that $a = a(x)$ and $c_c = c_c(x)$ are slowly varying functions of x. In this case, an effective approximate solution of eq. (7.66) may be obtained by means of the Steklov-Liouville method [173]. Another name for this method is the WKB-method [48]. Using this method, we introduce the new variables

$$U = c_c^{-1/4} v(y), \quad y = \int_0^x \sqrt{\frac{m}{c_c}} dx, \tag{7.67}$$

and equation (7.66) takes the form

$$\frac{d^2 v}{dy^2} + \left(\omega^2 - c_c^{-1/4} \frac{d^2}{dy^2} c_c^{-1/4}\right) v = 0. \tag{7.68}$$

Provided that the rigidity c_c changes rather slowly along the rod, the second term in the parentheses can be neglected, and eq. (7.68) is easily integrated, to give

$$v = De^{-i\omega y} + Fe^{i\omega y}, \tag{7.69}$$

where D and F are some integration constants.

In order to ensure that the solution remains bounded as $x \to \infty$, we put $F = 0$. Returning to the original variables yields

$$u = Dc_c^{-1/4} e^{-i\omega y} e^{i\omega t}. \tag{7.70}$$

Evaluation of the strain due to the assumption of slowly varying c_c gives

$$\varepsilon = i\omega D\sqrt{m} c_c^{-3/4} e^{-i\omega y} e^{i\omega t}. \tag{7.71}$$

The square of the strain amplitude is easy to obtain

$$a^2 = \omega^2 D^2 m \left|c_c\right|^{-3/2} e^{2\omega \operatorname{Im} y}. \tag{7.72}$$

As a matter of fact, this is an integral equation for a, which reduces to a differential equation by means of taking a logarithmic derivative of the both sides of eq. (7.72), to give

$$a' = -\frac{a\omega \operatorname{Im} \sqrt{m/c_c}}{1 + \dfrac{3}{4} \dfrac{a}{\left|c_c\right|} \dfrac{\partial \left|c_c\right|}{\partial a}}. \tag{7.73}$$

The assumption on a slow change of a has been made above. As seen from eq. (7.73) this is possible if the imaginary part of c_c is small compared to its real part, i.e.

$$h \ll q.$$

This fact allows us to use the following asymptotic formulae

$$\left|c_c\right| = q, \quad \operatorname{Im} \sqrt{\frac{m}{c_c}} = -\frac{h}{2q}\sqrt{\frac{m}{q}}. \tag{7.74}$$

Thus eq. (7.73) takes the form

$$a' = -\frac{ah\sqrt{m\omega^2}}{2q^{3/2}\left[1+\frac{3}{4}\frac{a}{q}\frac{\partial q}{\partial a}\right]}. \tag{7.75}$$

This coincides with eq. (7.64), so the method of slowly varying rigidity does not yield a new result.

2. The second method should be referred to as an inverse method. Its essence is as follows, cf. [123]. Given a particular dependence

$$c_c = c_c(x), \tag{7.76}$$

for which eq. (7.66) has a relatively simple closed form solution. Satisfying the boundary conditions renders the following dependence

$$a = a(x). \tag{7.77}$$

By means of the latter equation we eliminate x from (7.76) and obtain an expression

$$c_c = c_c(a). \tag{7.78}$$

A particular form of dependence (7.78) helps us to guess the character of nonlinearity Q.

Equations of the method of slowly varying waves have also been derived by Mironov in [107] where a two-frequency wave was studied. These equations were obtained by means of a modification of the Mitropolsky method. Provided that only a single-frequency wave is studied, they take the form of (7.63) and (7.64).

It is worth noting that the original equations for the wave parameters a, φ, B and ψ, e.g. eqs. (7.49), (7.55) and (7.66), are suitable not only for the analysis of slowly varying waves, but also for waves whose parameters admit a rapid change along the rod. Some examples of such waves can be found in the next Section.

7.3 Propagation of vibration in plastic media

In the present Section the laws of propagation of stationary vibration in plastic bodies with various constitutive equations are studied. A rigid-plastic material with linear hardening, an elastoplastic material with a linear hardening, and an elastoplastic material with an arbitrary hardening law, but a monotonic convex loading curve, are analysed consequently. The latter problem presents some interest in the theory of internal damping in materials under intensive stress.

7.3.1 Rigid-plastic materials with linear hardening

Let the constitutive equation have the form, cf. [121]

$$Q = c\varepsilon + H \operatorname{sgn} \dot{\varepsilon}. \tag{7.79}$$

One can easily recognize the deformation law of a rigid-plastic material with linear hardening and the Bauschinger effect being taken into account.

The linearisation factors are as follows

$$q = c,\ h = H_0/a,\ H_0 = 4H/\pi. \tag{7.80}$$

Equations (7.49) take the form

$$ca'' - ca\left(\varphi'\right)^2 + H_0\varphi'' + m\omega^2 a = 0, \tag{7.81}$$

$$H_0\left(\varphi'\right)^2 + 2ca'\varphi' + ca\varphi'' = 0. \tag{7.82}$$

Although we have not succeeded in constructing a general solution with four integration constants, we have obtained a solution which has two integration constants and satisfies conditions (7.50). This solution has the form

$$a = a_0 - \beta x,\ \varphi = \varphi_0 + \alpha x. \tag{7.83}$$

Direct substitution into (7.81) and (7.82) shows that the constants α and β have the following values

$$\alpha = \left(\frac{m\omega^2}{c}\right)^{1/2},\ \beta = \frac{\alpha H_0}{2c}. \tag{7.84}$$

The constants a_0 and φ_0 are still arbitrary and must be determined from the boundary condition at $x = 0$.

Solution (7.82) is valid only for those values of x for which $a > 0$, i.e.

$$0 < x < x_* = a_0/\beta. \tag{7.85}$$

For $x > x_*$ it should be taken that

$$a = 0,\ \varphi = \varphi^* = \varphi_0 + \alpha x_*. \tag{7.86}$$

A direct substitution of the latter solution into the system of eqs. (7.81) and (7.82) shows that this system is satisfied. Besides, the solutions (7.83) and (7.86) meet.

Inserting the obtained expressions for amplitude and phase into (7.83) we get the following expression for the deformation

$$\varepsilon = \left[\begin{array}{ll} (a_0 - \beta x)\cos(\omega t - \varphi_0 - \alpha x), & 0 < x < x_*, \\ 0, & x > x_*. \end{array}\right. \tag{7.87}$$

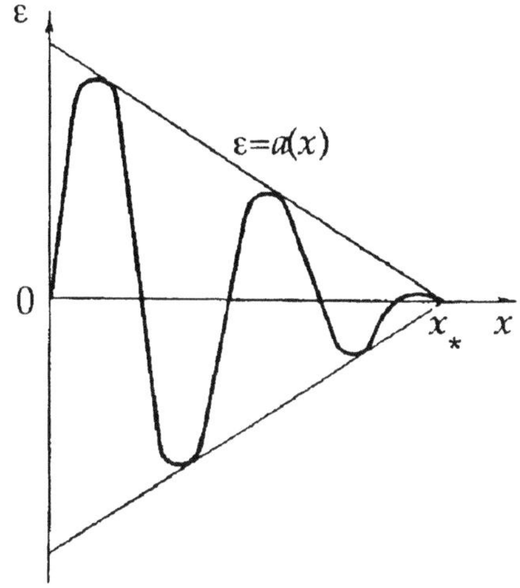

FIGURE 7.4. Strain amplitude ε versus distance x.

This dependence for t =const is shown in Fig. 7.4.

Let the first equation in (7.43) be the boundary condition at $x = 0$. Satisfying it by means of a linearised expression for force Q, we obtain the following values for the integration constants

$$a_0 = \frac{1}{c}\sqrt{B_0^2 - H_0^2}, \; \varphi_0 = \arctan\frac{H_0}{a_0 c}. \tag{7.88}$$

It follows from this equation that the constructed solution makes sense only if the amplitude of the force in cross-section $x = 0$ is larger that H_0. From a physical perspective, this means that force B_0 must exceed the yield stress H in the law (7.79). If force B_0 is smaller than H, then no motion in the rod occurs.

Displacement in this problem is obtained by integration of the equation for deformation (7.87), to give

$$\left[\begin{array}{ll} u = -\dfrac{a_0 - \beta x}{\alpha}\sin(\omega t - \varphi_0 - \alpha x) + & \\ +\dfrac{H_0}{2\alpha c}\left[\cos(\omega t - \varphi_0 - \alpha x) - \cos(\omega t - \varphi_0 - \alpha x_*)\right], & x < x_*, \\ u = 0, & x > x_*. \end{array}\right. \tag{7.89}$$

It is seen that in the region near the driven end the displacement is a superposition of running and standing waves. For $x > x_*$ there is no wave motion.

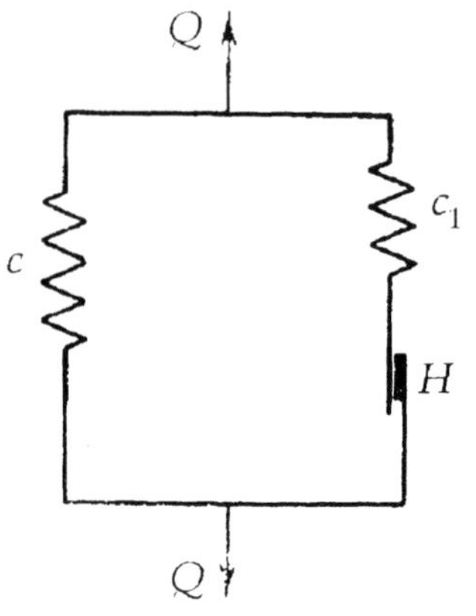

FIGURE 7.5. Rheological model for elastoplastic material with a linear hardening.

The region of wave motion x_* is obtained by substitution of eqs. (7.88) and (7.84) into (7.85) and is given by

$$x_* = \frac{2\sqrt{B_0^2 - H_0^2}}{H_0}\left(\frac{c}{m\omega^2}\right)^{1/2}. \tag{7.90}$$

The lower the material hardening and the higher the frequency of vibration the smaller is the zone of motion. Further, decreasing the yield stress H or increasing the amplitude B_0 enlarges the region of motion.

In addition, $x_* \to 0$ as $c \to 0$. Hence, in a rod made of a rigid-plastic material without hardening the vibration localizes in cross-section $x = 0$, where the source of vibration is located.

7.3.2 *Elastoplastic materials with linear hardening*

Let the rod's material be an elastoplastic material with a linear hardening law, see [121]. For developed plastic deformations the rod behaviour is governed by the following system of equations

$$Q = c\varepsilon + c_1(\varepsilon - \varepsilon_1), \; c_1(\varepsilon - \varepsilon_1) = H \operatorname{sgn}\dot{\varepsilon}_1. \tag{7.91}$$

Here ε_1 denotes plastic strain in the element H of the rheological model, cf. Fig. 7.5.

Function $\operatorname{sgn}\dot{\varepsilon}_1 = 1$ for a positive velocity of plastic strain and $\operatorname{sgn}\dot{\varepsilon}_1 = -1$ for a negative velocity. Let us agree that if the velocity of plastic strain is zero, then $\operatorname{sgn}\dot{\varepsilon}_1$ takes on that value from the interval $[-1, 1]$ which is prescribed by the second equation in (7.91). Under such a definition of the function $\operatorname{sgn}\dot{\varepsilon}_1$ the system of equations (7.91) is applicable for modelling the processes of loading and unloading of the rod, both with plastic de-

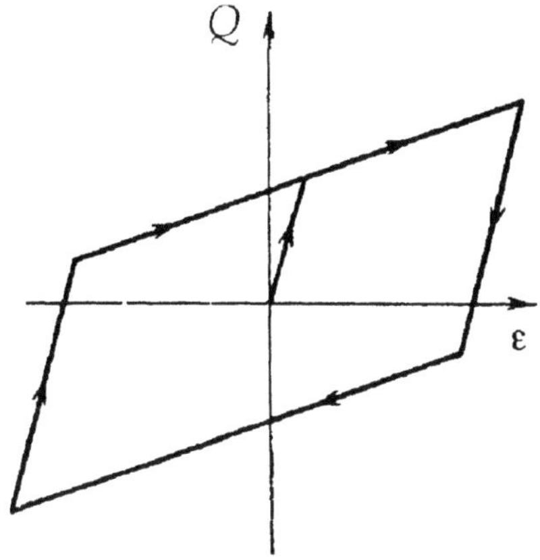

FIGURE 7.6. A typical diagram of cyclic deformation.

formations and without them, the Bauschinger effect being automatically taken into account.

A typical loop of a cyclic deformation of material is depicted in Fig. 7.6.

Analysis of eqs. (7.91) leads to the following conclusions. When plastic deformation is absent the rod rigidity is $c+c_1$. A plastic deformation begins when the tensile force achieves the value of $H(c+c_1)/c_1$, the hardening being determined by parameter c.

The system of equations (7.91) is an example of the constitutive equations which cannot be represented in a functional form (7.42). The method of harmonic linearisation can be applied directly to eq. (7.91). Following this method, we put

$$\varepsilon_1 = b\cos(\omega t - \psi), \tag{7.92}$$

where b and ψ are the amplitude and phase of plastic strain, respectively. Next, the nonlinear function in (7.91) is approximated by a linear one

$$H\mathrm{sgn}\dot{\varepsilon}_1 \approx \frac{H_0}{b\omega}\dot{\varepsilon}_1,\ H_0 = \frac{4H}{\pi}. \tag{7.93}$$

Inserting this approximation into (7.91) we find the following expressions for ε and Q in terms of ε_1

$$\varepsilon = \varepsilon_1 + \frac{H_0}{c_1 b\omega}\dot{\varepsilon}_1,\ Q = c\varepsilon_1 + \left(\frac{c}{c_1}+1\right)\frac{H_0}{b\omega}\dot{\varepsilon}_1. \tag{7.94}$$

Differentiating the equation of dynamics (7.40) with respect to x and using eqs. (7.41) and (7.94), we obtain an equation for the plastic strain

$$\frac{\partial^2}{\partial x^2}\left[c\varepsilon_1 + \left(\frac{c}{c_1}+1\right)\frac{H_0}{b\omega}\dot{\varepsilon}_1\right] - m\frac{\partial^2}{\partial t^2}\left[\varepsilon_1 + \frac{H_0}{c_1 b\omega}\dot{\varepsilon}_1\right] = 0. \tag{7.95}$$

Substituting the expression for ε_1 due to (7.92) into (7.95) and combining coefficients in sine and cosine, we arrive at the following system of two differential equations for the amplitude and phase of plastic strain

$$c\left[b'' - b\left(\psi'\right)^2\right] + H_0\left(\frac{c}{c_1}+1\right)\psi'' + m\omega^2 b = 0, \tag{7.96}$$

$$-c\left[b\psi'' + 2b'\psi'\right] - \left(\frac{c}{c_1}+1\right)H_0\left(\psi'\right)^2 + \frac{m\omega^2 H_0}{c_1} = 0. \tag{7.97}$$

This system of equations has a solution

$$b = b_0 - \beta x,\ \psi = \psi_0 + \alpha x, \tag{7.98}$$

where b_0 and ψ_0 are integration constants, and α and β are easily obtained by substitution of (7.98) into the system of equations (7.96) and (7.97). These are given by

$$\alpha = \left(\frac{m\omega^2}{c}\right)^{1/2},\quad \beta = \frac{H_0\alpha}{2c}. \tag{7.99}$$

Solution (7.98) makes sense only in that part of the rod where the amplitude b is positive, i.e. only for x satisfying the following inequality

$$0 < x < x_* = b_0/\beta. \tag{7.100}$$

For $x > x_*$ one must take

$$\begin{aligned} &b = 0,\ \psi = \psi_* + \gamma\left(x - x_*\right), \\ &\psi_* = \psi_0 + \alpha x_*,\ \gamma = \left(\frac{m\omega^2}{c + c_1}\right)^{1/2}. \end{aligned} \tag{7.101}$$

Direct substitution of solution (7.101) into the system of equations (7.96) and (7.97) shows that these equations are satisfied. In addition, the solutions (7.98) for $0 < x < x_*$ and (7.101) for $x > x_*$ meet. Indeed, at $x = x_*$ the amplitude and phase of the plastic strain are continuous. By virtue of (7.94) the total strain and the force in the rod are continuous as well. Therefore, the plastic strain is equal to zero for $x > x_*$. The solution constructed satisfies the Sommerfeld radiation condition at infinity, as $\psi' > 0$ as $x \to \infty$.

Let us determine the integration constants and find the main variables of the problem, namely the total strain ε and the displacement u. A convenient way to do this is to transform to a complex exponential form in eq. (7.92)

$$\varepsilon_1 = be^{i(\omega t - \psi)} \tag{7.102}$$

which is allowed since eqs. (7.94) are linear in ε_1. It goes without saying that only the real parts of (7.102) and others have a physical meaning.

Inserting (7.102) into (7.94) yields the equations

$$\begin{aligned} Q &= \left[cb + iH_0\left(\frac{c+c_1}{c}\right)\right] e^{i(\omega t-\psi)}, \\ \varepsilon &= \left(b + i\frac{H_0}{c_1}\right) e^{i(\omega t-\psi)}. \end{aligned} \tag{7.103}$$

Using the first equation in (7.103) and the first boundary condition in (7.43) at $x = 0$, we easily obtain the amplitude and phase of the plastic strain in the plastic zone

$$b_0 = \frac{1}{c}\left[B_0^2 - H_0^2\left(\frac{c+c_1}{c}\right)^2\right]^{1/2}, \quad \psi_0 = \arctan\frac{H_0\,(c+c_1)}{cc_1 b_0}. \tag{7.104}$$

The solution obtained is seen to be meaningful only if the amplitude of the external force satisfies the following inequality

$$B_0 > H_0\,(c+c_1)\,/c_1. \tag{7.105}$$

This inequality is a condition of initiation of plastic deformations in the rod. This condition is an approximate one and differs from the exact condition only in an inessential factor $4/\pi$ in H_0.

As seen from the second equation (7.103) the strain amplitude is

$$a = \left(b^2 + H_0^2/c_1^2\right)^{1/2}, \tag{7.106}$$

or accounting of an exact expression for b

$$\left[\begin{aligned} a &= \left[(b_0 - \beta x)^2 + H_0^2/c_1^2\right]^{1/2}, && 0 < x < x_*, \\ a &= H_0/c_1, && x > x_*. \end{aligned}\right. \tag{7.107}$$

It follows from these equations that the strain amplitude decreases from its maximum value at $x = 0$ to the value of H_0/c_1, and then remains constant and is equal to the maximal elastic strain.

Let us determine the displacement u. To this end, we insert expressions for b and ψ from (7.98) and (7.101) into (7.103), and integrate over x taking into account the continuity of u at $x = x_*$ and the radiation condition as $x \to \infty$. The result is

$$\left[\begin{aligned} u &= \left(i\frac{b}{\alpha} - \frac{H_0}{c_1\alpha} - \frac{\beta}{\alpha^2}\right) e^{i(\omega t-\psi)} + \\ &\quad \left[\frac{H_0}{c_1}\left(\frac{1}{\alpha} - \frac{1}{\gamma}\right) + \frac{\beta}{\alpha^2}\right] e^{i(\omega t-\psi_*)}, && x < x_*, \\ u &= -\frac{H_0}{c_1\gamma} e^{i(\omega t-\psi)}, && x > x_*. \end{aligned}\right. \tag{7.108}$$

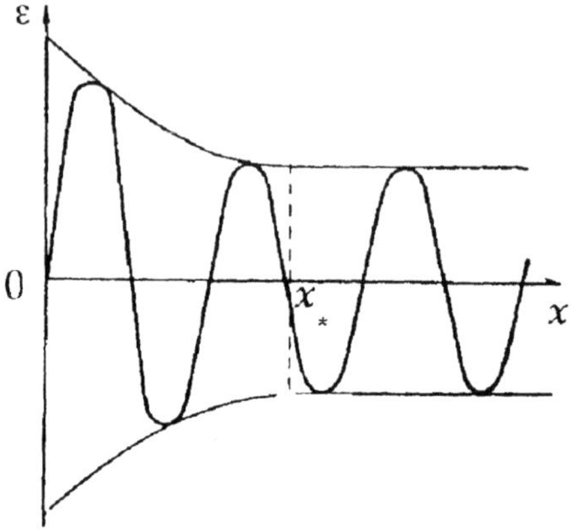

FIGURE 7.7. The total strain ε versus x.

The parameters which appear in this equation have been determined above. As seen from (7.108) a wave of constant amplitude is outgoing at infinity for $x > x_*$. For $0 < x < x_*$ the rod vibration is a superposition of propagating and standing waves, where the amplitude of the running wave decreases with increasing distance from the source of vibration.

Let us draw some conclusions. The cross-section $x = x_*$ is a border between a part of the rod with plastic deformation and a part with pure elastic deformation. An expression for x_* is obtained by substitution of the expressions for b_0 (7.104) and β and α (7.99) into (7.100)

$$x_* = \frac{2}{H_0}\left[B_0^2 - H_0^2\left(\frac{c+c_1}{c}\right)^2\right]^{1/2}\left(\frac{c}{m\omega^2}\right)^{1/2}. \tag{7.109}$$

Thus, the higher the material hardening is and the lower the mass per unit length and vibration frequency are, the larger the zone of plastic strain is. In particular, when the rod material is ideal-plastic ($c = 0$), then as follows from (7.109) $x_* = 0$, i.e. the plastic strain zone is localised in that cross-section to which the vibration source is attached.

It should be noted that the plastic strain zone is a region of the most intensive vibration, i.e. a zone of large displacements. In the elastic zone the amplitude of vibration is minimal and due to (7.108) does not depend upon the shaker force but depends on the properties of the rod material and frequency (see equation for γ), namely the higher the frequency, the lower the amplitude of vibration.

Figure 7.7 displays schematically the total strain versus x at an arbitrary time instant. The plastic strain zone $[0, x_*]$ is seen to be a sequence of pieces of plastic strain of alternating sign. A number of these pieces is equal to the number of extrema of ε_1. The latter is easy to obtain from eq. (7.92) and is given by

$$n = \frac{1}{\pi}\left(\psi_* - \psi_0\right). \tag{7.110}$$

Substituting expressions for ψ_* and ψ_0 into (7.110) yields

$$n = \frac{2}{\pi H_0}\left[B_0^2 - H_0^2\left(\frac{c+c_1}{c}\right)^2\right]^{1/2}. \tag{7.111}$$

If the excitation amplitude B_0 exceeds considerably the yield stress and the material hardening is not very large, the second term in the square brackets may be neglected as being small. This gives

$$n = \frac{2B_0}{\pi H_0}. \tag{7.112}$$

One sees that the number of pieces may be considerable. This fact complicates construction of an exact solution since one has to deal not only with a sequence of waves of loading and unloading ([155] and [164]) but also with an unknown *a priori* initial strain distribution as the problem is periodic in time.

Another approximate solution of the problem under consideration has been reported in [25]. The method of equivalent linearisation has been used to construct the solution, cf. [27]. As the methods of equivalent linearisation and harmonic linearisation coincide it may seem that the solution of [25] should coincide with the above solution. However this does not occur for a number of reasons. Firstly, a linearisation of the second equation was not used in (7.91), but an equivalent linearisation of the constitutive law $Q = Q(\varepsilon)$ depicted in Fig. 7.6 was utilized in [25]. Secondly, in contrast to the exact solution obtained here, a linearised equation was solved in [25] approximately by the method of slowly varying waves. The formulae derived in [25] turned out to be more complicated. Although similar to eq. (7.107), they predict a nearly linear decrease in the strain amplitude as x increases.

Thirdly, in [25] the amplitude tends asymptotically to the maximal amplitude of elastic waves, and any quantitative discrepancy between the solutions is negligible. There is, however, an important qualitative difference between the solutions, namely, for any sufficiently large x the amplitude in [25] is very close to the maximal amplitude of elastic waves, but it exceeds this maximal amplitude. This means that plastic strains take place in any cross-section even if the distance from the source of vibration goes beyond any limit. This fact contradicts the present solution which allows plastic strains only in a finite region $0 < x < x_*$. This contradiction raises the question of which solution is closer to reality. Since two approximate and quantitatively close solutions are compared, the problem cannot be solved by means of their comparison. In order to answer the question some exact solutions must be considered.

Historically, periodic elastoplastic waves in rods were first studied by Lensky [95]. A semi-infinite rod with a deformation curve of Fig. 7.6 and a periodic loading at its end was analysed. It was shown that this problem reduced to a system of seven differential-functional equations. It was pointed out that there was no steady-state periodic solution for a jump loading. A periodic solution proved to exist only for a rod of finite length excited at the second end by a periodic load with a specially chosen period in such a way that no reflected waves occur. The paper gives, however, no direct answer on the posed question about the size of the zone of plastic strains.

The same problem has been analysed in [116]. The study was limited to the case in which the excitation slightly exceeds the limit of the plastic stress. An exact solution was constructed by means of the inverse method for a piecewise-linear serrated load. It turned out that the plastic strains were localised in a finite region adjoining the loaded end. Further, elastic waves propagated along the rest of the rod. This corresponds exactly to the conclusions of the present approximate analysis.

Note in passing that a number of exact solutions on cyclic loading of elastoplastic rods of finite length with piecewise-linear characteristics have been reported, cf. [72], [73] and [74]. Paper [72] is of most interest to the present investigation. A closed form solution was derived there under very restrictive assumptions. The external force was assumed to be a rectangular sinusoid with amplitude equal to the yield stress of the material. The Bauschinger effect was also ignored and the period of external excitation was specially chosen in such a way that a resonance occurred.

7.3.3 Elastoplastic materials with an arbitrary hardening law

Consider an elastoplastic material with a rheological equation (2.3). This material has been studied in detail in Chapter 2. The laws of vibration propagation in a rod made of this material have been investigated in [127]. Apparently

$$Q = F\sigma, \tag{7.113}$$

where F is a cross-sectional area of the rod.

As shown in Chapter 2 a harmonic linearisation of the nonlinearity in the constitutive equation and the complex form of the variables lead to eq. (2.26), which due to (7.113), gives

$$\begin{aligned} &Q = c_c\varepsilon, \; c = EF, \; n = 4/\pi, \\ &c_c = c\left[1 - \int_0^1 \left(1 - \eta^2 - i\eta\sqrt{1-\eta^2}\right) p\left(\frac{a\eta}{n}\right)\frac{a}{n}\,d\eta\right]. \end{aligned} \tag{7.114}$$

Substituting the expression for the complex rigidity given by (7.114) into (7.75) and retaining only terms of the first order in effects of plastic deformation yields

$$a' = -\frac{\omega a}{2}\sqrt{\frac{m}{c}}\int_0^1 \eta\sqrt{1-\eta^2 p\left(\frac{a\eta}{n}\right)\frac{a}{n}}d\eta. \qquad (7.115)$$

Integration of this equation gives $a = a(x)$. Inserting this into the previous equations one can investigate the laws of vibration propagation of interest.

For a power function of defects distribution (2.28), equation (7.115) takes the form

$$a' = -\frac{\omega g}{2}\sqrt{\frac{m}{c}}a^{\alpha+1}, \qquad (7.116)$$

where g is given by (2.30). Integrating (7.116) we obtain

$$a = a_0\left(1+\frac{\omega\alpha g}{2}\sqrt{\frac{m}{c}}a_0^{\alpha}x\right)^{-1/\alpha}. \qquad (7.117)$$

It is seen that vibration occupies the whole rod decaying from the value $a = a_0$ at $x = 0$ to zero. The following peculiarity of the vibration field is of interest. For large values of x we can neglect the value of unity in the parentheses of (7.117) to obtain

$$a < \left(\frac{\omega\alpha g}{2}\sqrt{\frac{m}{c}}x\right)^{-1/\alpha}. \qquad (7.118)$$

The strain amplitude at the loaded end a_0 is absent in this equation. Hence, it indicates an upper limit of vibration at any point. This limit does not depend upon on the power of the vibration source. Figure 7.8 displays the curves $a = a(x)$, due to eq. (7.117), for various a_0. The upper curve with a shading corresponds to an universal hyperbola

$$a = \left(\frac{\omega\alpha g}{2}\sqrt{\frac{m}{c}}x\right)^{-1/\alpha}. \qquad (7.119)$$

In accordance with inequality (7.118), all curves $a = a(x)$ lie beneath the hyperbola (7.119) approaching it at large x. Therefore, the intensity of vibration does not depend on the power of the source provided that the latter exceeds some level. In particular, it means that the amplitude of the excitation cannot be reliably determined by measurements from the far field of elastoplastic vibration.

7.3.4 A viscoelastoplastic material

Consider now a more complicated rheological model of a material which is depicted in Fig. 7.9. The model consists of the elastoplastic element of

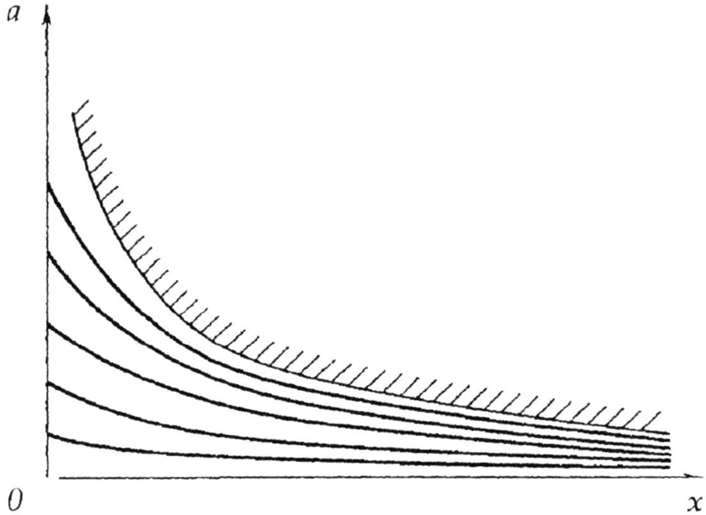

FIGURE 7.8. The strain amplitude a versus x. The upper curve is an universal hyperbola due to eq. (7.119).

Fig. 2.1 in parallel with a generalised viscoelastic element. The latter is introduced as an adequate description of the energy dissipation in a material under small strain amplitudes. An influence of the viscoelastic element becomes insignificant at large amplitudes since the energy dissipation is determined by the effects of plastic deformations in the elastoplastic element. A similar rheological model has been applied in [9] to the description of vibration in complex structures. The elastoplastic part of the model describes material damping and interface friction between the structural members during their relative motion, while the dashpot models a linear resonant absorption of vibration by secondary systems, see the example of Section 8.2 for details.

Repeating the derivation of the previous part of this Section we find that the strain amplitude $a(x)$ decreases and is governed by the following ordinary differential equation

$$a' = -\gamma\sqrt{\frac{m}{c}}a^{\alpha+1} - \beta a, \tag{7.120}$$

where

$$\gamma = \frac{\omega g}{2}\sqrt{\frac{m}{c}}, \ \beta = \frac{\omega f}{2}\sqrt{\frac{m}{c}}. \tag{7.121}$$

Here f characterises energy dissipation due to viscosity, where

$$f = f(\omega).$$

The other variables coincide with those of the previous part.

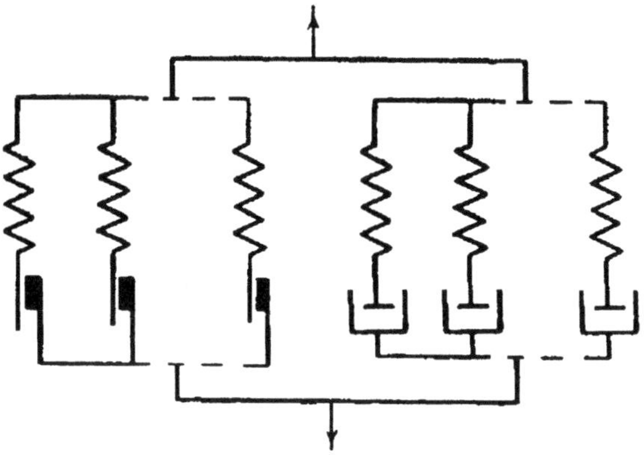

FIGURE 7.9. The rheological model of a viscoelastoplastic material.

The integral of eq. (7.120) is given by

$$a\left(x\right)=a_{0}\left[1+\frac{1}{\beta}\left(e^{\alpha\beta x}-1\right)\left(\beta+\gamma a_{0}^{\alpha}\right)\right]^{-1/\alpha}, \qquad (7.122)$$

where a_0 is the strain amplitude at $x = 0$.

As seen from eq. (7.122), vibration occupies the whole rod decaying from the value $a = a_0$ at $x = 0$ to zero.

The vibration distribution has the same qualitative feature as in the previous example. Indeed, increase of the strain amplitude at $x = 0$ leads to an increase in the amplitude a at any x. However, this increase at a given x has a limit which is formally obtained from eq. (7.122) as $a_0 \to \infty$

$$a_{*}\left(x\right)=\left[\frac{\beta}{\gamma\left(e^{\alpha\beta x}-1\right)}\right]^{1/\alpha}. \qquad (7.123)$$

Hence, the following inequality holds

$$a\left(x\right)<a_{*}\left(x\right) \qquad (7.124)$$

which is similar to inequality (7.118).

Again this indicates the existence of an upper limit of vibration which does not depend on the power of the source of vibration, but is a function of frequency, the rod parameters, and the distance from the source.

The effect of vibration saturation observed in the two last examples may be explained as follows. As the power of the source of vibration increases, the density of vibrational energy increases in the cross-sections near the loaded end. The energy dissipation in this zone increases even more intensively. This results in an absorption of vibration in a region adjoining the loaded end, while the vibrational energy in the remote cross-sections increases negligibly.

7.4 Vibration in a rod with power law elastic and dissipative characteristics

The goal of this Section is to show that the nonlinear effects discovered in the previous Section, namely, vibration saturation (7.3.2, 7.3.3 and 7.3.4) and a finite distance which the vibration travels down the rod (7.3.1) occur in the non-plastic rods, too.

7.4.1 First example

Let the governing equation for a rod have the hypothetical form, cf. [126]

$$Q = k_1 |\varepsilon|^{\kappa} \operatorname{sgn}\varepsilon + k_2 |\dot{\varepsilon}|^{\mu} \operatorname{sgn}\dot{\varepsilon}. \tag{7.125}$$

The linearisation factors are as follows

$$q = \alpha a^{\kappa-1}, \; h = \beta a^{\mu-1}, \tag{7.126}$$

where

$$\begin{aligned} \alpha &= k_1 \frac{2^{\kappa+2}}{\pi} \mathrm{B}\left(\frac{\kappa+1}{2}, \frac{\kappa+1}{2}\right), \\ \beta &= k_2 \omega^{\mu} \frac{2^{\mu+2}}{\pi} \mathrm{B}\left(\frac{\mu+1}{2}, \frac{\mu+1}{2}\right), \end{aligned} \tag{7.127}$$

B being the Eulerian beta-function.

In order to solve the problem we make use of the method of slowly varying parameters since the other methods of Section 7.2 do not provide us with a closed form solution.

Inserting expressions (7.126) into eqs. (7.63) and (7.64) we arrive at the following equations

$$a' = -\gamma a^{s+1}, \; \varphi' = \sqrt{\frac{m\omega^2}{\alpha}} a^{(1-\kappa)/2}. \tag{7.128}$$

Their integration due to the boundary conditions yields

$$a = a_0 \left(1 + s\gamma a_0^s x\right)^{-1/s}, \tag{7.129}$$

$$\varphi = \frac{\alpha}{\beta} \frac{3\kappa+1}{\kappa-\mu} \left(a_0^{(\kappa-\mu)/2} - a^{(\kappa-\mu)/2}\right) + \varphi_0. \tag{7.130}$$

Here a_0 and φ_0 are the values of a and φ at $x = 0$, respectively. The other denotations are

$$\gamma = \sqrt{\frac{m\omega^2}{\alpha}} \frac{2\beta}{\alpha(3\kappa+1)}, \; s = \frac{1+2\mu-3\kappa}{2}. \tag{7.131}$$

Let us next consider some particular cases.

Let $s < 0$, i.e.

$$3\kappa > 2\mu + 1. \tag{7.132}$$

As follows from (7.129) the amplitude of vibration vanishes at $x = x_*$, where

$$x_* = \frac{a_0^{-s}}{\gamma |s|}. \tag{7.133}$$

Equation (7.129) is meaningless for $x > x_*$. On this account, it should be taken that the strain is equal to zero for $x > x_*$.

Let us consider now $s > 0$, i.e.

$$3\kappa < 2\mu + 1. \tag{7.134}$$

The solution (7.129) is then valid for any x. In particular, the following inequality holds

$$a < \left(\frac{1}{s\gamma x}\right)^{1/s}, \tag{7.135}$$

expressing a condition of vibration saturation.

Finally, let $s = 0$, i.e.

$$3\kappa = 2\mu + 1. \tag{7.136}$$

Passing to the limit at $s \to 0$ renders the result

$$a = a_0 e^{-\gamma x}. \tag{7.137}$$

This particular case shows that an exponential decay of vibration is possible not only in a linear elastic rod with linear damping, but also in a nonlinear case.

7.4.2 Second example

Consider now a particular case of dependence (7.125), namely $\kappa = \mu$. The complex rigidity is then given by

$$c = (\alpha + i\beta)\, a^{\mu - 1}. \tag{7.138}$$

It turns out that this particular case can be analysed more accurately, i.e. without utilization of the assumption of a slowly varying amplitude. It is possible to do by means of the inverse method outlined in Section 7.2. Assume [123]

$$c = c_0 (1 + \gamma x)^2, \tag{7.139}$$

where c_0 is a complex value while γ is a real value. Insertion of the complex rigidity (7.139) into eq. (7.66) leads to the Euler equation

$$(1 + \gamma x)^2 U'' + 2\gamma (1 + \gamma x) U' + \frac{m\omega^2}{c_0} U = 0. \tag{7.140}$$

A solution which satisfies the kinematic boundary condition (7.43) takes the form

$$U = A_0 (1 + \gamma x)^n , \tag{7.141}$$

where n is a root of the quadratic equation

$$n^2 + n + \frac{m\omega^2}{c_0 \gamma^2} = 0. \tag{7.142}$$

By using eq. (7.141) we find the strain amplitude

$$a(x) = \left| \frac{dU}{dx} \right| = a_0 (1 + \gamma x)^s , \tag{7.143}$$

where

$$s = \operatorname{Re} n - 1, \; a_0 = A_0 |\gamma n| . \tag{7.144}$$

Eliminating x from (7.139) by means of (7.143), we obtain the following dependence for c on a

$$c = D a^{2/s}, \;\; D = c_0 a_0^{-2/s}. \tag{7.145}$$

Choosing appropriate values of s and using the previous equations we can construct the solution of the problem under consideration.

The question of the boundary condition at infinity is, as yet, unanswered. To do this, we have to consider two cases subject to the sign of γ.

a) $\gamma > 0$. In order to satisfy the Sommerfeld radiation condition it is necessary to require a solution that decays as $x \to \infty$. For this reason, n is that root of the quadratic equation (7.142) which has a negative real part. From the first equation in (7.144) we see that s satisfies the following inequality

$$s < -1. \tag{7.146}$$

b) $\gamma < 0$. In order to ensure that the solution (7.141) is bounded at $x = x_* = 1/|\gamma|$ it is necessary to take that root of the quadratic equation (7.142) which has a positive real part. By virtue of eq. (7.144) we get

$$s > -1. \tag{7.147}$$

If it is the case the displacement amplitude U vanishes at $x = x_*$, i.e. for $x \geq x_*$

$$U \equiv 0. \tag{7.148}$$

Equation (7.66) is satisfied by this solution for any dependence $c = c(a)$, apparently for the dependence (7.145). Further, this solution meets the solution constructed for $x \leq x_*$ in displacement and force.

Thus, for $s > -1$ we have $\gamma < 0$ and

$$U = \left[\begin{array}{ll} A_0 (1 + \gamma x)^n , & x \leq x_*, \\ 0, & x \geq x_*. \end{array} \right. \tag{7.149}$$

Comparing eqs. (7.138) and (7.145) we conclude that

$$\frac{2}{s} = \mu - 1, \; c_0 = (\alpha + i\beta)\, a_0^{\mu-1}. \tag{7.150}$$

Further, by means of eq. (7.144) we obtain

$$n = \lambda + i\nu, \; \lambda = \frac{\mu+1}{\mu-1}. \tag{7.151}$$

In order to find ν we substitute n, due to (7.151), and c_0, due to (7.150), into eq. (7.142) to obtain

$$\lambda^2 + \lambda - \nu^2 + (2\lambda+1)\, i\nu + \frac{m\omega^2 (\alpha - i\beta)\, a_0^{1-\mu}}{\gamma^2 \left(\alpha^2 + \beta^2\right)} = 0. \tag{7.152}$$

This equation with complex coefficients possesses real roots only for a specially chosen γ. These roots must satisfy the following equation

$$\nu^2 - \frac{\alpha}{\beta}(2\lambda+1)\,\nu - \lambda^2 + \lambda = 0. \tag{7.153}$$

where γ^2 is given by

$$\gamma^2 = \frac{\beta m\omega^2 a_0^{1-\mu}}{\gamma(2\lambda+1)\left(\alpha^2+\beta^2\right)}. \tag{7.154}$$

As γ^2 must be positive, ν is that root of eq. (7.153) which has the same sign as the value of $(2\lambda+1)$.

To obtain the strain amplitude a_0 at $x = 0$ we use the second equation in (7.144). After the substitution of the obtained values of γ and n into it we obtain the following equation

$$a_0^{1+\mu} = \frac{\lambda^2+\mu^2}{\nu(2\lambda+1)} \frac{\beta m\omega^2}{\left(\alpha^2+\beta^2\right)} A_0^2. \tag{7.155}$$

Inserting this expression for a_0 into eq. (7.154) and taking account of the fact that γ and λ have opposite signs, we obtain

$$\begin{aligned} \gamma &= (\Gamma A_0)^{-1/\lambda}\, \mathrm{sgn}\,(-\lambda), \\ \Gamma &= \left(\lambda^2+\mu^2\right)^{1/2} \left[\frac{\beta m\omega^2}{\nu(2\lambda+1)\left(\alpha^2+\beta^2\right)}\right]^{(1-\lambda)/2}, \end{aligned} \tag{7.156}$$

which give an expression for γ in terms of the main parameters of the problem.

Let us consider some possible particular cases.

a) $\mu < 1$. According to eq. (7.151), $\lambda < 0$ and

$$2\lambda + 1 < 0,\ \nu < 0,\ \gamma > 0. \tag{7.157}$$

An expression for the displacement is obtained by substituting (7.141) into (7.65). Accounting for n, due to (7.151), we find

$$u = \frac{A_0}{(1+\gamma x)^{|\lambda|}} \cos\left[\omega t - |\nu| \ln(1+\gamma x)\right]. \tag{7.158}$$

Hence, a propagating heterogeneous wave travels down the whole rod in this case. The amplitude of the wave obeys the law

$$A(x) = \frac{A_0}{(1+\gamma x)^{|\lambda|}}. \tag{7.159}$$

Omitting the value of unity in the parentheses and using the expression for γ due to (7.156) leads to the inequality

$$A(x) < \frac{A_0}{\Gamma x^{|\lambda|}}, \tag{7.160}$$

which is the known property of vibration saturation.

b) $\mu > 1$. In accordance with the equations of this Section we have

$$\lambda > 0,\ \nu > 0,\ \gamma < 0. \tag{7.161}$$

The displacement is then given by

$$u = \begin{cases} A_0 (1-|\gamma|\, x)^{\lambda} \cos\left[\omega t + \nu \ln(1-|\gamma|\, x)\right], & 0 < x < x_*, \\ 0, & x > x_*. \end{cases} \tag{7.162}$$

The coordinate of the border of the motion zone has the following expression

$$x_* = (\Gamma A_0)^{1/\lambda}. \tag{7.163}$$

The border of the motion zone enlarges as the vibration amplitude of a shaker A_0 increases.

The results obtained confirm completely the general conclusions which were drawn at the beginning of the Section.

The material of this Section is a specific particular case of a viscoelastic material. Another particular case has been studied in [4]. The rheological equation has been written there by utilizing an inheritance integral, where terms up to second order in strain were retained. A solution of the vibration problem in a semi-infinite rod has been obtained by means of the method of slowly varying waves.

7.5 Propagation of vibration in a nonlinear dissipative rod of finite length

The methods of Section 7.2 fail to analyse the vibration in a rod of finite length. One of the most general methods, namely the method of slowly varying amplitude is not applicable in this case. This follows from the fact that in this linear problem, the strain amplitude can change as fast as the solution itself. Consequently, the basic assumption of the method is not fulfilled. Other, more sophisticated methods, like the method of equations (7.49) or the inverse method cannot be used either since it is impossible to construct solutions with a sufficient number of integration constants. Further, it is not expedient to return to the modal analysis (as used in Chapter 6) as the solution becomes cumbersome at higher frequencies. It appears that the only way out of this situation is to use a slight modification of the method of harmonic linearisation.

As in the method of harmonic linearisation, the nonlinear function (7.42) is approximated by a linear one

$$Q(\varepsilon, \dot{\varepsilon}) \approx q\varepsilon + \frac{h}{\omega}\dot{\varepsilon}. \tag{7.164}$$

The linearisation factors are found by minimising the square of error inherent in approximation (7.164)

$$\int_{x}^{x+\Delta x}\int_{0}^{T}\left[Q(\varepsilon, \dot{\varepsilon}) - q\varepsilon - \frac{h}{\omega}\dot{\varepsilon}\right]^2 dtdx = \min_{q,\,h} \tag{7.165}$$

due to harmonic deformation (7.44)

$$\varepsilon = a(x)\cos\left[\omega t - \varphi(x)\right].$$

Contrary to the standard method of harmonic linearisation an additional averaging over the spatial coordinate is introduced here. An averaging scale should be sorted out in such a way that some smooth functions $q(x)$ and $h(x)$ will be obtained even for an oscillating function $a(x)$. This is important for the forthcoming integration of equations obtained after the linearisation since it allows us to apply the effective method of slowly varying parameters.

Realising requirement (7.165) and assuming that q and h vary insignificantly in the averaging interval we obtain

$$q = \frac{\int_{x}^{x+\Delta x} q^*(a)\,a^2 dx}{\int_{x}^{x+\Delta x} a^2 dx}, \quad h = \frac{\int_{x}^{x+\Delta x} h^*(a)\,a^2 dx}{\int_{x}^{x+\Delta x} a^2 dx}, \tag{7.166}$$

where q^* and h^* are the linearisation factors, cf. eq. (7.46)

$$q^*(a) = \frac{1}{\pi a}\int_0^{2\pi} Q(a\cos\kappa, -a\omega\sin\kappa)\cos\kappa d\kappa,$$

$$h^*(a) = -\frac{1}{\pi a}\int_0^{2\pi} Q(a\cos\kappa, -a\omega\sin\kappa)\sin\kappa d\kappa.$$

It is seen from (7.166) that q and h tend to q^* and h^* as $\Delta x \to 0$ and then the advantage of the approach disappears. On the other hand, it is not advisable to take a large value of Δx because the information about $a(x)$ is lost. The question of selection of the averaging scale will be discussed in detail later on. However, it is essential to realize that this choice ensures that q and h are slowly varying factors.

In what follows we make use of the complex rigidity

$$c_c = q + ih, \tag{7.167}$$

which, due to eqs. (7.166) is given by

$$c_c = \frac{\int_x^{x+\Delta x} a^2 c_c^*(a)\,dx}{\int_x^{x+\Delta x} a^2 dx}, \quad c_c^*(a) = q^*(a) + ih^*(a). \tag{7.168}$$

We see from this equation that c_c is a functional of a.

By means of approximation (7.164) and eqs. (7.40) and (7.41) we arrive at eq. (7.47). Separating the variables by means of (7.65) we obtain eq. (7.66)

$$\left(c_c U'\right)' + m\omega^2 U = 0.$$

Since the complex rigidity is a slowly varying function of the coordinate it is reasonable to apply the Steklov-Liouville method or the WKB-method to solve the above equation. This method, due to (7.67) and (7.69), gives the following approximate solution

$$u = c_c^{-1/4}\left[D_1\cos\omega y(x) + D_2\sin\omega y(x)\right]e^{i\omega t}. \tag{7.169}$$

Let us satisfy the boundary conditions. Let the end of the rod $x = 0$ be free and the end $x = L$ be subject to a harmonic force of frequency ω and amplitude B_0. Satisfying the boundary conditions, evaluating the strain and neglecting the terms of higher order we obtain

$$\varepsilon = B_0 c_c^{-3/4}(x)\, c_c^{-1/4}(L)\frac{\sin\omega y(x)}{\sin\omega y(L)}e^{i\omega t}. \tag{7.170}$$

By means of this expression we easily find the strain amplitude

$$a^2 = \frac{B_0^2}{|c_c(x)|^{3/2} |c_c(L)|^{1/2}} \frac{\cosh\beta(x) - \cos\gamma(x)}{\cosh\beta(L) - \cos\gamma(L)}, \tag{7.171}$$

where the following notation has been introduced

$$2\omega y(x) = \gamma(x) - i\beta(x). \tag{7.172}$$

It is evident that γ and β depend also on frequency, however this argument is not shown since the frequency is fixed. Argument x may take either a current value or $x = L$ as is seen from (7.171).

The physical background to the problem suggests that the assumption of a slowly varying c_c is fulfilled if the material damping is not large. Use of the corresponding asymptotics leads to the following approximate results

$$|c_c| = q,\ \gamma = 2\omega \int_0^x \sqrt{\frac{m}{q(x)}} dx,\ \beta = \omega \int_0^x \sqrt{\frac{m}{q(x)}} \frac{h(x)}{q(x)} dx, \tag{7.173}$$

which are valid up to values of the first order in h/q. It is seen from these equations that the inequality

$$|\beta(x)| \ll |\gamma(x)|. \tag{7.174}$$

holds for $h \ll q$.

Inserting the amplitude (7.171) into (7.166) and averaging over x we obtain a system of integral equations for two unknown functions q and h. After q and h have been found, the solution of the problem can be reconstructed by means of the above equations.

Let us briefly consider the question of averaging over x. As it has been pointed out, this averaging should be performed in such a way that its result would be a slowly varying function of the spatial coordinate. The presence of trigonometric functions causes a rapid change in a, while other functions vary slowly. Indeed, the rigidity $c(x)$ varies slowly due to the above assumption. Further, $\cosh\beta(x)$ varies slower than $\cos\gamma(x)$ because $\beta(x)$ is asymptotically small compared to $\gamma(x)$ due to (7.174). Therefore, an averaging must be performed over parameter γ. While averaging one must try to keep the averaging interval as small as possible in order to get a detailed dependence q and h on x.

One or more waves of oscillation of amplitude a exist in the rod when $\gamma(L) > 2\pi$. In this case the period of this oscillation should be chosen as a minimal averaging interval. In other words, the averaging over γ must be performed within the interval $[0, 2\pi]$. This interval, however, can be halved because of the cosine function. If $\gamma(L) < \pi$ then no full half-wave of oscillation of amplitude a exists within the length of the rod. In this case, the averaging must be performed over the length of the rod, or, in other words, over the following interval of γ $[0, \gamma(L)]$. This allows eq. (7.166) to be written in the following form

$$q = \frac{\int_0^\Pi a^2 q^*(a)\, d\gamma}{\int_0^\Pi a^2 d\gamma}, \; h = \frac{\int_0^\Pi a^2 h^*(a)\, d\gamma}{\int_0^\Pi a^2 d\gamma}, \tag{7.175}$$

where Π denotes the averaging interval

$$\Pi = \min[\pi, \gamma(L)]. \tag{7.176}$$

The result of evaluating the integral in the denominator of (7.175) is

$$\frac{1}{\Pi}\int_0^\Pi a^2 d\gamma = \frac{B_0^2}{q^{3/2}(x)\, q^{1/2}(L)} \frac{\cosh\beta(x) - \sin(\Pi)/\Pi}{\cosh\beta(L) - \cos\gamma(L)}. \tag{7.177}$$

Evaluation of the integrals appearing in the numerators of (7.175) is not possible in general, since they depend essentially on the character of the nonlinearity.

Consider now a rod made of the elastoplastic material analysed in Chapter 2. By virtue of (2.29) its complex rigidity is given by

$$c_c^* = c(1 - ra^\alpha + iga^\alpha), \; c = EF, \tag{7.178}$$

where E is the Young's modulus and F is the cross-sectional area. We then obtain

$$q^* = c(1 - ra^\alpha), \; h^* = cga^\alpha. \tag{7.179}$$

Inserting these expressions into (7.175) gives

$$q = c\left[1 - r\int_0^\Pi a^{2+\alpha} d\gamma \left(\int_0^\Pi a^2 d\gamma\right)^{-1}\right], \tag{7.180}$$

$$h = cg\int_0^\Pi a^{2+\alpha} d\gamma \left(\int_0^\Pi a^2 d\gamma\right)^{-1}. \tag{7.181}$$

The first of these equations may be simplified by means of the second one, to give

$$q = c(1 - \mu h/c), \; \mu = r/g. \tag{7.182}$$

Let $\gamma(L) > \pi$. Then $\Pi = \pi$. Substituting the amplitude (7.171) into (7.181) leads to an integral which is reduced to the first Laplace integral for the Legendre functions, cf. [93]

$$\mathrm{P}_\nu(z) = \frac{1}{\pi}\int_0^\pi \left[z + \left(z^2 - 1\right)^{1/2}\cos\gamma\right]^\nu d\gamma. \tag{7.183}$$

Using the substitution

$$z = \coth\beta,$$

we transform eq. (7.183) to the following form

$$\frac{1}{\pi}\int_0^{\pi}(\cosh\beta - \cos\gamma)^{\nu}\, d\gamma = \sinh^{\nu}\beta \mathrm{P}_{\nu}(\coth\beta), \tag{7.184}$$

which is required to evaluate the integral in (7.181). The result of this evaluation is as follows

$$\frac{h}{cg} = \left[\frac{B_0}{q^{3/4}(x)\, q^{1/4}(L)}\right]^{\alpha} \frac{\sinh^{1+\alpha/2}\beta(x)\, \mathrm{P}_{1+\alpha/2}(\coth\beta(x))}{[\cosh\beta(L) - \cos\gamma(L)]^{\alpha/2}\cosh\beta(x)}. \tag{7.185}$$

Introducing the non-dimensional variables

$$\rho = \frac{h}{c},\ \kappa = 2\omega L\sqrt{\frac{m}{c}},\ \zeta = \frac{\kappa}{2}\frac{x}{L}, \tag{7.186}$$

and using expression (7.182), we rewrite eq. (7.185) in the following form

$$\frac{\rho}{g} = \left[\frac{B_0/c}{(1-\mu\rho)^{3/4}(1-\mu\rho_1)^{1/4}}\right]^{\alpha} \frac{\sinh^{1+\alpha/2}\beta \mathrm{P}_{1+\alpha/2}(\coth\beta)}{[\cosh\beta_1 - \cos\gamma_1]^{\alpha/2}\cosh\beta}. \tag{7.187}$$

For small ρ the parameters in the latter equation have the following asymptotic representations

$$\beta = \int_0^{\zeta}\rho d\zeta,\ \beta_1 = \int_0^{\kappa/2}\rho d\zeta,\ \gamma_1 = \kappa + \mu\beta_1. \tag{7.188}$$

Further, ρ_1 is the value of ρ at $x = L$ or, due to (7.186), at $\zeta = \kappa/2$. Equation (7.187) is the required nonlinear integral equation for the unknown function $\rho = \rho(\zeta)$. After this equation has been solved, q is obtained by means of (7.182). Determination of the variables of interest now presents no problem.

Let us analyse eq. (7.187). Putting $\zeta = 0$ in it yields

$$\frac{\rho_0}{g} = \left[\frac{B_0/c}{(1-\mu\rho_0)^{3/4}(1-\mu\rho_1)^{1/4}}\right]^{\alpha} \frac{D}{[\cosh\beta_1 - \cos\gamma_1]^{\alpha/2}}, \tag{7.189}$$

where

$$D = \lim_{\beta\to 0}\sinh^{1+\alpha/2}\beta \mathrm{P}_{1+\alpha/2}(\coth\beta) = \frac{2^{3(1+\alpha/2)}}{\pi}\mathrm{B}\left(\frac{3+\alpha}{2}, \frac{3+\alpha}{2}\right),$$

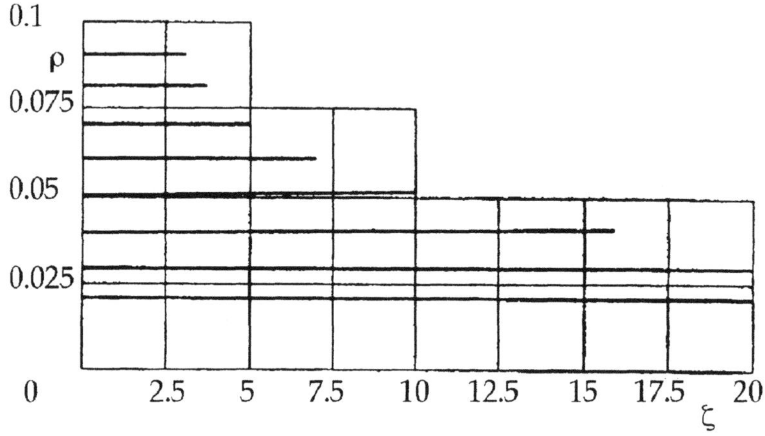

FIGURE 7.10. Dependence ρ versus ζ for $\alpha = 1$.

B being the Eulerian beta-function.

Dividing both sides of eq. (7.187) by the corresponding sides of eq. (7.189) we get

$$\frac{\rho}{\rho_0}\left(\frac{1-\mu\rho}{1-\mu\rho_0}\right)^{\frac{3}{4\alpha}} = \frac{\sinh^{1+\alpha/2}\beta}{D\cosh\beta}\mathrm{P}_{1+\alpha/2}\left(\coth\beta\right). \tag{7.190}$$

The latter equation may be easily reduced to an ordinary differential equation by taking a logarithmic derivative. As the Cauchy problem for this equation has an unique solution, there exists a unique function $\rho = \rho(\zeta)$ for each value ρ_0. This function can be obtained, for example, by means of the method of successive approximations using integral equation (7.190). The solutions are displayed in Fig. 7.10 for $\alpha = 1$.

Let us now study eq. (7.189). Solving it for γ_1 gives

$$\gamma_1 = \kappa + \mu\beta_1 = 2\pi n \pm \arccos\left[\cosh\beta_1 - \frac{A}{\rho_0^{2/\alpha}\left(1-\mu\rho_0\right)^{3/2}\left(1-\mu\rho_1\right)^{1/2}}\right], \tag{7.191}$$

where

$$A = (gD)^{2/\alpha}\left(\frac{B_0}{c}\right)^2, \; n = 1, 2, 3, \ldots, \infty. \tag{7.192}$$

Equation (7.191) allows us to make some conclusions about the relationship between ρ_0 and κ, where κ is the non-dimensional frequency of vibration. Positive values of κ appear only when the following inequality holds

$$-1 \leq \cosh\beta_1 - \frac{A}{\rho_0^{2/\alpha}\left(1-\mu\rho_0\right)^{3/2}\left(1-\mu\rho_1\right)^{1/2}} \leq 1. \tag{7.193}$$

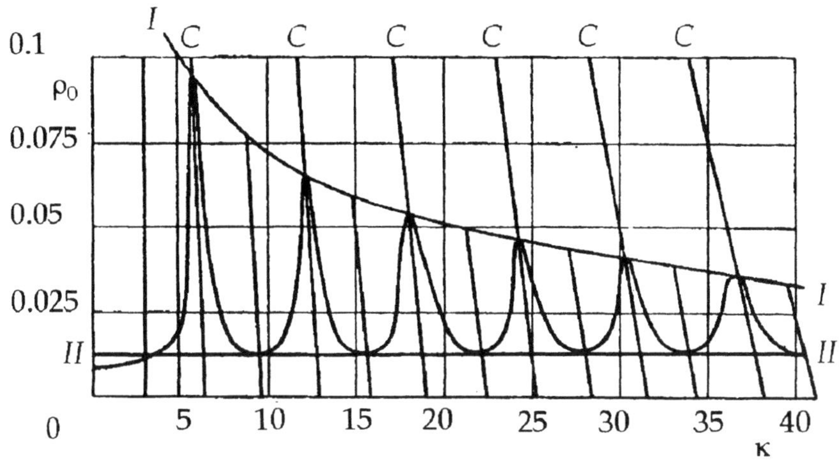

FIGURE 7.11. Dependence ρ_0 versus ζ.

This inequality gives, in the plane of the parameters κ and ρ_0, some domain in which the whole curve (7.191) is located. Given the integrals of eq. (7.190) the boundaries of this domain are easily obtained. These are depicted in Fig. 7.11 by lines I-I and II-II (A is taken to be $3.4 \cdot 10^{-4}$). The above domain lies between these lines. It is also possible to get an approximate closed form expression for these boundaries. Figure 7.10 plots the dependences $\rho = \rho(\zeta)$ only for those values of parameters which satisfy inequality (7.193). With great accuracy we can take $\rho = \rho_0 = \text{const}$. Then, equation (7.188) gives

$$\beta_1 = \rho_0 \kappa / 2, \tag{7.194}$$

and the boundaries (7.193) are given by the following equation

$$\kappa = \frac{2}{\rho_0} \operatorname{arg\,cosh} \left[\frac{A}{\rho^{2/\alpha} (1 - \mu\rho_0)^2} \pm 1 \right], \tag{7.195}$$

where plus corresponds to the line I-I and minus corresponds to the line II-II.

Consider the curves

$$\gamma_1 = \kappa + \mu\beta_1 = 2\pi n, \tag{7.196}$$

$$\gamma_1 = \kappa + \mu\beta_1 = 2\pi n - \pi. \tag{7.197}$$

These can be easily plotted in the plane ρ_0 and κ by using the integrals of eq. (7.190). It is also possible to derive an approximate closed form expression for κ. By means of the approximate equality (7.194) we find, respectively

$$\kappa = \frac{2\pi n}{1+\mu\rho_0/2}, \quad \kappa = \frac{2\pi n - \pi}{1+\mu\rho_0/2}. \tag{7.198}$$

These curves are also shown in Fig. 7.11. The first group of these curves, denoted by C is analogous to the backbone curves in nonlinear vibration theory.

The domain between the lines I-I and II-II is cut into sections by curves (7.196) and (7.197). As seen from eq. (7.191) only one branch of curve (7.191) is located within each section passing through the points where curve (7.196) crosses I-I and curve (7.197) crosses II-II.

An actual construction of curve (7.191) within each section is easily performed by the method of successive approximations using the following scheme

$$\kappa = 2\pi n - \mu\beta_1 \pm \arccos\left[\cosh\beta_1 - \frac{A}{\rho_0^{2/\alpha}\left(1-\mu\rho_0\right)^{3/2}\left(1-\mu\rho_1\right)^{1/2}}\right]. \tag{7.199}$$

To this end, one must choose a value of n and a sign in the latter equation which corresponds to the section under consideration. Next, some values of κ and ρ_0 from this section should be taken and β_1 and ρ_1 should be computed by means of the integrals of eq. (7.190). Substituting these values into the right hand side of eq. (7.199) we obtain a more accurate value of κ corresponding to the chosen ρ_0 etc. The process converges very quickly and one or two approximations are sufficient in most cases. The results of such computations are shown in Fig. 7.11. This completes construction of the dependences ρ_0 versus κ for the case $\gamma_1 > \pi$.

Let us consider now the case $\gamma_1 < \pi$. Then $\Pi = \gamma(L)$. Note that this case consists of very low frequencies, i.e. κ is small. As ρ is small, β and β_1 are small too, i.e. $\beta \ll 1$ and $\beta_1 \ll 1$ and therefore $\cosh\beta \approx 1$ and $\rho \approx \rho_0$. Accounting for this fact and inserting a^2, due to (7.171), into (7.181) we obtain by virtue of (7.182), (7.186) and (7.188)

$$\rho = \rho_0 = g\left[\frac{B_0}{c\left(1-\mu\rho_0\right)}\right]^{\alpha} \frac{\frac{1}{\kappa}\int\limits_0^{\kappa}\left(1-\cos\gamma\right)^{1+\alpha/2}d\gamma}{\left(1-\cos\kappa\right)^{\alpha/2}\left[1-\left(\sin\kappa\right)/\kappa\right]}. \tag{7.200}$$

The integral in this equation cannot be evaluated in terms of elementary functions, but can be computed digitally. Then, eq. (7.200) yields the required relation between ρ_0 and κ for $\kappa < \pi$. The curve (7.200) at $\kappa = \pi$ and the curve (7.189) meet as value of β is small for this value of κ. By means of (7.200) it is easy to analyse the behaviour of the curve $\rho_0(\kappa)$ for small κ. Using the asymptotic equations for trigonometric functions we obtain

$$\rho_0\left(1-\mu\rho_0\right)^{\alpha} = g\left(\frac{B_0}{c}\right)^{\alpha}\frac{3}{3+\alpha}. \tag{7.201}$$

Thus, the curve does not pass through the origin of the coordinates.

The construction of the entire curve $\rho_0(\kappa)$ is now completed. Utilizing it for a given value of frequency, we obtain ρ_0 and then $h = c\rho$. The vibration parameters will be found by means of the above equations. For example, the average squared strain at $x = 0$ due to (7.177) is given by

$$a_{av}^2 = \frac{B_0^2}{q^{3/2}(x)\, q^{1/2}(L)} \frac{1 - (\sin \Pi)/\Pi}{\cosh \beta(L) - \cos \gamma(L)}. \tag{7.202}$$

In the case of $\gamma(L) > \pi$ this formula yields

$$a_{av}^2 = \frac{(B_0/c)^2}{(1-\mu\rho_0)^{3/2}(1-\mu\rho_1)^{1/2}} \frac{1}{\cosh \beta_1 - \cos \gamma_1}, \tag{7.203}$$

where the previous notation was used. Combining this equation with (7.189) we obtain

$$a_{av} = \left(\frac{\rho_0}{gD}\right)^{1/\alpha}. \tag{7.204}$$

In the present example $\alpha = 1$. Hence, in this case the diagram of Fig. 7.11 gives us the dependence $a_{av}(\kappa)$ up to the scale of the ordinate axis.

Consider now the case $\gamma(L) < \pi$ which is the case of low frequencies. Thus, $\beta \ll 1$ and $\cosh \beta \approx 1$. Equation (7.202) is transformed to the following form

$$a_{av}^2 = \frac{(B_0/c)^2}{(1-\mu\rho_0)^2} \frac{1-(\sin\kappa)/\kappa}{1-\cos\kappa}, \quad \kappa < \pi. \tag{7.205}$$

As $\kappa \to 0$ we have the equality

$$a_{av} = \frac{B_0/c}{(1-\mu\rho_0)\sqrt{3}}, \tag{7.206}$$

from which, by means of (7.201), we find

$$a_{av} = \frac{\rho_0}{gD}\frac{D(3+\alpha)}{3\sqrt{3}}.$$

At $\alpha = 1$ this gives

$$a_{av} = 0.97\,\frac{\rho_0}{gD},$$

which actually coincides with eq. (7.204). Exact agreement between these curves also takes place at another values of κ. Hence, the entire curve $\rho_0(\kappa)$ is similar to the curve $a_{av}(\kappa)$. The resonance curve also bends off to the left, which is well-known phenomenon for mechanical systems with internal damping. In this example this effect is not considerable and does not cause any multiple-valued amplitude-frequency characteristics.

Other amplitude-frequency dependences may be plotted by analogy.

Let us consider how close a mode of resonant vibration is to a mode of free elastic vibration. As shown in Section 7.1 this difference is determined by the value of cosh $\beta_1 - 1$. For the extreme left peak this value is 0.04 while for the extreme right peak it is 0.25. In other words, the difference is negligible at low frequencies, but it is essential for the extreme right peak.

Note, finally, that the construction of the whole amplitude-frequency dependence is done by utilizing an uncomplicated transcendental equation. This is an advantage of the method of harmonic linearisation over other methods of nonlinear mechanics used for the analysis of vibrations in systems with distributed parameters.

The problem of longitudinal vibration of elastoplastic rods subject to a harmonic distributed loading has been considered in [61]. The case in which the loading curve is a cubic parabola was studied there. While constructing an approximate solution it was assumed that the motion of the rod was nearly harmonic. A vibration mode of the vibrating rod was sought in the form of a expansion in terms of the normal modes of the free elastic rod. The Galerkin method was used to obtain the parameters of the expansion. The amplitude-frequency characteristic was computed in the frequency domain for only the first three resonance peaks. It was shown that the influence of the higher modes of free elastic vibrations on the first resonance is very small. Substantial numerical difficulties associated with the solution of the equations obtained by Galerkin's method have been reported.

Vibrations in a half-wavelength rod, i.e. a rod of length equal to half the wavelength of the travelling wave, with amplitude-dependent losses have been studied in [92]. The losses were computed using an elastic mode of forced vibration of the rod. Then, by means of the WKB-method, the vibration mode for the rod with losses was obtained. In such a way, the frequency domain near the first resonance was investigated.

7.6 Propagation of random vibration in a rod with nonlinear properties

Consider a longitudinal vibration in a rod with a nonlinear constitutive equation. Let us write down the equations of the rod motion

$$Q' - m\ddot{u} = 0, \quad \varepsilon = u', \quad Q = Q(\varepsilon, \dot{\varepsilon}). \tag{7.207}$$

As before we assume that Q is an odd function of its arguments. Assume that the rod has a finite length L, one end ($x = 0$) being free and the other end ($x = L$) being subject to a force $p(t)$. The latter is assumed to be a stationary random function of time with a zero mean value.

It is known that the stated problem has no exact solution. In what follows we offer its approximate solution by means of the method of statistical linearisation, cf. [124]. According to this method, the third nonlinear equation (7.207) is replaced by an approximated linear equation

$$Q \approx h_1 \varepsilon + h_2 \dot{\varepsilon}. \tag{7.208}$$

The linearisation factors are conventionally chosen from the condition that the linear function (7.208) optimally approximates the original nonlinear relation due to the criterion of minimising the mean-square error. These factors are as follows, see [141]

$$\begin{aligned} h_1 &= \frac{1}{\sigma_1^2} \iint\limits_{-\infty}^{\infty} Q(\varepsilon, \dot{\varepsilon})\, \varepsilon w(\varepsilon, \dot{\varepsilon})\, d\varepsilon d\dot{\varepsilon}, \\ h_2 &= \frac{1}{\sigma_2^2} \iint\limits_{-\infty}^{\infty} Q(\varepsilon, \dot{\varepsilon})\, \dot{\varepsilon} w(\varepsilon, \dot{\varepsilon})\, d\varepsilon d\dot{\varepsilon}. \end{aligned} \tag{7.209}$$

Here σ_1 and σ_2 are the standard deviations of strain and its velocity in cross-section x, respectively, and $w(\varepsilon, \dot{\varepsilon})$ denotes a joint probability density function. As the distribution law for ε and $\dot{\varepsilon}$ is unknown until the whole problem is solved, we assume a normal distribution [141], i.e.

$$w(\varepsilon, \dot{\varepsilon}) = \frac{1}{2\pi\sigma_1\sigma_2} \exp\left(-\frac{\varepsilon^2}{2\sigma_1^2} - \frac{\dot{\varepsilon}^2}{2\sigma_2^2} \right). \tag{7.210}$$

Equations (7.209) and (7.210) indicate that the linearisation factors depend on the still unknown root-mean-squares of strain and its velocity, i.e.

$$h_i = h_i(\sigma_1, \sigma_2). \tag{7.211}$$

Combining eq. (7.208) and the first two equations in (7.207) we obtain a single equation for the displacement

$$\left(h_1 u' + h_2 \dot{u}' \right)' - m\ddot{u} = 0. \tag{7.212}$$

This is the governing equation for longitudinal vibrations in a rod for the Kelvin-Voigt material. It should, however, be kept in mind that its coefficients depend on the unknown characteristics of the final solution σ_1 and σ_2. It is assumed that σ_1 and σ_2 are time-independent as the steady-state vibrations are analysed.

Applying the spectral decomposition of the load on the end $x = L$

$$p(t) = \int\limits_{-\infty}^{\infty} e^{i\omega t} V(\omega)\, d\omega$$

and looking for a steady-state solution of eq. (7.212) by the method of separation of variables, we arrive at the integral canonical representation of the displacement

$$u = \int_{-\infty}^{\infty} e^{i\omega\tau} \Phi(\omega, x) V(\omega) d\omega, \tag{7.213}$$

where function Φ is governed by the following equation

$$\left(c_c \Phi'\right)' + m\omega^2 \Phi = 0, \; c_c = h_1 + i\omega h_2. \tag{7.214}$$

As before, c_c is referred to as the complex rigidity. Assuming slow variations of σ_1 and σ_2 and thus slow variations of $c_c(x)$, we will construct the solution of eq. (7.214) by the Steklov-Liouville method described above. The solution is given by

$$\begin{aligned} \Phi &= -\frac{\cos \omega y(x)}{\omega \sqrt{m} \left[c_c(x) c_c(L)\right]^{1/4} \sin \omega y(L)}, \\ y &= \int_0^x \sqrt{\frac{m}{c_c(x)}} dx . \end{aligned} \tag{7.215}$$

In order to satisfy the boundary conditions, the derivatives of c_c with respect to x have been neglected as they are asymptotically small compared to wc_c/a.

Substituting (7.215) into (7.213) renders the following canonical integral representation for ε and $\dot{\varepsilon}$

$$\begin{aligned} \varepsilon &= \int_{-\infty}^{\infty} e^{i\omega\tau} \Psi(\omega, x) V(\omega) d\omega, \\ \dot{\varepsilon} &= \int_{-\infty}^{\infty} i\omega e^{i\omega\tau} \Psi(\omega, x) V(\omega) d\omega, \end{aligned} \tag{7.216}$$

where

$$\Psi(\omega, x) = \frac{\sin\left[\omega y(x)\right]}{c_c^{3/4}(x) c_c^{1/4}(L) \sin\left[\omega y(L)\right]}. \tag{7.217}$$

By means of (7.216) due to the standard technique, we find the variations of ε and $\dot{\varepsilon}$

$$\sigma_1^2 = \int_{-\infty}^{\infty} \left|\Psi(\omega, x)\right|^2 S(\omega) d\omega,$$

$$\sigma_2^2 = \int_{-\infty}^{\infty} \omega^2 \left|\Psi(\omega, x)\right|^2 S(\omega)\, d\omega. \tag{7.218}$$

Expressions (7.218) are actually equations for functions σ_1 and σ_2. The unknown functions $\sigma_1(x)$ and $\sigma_2(x)$ appear in the integrands on the right hand sides nonlinearly as y is an integral over x. Therefore, equations (7.218) form a system of two nonlinear integral equations. After this system has been solved one can easily find the statistical characteristics of the vibrational field by means of the above formulae. On this account, the main problem at the moment is to solve the system of integral equations (7.218). In the most general case, it can be solved only numerically, say by the method of successive approximations. Below we will consider some cases which enable approximate analytical solutions. In order to find an analytical solution of (7.218) a maximal simplification of this system is desirable. By means of the notation

$$2y(x) = A(x) - i\omega B(x), \tag{7.219}$$

the square of the absolute value of Ψ is written in the form

$$|\Psi|^2 = |c_c(L)|^{-1/2} |c_c(x)|^{-3/2} \frac{\cosh\left[\omega^2 B(x)\right] - \cos\left[\omega A(x)\right]}{\cosh\left[\omega^2 B(L)\right] - \cos\left[\omega A(L)\right]}. \tag{7.220}$$

The assumption of a slowly varying complex rigidity c_c has been used. For physical reasons, it is clear that slowly varying σ_1 and σ_2, as well as a slowly varying complex rigidity c_c (which depends on σ_1 and σ_2), are possible only for not too large values of damping. This means that the imaginary part in eq. (7.219) is small compared to the real part for each frequency of the load spectrum. This allows us to use the following approximate equations

$$|c_c| = h_1,\ A(x) = 2\int_0^x \sqrt{\frac{m}{h_1}}\, dx,\ B(x) = \omega \int_0^x \sqrt{\frac{m}{h_1}}\frac{h_2}{h_1}\, dx. \tag{7.221}$$

These are valid with an accuracy up to the first order of smallness.

The second simplification is more restrictive. When the damping is small, expression (7.212) is a rapidly changing function of frequency with sharp peaks (they would be resonance peaks if the system were linear). For this reason it is necessary to average $|\Psi|^2$ over a frequency band of the order of the distance between adjacent peaks and then integrate as equations (7.218) require. Averaging as in Section 7.1 we obtain

$$|\Psi|^2_{av} = h_{1L}^{-1/2} h_{1x}^{-3/2} \frac{\cosh\left[\omega^2 B(x)\right]}{\sinh\left[\omega^2 B(L)\right]}. \tag{7.222}$$

Here and in what follows, the subindex denotes the value of the argument, e.g. $h_{1x} = h_1(x)$ and $h_{1L} = h_1(L)$. Inserting eq. (7.222) into (7.218) we obtain a simplified version of the system of integral equations

$$\sigma_{1x}^2 = \int_{-\infty}^{\infty} |\Psi(\omega, x)|_{av}^2 S(\omega)\, d\omega,$$

$$\sigma_{2x}^2 = \int_{-\infty}^{\infty} \omega^2 |\Psi(\omega, x)|_{av}^2 S(\omega)\, d\omega. \tag{7.223}$$

The above averaging procedure assumes the spectral densities of the load to be smooth.

Let us next consider some particular cases.

1. Let the argument of the hyperbolic functions be small for the main frequencies of the frequency spectrum of the load. Using the asymptotic formulae we arrive at the system of equations which predicts that σ_1 and σ_2 are constant. The equations for their determination are

$$\sigma_1^2 h_1^2 B(L) = \int_{-\infty}^{\infty} \frac{S(\omega)}{\omega^2} d\omega,$$

$$\sigma_2^2 h_1^2 B(L) = \int_{-\infty}^{\infty} S(\omega)\, d\omega. \tag{7.224}$$

2. Another important limiting case is that in which the argument of the hyperbolic function is large. Utilizing the corresponding asymptotic formulae in (7.222), the system (7.223) becomes

$$\sigma_{1x}^2 = \int_{-\infty}^{\infty} e^{\omega^2[B(x)-B(L)]} \frac{S(\omega)}{h_{1L}^{1/2} h_{1x}^{3/2}} d\omega,$$

$$\sigma_{2x}^2 = \int_{-\infty}^{\infty} e^{\omega^2[B(x)-B(L)]} \frac{\omega^2 S(\omega)}{h_{1L}^{1/2} h_{1x}^{3/2}} d\omega. \tag{7.225}$$

It is instructive to note that exactly this system of equations is obtained when a semi-infinite rod $-\infty < x < L$ subject to a random load at $x = L$ is considered. From a physical point of view, the transition from the system (7.225) to a new system (7.218) means neglecting the reflection at the free end of rod.

The first difficulty that arises when one tries to solve the system (7.225) is an evaluation of the integral over frequency. For some classes of spectral density of the load such evaluation is possible in a general form. For instance, if

$$S(\omega) = D\,|\omega|^s e^{-p\omega^2},\ s > 0,\ p > 0,\ D > 0, \tag{7.226}$$

then the evaluation reduces to a known integral

$$\int_0^\infty z^{s-1} e^{-pz^2} dz = \frac{1}{2} p^{-s/2} \Gamma\left(\frac{s}{2}\right), \tag{7.227}$$

with Γ being the Eulerian gamma-function. In this case, the system of equations (7.225) takes the form

$$\begin{aligned} \left[\frac{\sigma_{1x}}{\sigma_{1L}}\left(\frac{h_{1x}}{h_{1L}}\right)^{3/4}\right]^{-4/(s+1)} &= \left[\frac{\sigma_{2x}}{\sigma_{2L}}\left(\frac{h_{1x}}{h_{1L}}\right)^{3/4}\right]^{-4/(s+3)} = \\ &= 1 - \frac{B(x) - B(L)}{\rho}. \end{aligned} \tag{7.228}$$

The left equation in (7.228) links σ_{1x} and σ_{2x}, while the right equation in (7.228) yields the following differential equation

$$\frac{\partial}{\partial \sigma_{2x}} \left[\frac{\sigma_{2x}}{\sigma_{2L}}\left(\frac{h_{1x}}{h_{1L}}\right)^{3/4}\right]^{-4/(s+3)} \frac{d\sigma_{2x}}{dx} = -\frac{1}{\rho}\frac{dB(x)}{dx}.$$

The function on the right hand side of this equation is a known function of σ_{1x} and σ_{2x} due to the last formula in eq. (7.221). The boundary values σ_{1L} and σ_{2L} must be obtained directly from the system (7.225) at $x = L$

$$\sigma_{1L} h_{1L} = \sigma_p, \ \sigma_{2L} h_{1L} = \sigma_{\dot{p}}, \tag{7.229}$$

where σ_p and $\sigma_{\dot{p}}$ are referred to as the root-mean-squares of load and its rate, respectively.

For a rod's material with the constitutive equation

$$Q = k\varepsilon + r\left|\dot{\varepsilon}\right|^{\mu} \operatorname{sgn}\dot{\varepsilon}, \ k, \ r, \ \mu > 0 \tag{7.230}$$

we have

$$\begin{aligned} h_1 &= k, \ h_2 = d\sigma_2^{\mu-1}, \ \frac{dB}{dx} = \gamma\sigma_2^{\mu-1}, \\ d &= \frac{r}{\sqrt{\pi}} 2^{(1+\mu)/2} \Gamma\left(1 + \frac{\mu}{2}\right), \ \gamma = \left(\frac{m}{k}\right)^{1/2} \frac{d}{k}. \end{aligned} \tag{7.231}$$

In this case equation (7.228) is as follows

$$\frac{4}{s+3}\left(\frac{\sigma_{2x}}{\sigma_{2L}}\right)^{-1-4/(s+3)} \frac{1}{\sigma_{2L}}\frac{d\sigma_{2x}}{dx} = \frac{\gamma}{\rho}\sigma_{2x}^{\mu-1}. \tag{7.232}$$

Its solution takes the form

$$\sigma_{2x} = \sigma_{2L}\left[1 + \frac{(s+3)\gamma\lambda}{4\rho}\sigma_{2L}^{\mu-1}(L - x)\right]^{-1/\lambda}, \ \lambda = \mu - 1 + \frac{4}{s+3}. \tag{7.233}$$

This expression enables some general conclusions on the character of the vibrational field to be drawn. For definiteness consider a semi-infinite rod $-\infty < x < L$. It follows from (7.233) that vibration occupies the entire rod for $\lambda > 0$, whereas for $\lambda < 0$ the distance which vibration propagates down the rod is

$$(L - x)_* = \frac{4\rho}{(s+3)\,\gamma\,|\lambda|}\sigma_{2L}^{1-\mu}. \tag{7.234}$$

The rest of the rod does not move. When $\lambda = 0$ the vibration decays exponentially.

These conclusions are approximately valid for a rod of finite length provided that a considerable decrease of σ_2 is observed along the length L. Regretfully, the general case in which the arguments of the hyperbolic functions can also take intermediate values cannot be analysed analytically. The reason for this is a difficulty in integrating over frequency.

3. There exists, however, one particular case in which this integral can be approximately evaluated in a general form. This is the case of a narrow band loading in which the system of equations (7.223) takes the form

$$\sigma_1^2 = |\Psi(\Omega,\, x)|_{av}^2\,\sigma_p^2, \quad \sigma_2^2 = \Omega^2\,|\Psi(\Omega,\, x)|_{av}^2\,\sigma_p^2, \tag{7.235}$$

where Ω is a central frequency of the load and σ_p is the standard deviation of the load. The frequency spectrum of the load must be broad enough to ensure that the above frequency averaging takes place. The frequency averaging should include a few resonance peaks of the amplitude-frequency characteristic of the rod.

The system of equations (7.235) may be rewritten as follows

$$\sigma_2 = \Omega\sigma_1, \quad \sigma_1^2 = |\Psi(\Omega,\, x)|_{av}^2\,\sigma_p^2. \tag{7.236}$$

The second equation in (7.236) gives σ_1 provided that σ_2 is eliminated by means of the first equation.

An analysis of the vibrational field due to eq. (7.236) is relatively simple. A homogeneous field of vibration can be investigated by the methods of the first part of this Section, whilst an essentially inhomogeneous field can be investigated by the methods of the second part of the Section. An intermediate case, namely a case of a weakly heterogeneous field of vibration has not yet been studied. For this reason we begin with the latter case.

It is hardly possible to obtain an exact solution of the integral equation (7.236) in this intermediate case. Thus we should look for an approximate analytical solution. It is known that if the character of the solution can be predicted, the direct methods give satisfactory results. One of the direct methods, namely the collocation method is applied in what follows.

Let us consider the possible character of solution of (7.236). As σ_1 is a standard deviation it is necessarily non-negative and $\sigma_1 = \text{const}$ for a homogeneous field. If the vibration field is weakly heterogeneous, then σ_1

changes slightly decreasing with the distance from the loaded end. It is rather evident that the following dependence

$$\sigma_{1x} = \sigma_{1L} e^{-b(1-x/L)} \tag{7.237}$$

is a rather good approximation to the vibrational field provided that its parameters σ_{1L} and b are properly chosen. Let us determine them from the condition that the solution (7.237) satisfies equations (7.236) only at two points, namely at the rod ends. We then obtain

$$\sigma_{1L}^2 = \left|\Psi(\Omega,\, L)\right|_{av}^2 \sigma_p^2, \quad \sigma_{10}^2 = \left|\Psi(\Omega,\, 0)\right|_{av}^2 \sigma_p^2. \tag{7.238}$$

Because both sides of these equations contain σ_{1x}, due to (7.237), we conclude that (7.238) presents a system of algebraic or transcendental equations for σ_{1L} and b.

It is clear that the constructed approximate solution can describe well homogeneous and weakly inhomogeneous vibrational fields. However, this approximation is not suitable to deal with an essentially heterogeneous field of vibration. On the other hand, the approach of the second part of this Section works well in this case. Applying this approach yields the following equation

$$\sigma_{1x}^2 = \frac{\sigma_p^2}{h_{1L}^{1/2} h_{1x}^{3/2}} e^{\Omega^2 [B(x) - B(L)]}. \tag{7.239}$$

By taking the logarithmic derivative of both sides of this equation we reduce it to the following differential equation

$$\left[\frac{\partial}{\partial \sigma_{1x}} \ln\left(\sigma_{1x}^2 h_{1x}^{3/2}\right)\right] \frac{d\sigma_{1x}}{dx} = \Omega^2 \frac{dB(x)}{dx}. \tag{7.240}$$

While differentiating with respect to σ_{1x} one should keep in mind that σ_{2x} is dependent on σ_{1x} due to the first equation in (7.236).

The first condition in (7.229) is the boundary condition for (7.240) since it is valid for any spectral density of the load.

Thus, all possible cases of heterogeneity of the vibration field have been analysed.

Let us consider an example. Let a material with the rheological equation (7.230) be taken. The system (7.238) is written down as follows

$$\begin{aligned} \left(\frac{k\sigma_{1L}}{\sigma_p}\right)^2 &= \coth\left\{\gamma L \Omega^{\mu+1} \sigma_{1L}^{\mu-1} \frac{1 - e^{-b(\mu-1)}}{b(\mu-1)}\right\}, \\ \left(\frac{k\sigma_{1L} e^{-b}}{\sigma_p}\right)^2 &= 1/\sinh\left\{\gamma L \Omega^{\mu+1} \sigma_{1L}^{\mu-1} \frac{1 - e^{-b(\mu-1)}}{b(\mu-1)}\right\}. \end{aligned} \tag{7.241}$$

This can be transformed to a simpler form

$$R = \left(1 - e^{-4b}\right)^{-1/4}, \; e^{2b} = \cosh\left\{(BR)^{\mu-1} \frac{1 - e^{-b(\mu-1)}}{b(\mu-1)}\right\} \tag{7.242}$$

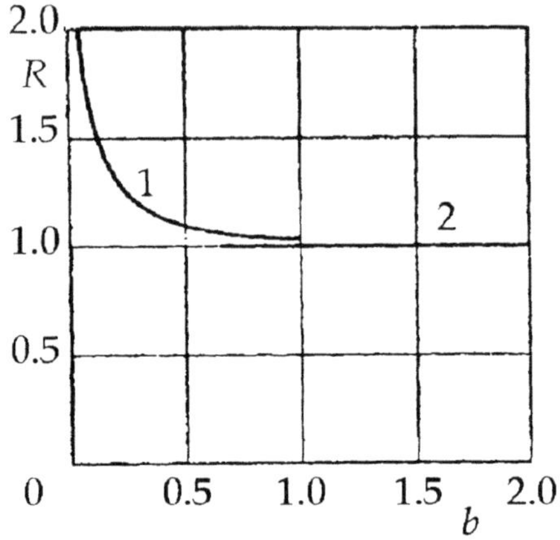

FIGURE 7.12. Dependence $R(b)$ for some values of μ.

by introducing the following notation

$$R = \frac{k\sigma_{1L}}{\sigma_p}, \; B = \frac{\sigma_p}{k}\left(\gamma L \Omega^{\mu+1}\right)^{1/(\mu-1)}. \tag{7.243}$$

The dependencies $R = R(b)$ and $B = B(b)$, due to eqs. (7.243), are plotted in Figs. 7.12 and 7.13 and are marked by a sign "1".

In this case the differential equation of the heterogeneous field (7.240) is given by

$$\frac{d\sigma_{1x}}{dx} = \gamma \frac{\Omega^{\mu+1}}{2} \sigma_{1x}^{\mu}. \tag{7.244}$$

Its solution is

$$\sigma_{1x} = \sigma_{1L}\left[1 + \frac{\mu-1}{2}\gamma\Omega^{\mu+1}\sigma_{1L}^{\mu-1}(L-x)\right]^{-1/(\mu-1)}, \tag{7.245}$$

where σ_{1L} is determined from the boundary condition (7.229).

This expression is suitable only for those values of x that ensure a positive and real right hand side of eq. (7.245). For $\mu > 1$ it is suitable for any x. If $\mu < 1$ then solution (7.245) becomes zero for a certain $x = x_*$, and makes no sense for $x < x_*$. For this reason, we must take $\sigma_{1x} = 0$ for $x < x_*$ because this solution satisfies eq. (7.244) and meets the solution (7.245) for $x > x_*$. We depict the solution of (7.245) in Figs. 7.12 and 7.13 in the planes of $b - R$ and $b - B$, respectively. Firstly, the boundary condition (7.229) gives

$$R = 1. \tag{7.246}$$

Further, let the heterogeneity degree b be the logarithm of the ratio of the standard deviation of strain at $x = L$ (the cross-section where the load is applied) to that at $x = 0$ (the free end). This definition of the heterogeneity

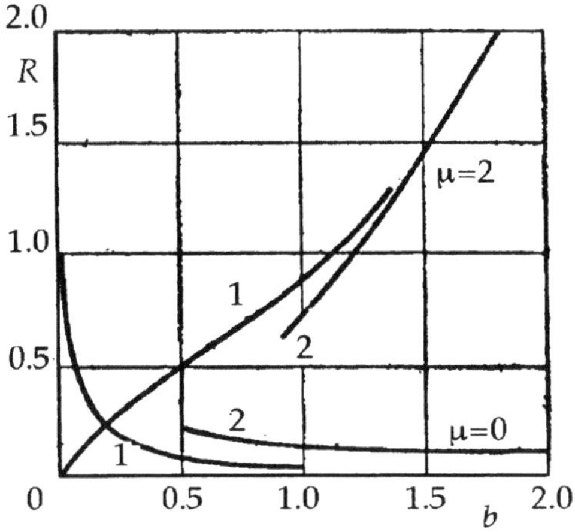

FIGURE 7.13. Dependence $B\,(b)$ for some values of μ.

degree is in agreement with approximation (7.237). According to formula (7.245) we have

$$b = \frac{1}{\mu - 1} \ln \left[1 + \frac{\mu - 1}{2} \gamma \Omega^{\mu+1} \sigma_{1L}^{\mu-1} L \right]. \tag{7.247}$$

By means of (7.243) this result is expressed in the following form

$$B = \frac{1}{R} \left\{ \frac{2}{\mu - 1} \left[e^{b(\mu-1)} - 1 \right] \right\}^{1/(\mu-1)}. \tag{7.248}$$

Figures 7.12 and 7.13 show the dependences $R = R\,(b)$ and $B = B\,(b)$ due to eqs. (7.246) and (7.248) for two typical values of μ, and are denoted by "2". As follows from (7.248) for $\mu < 1$ and large values of b, the value of B is nearly constant and is given by

$$B = \left(\frac{2}{1 - \mu} \right)^{1/(\mu-1)}. \tag{7.249}$$

Note that the value of B is proportional to the load σ_p. Consider a case in which the load is lower than the load corresponding to the limit in (7.249). By virtue of eqs. (7.243), (7.246) and (7.249) this requirement is equivalent to the following inequality

$$\frac{1 - \mu}{2} \gamma L \Omega^{\mu+1} \sigma_{1L}^{\mu-1} > 1. \tag{7.250}$$

Equation (7.245) shows that in this case vibration does not reach the free end, i.e. $\sigma_{10} = 0$. Hence, an infinitely large heterogeneity degree corresponds to this case.

As mentioned above, in the case of a weakly heterogeneous field of vibration (say $b < 1$) it is necessary to use curve 1, whereas in the case of an essentially heterogeneous field (say $b > 1$) we must take curve 2. It is worth mentioning that curves 1 and 2 converge naturally. Therefore the entire range of possible values of b is covered.

Note that the curve $\mu = 2$ is typical for the family of curves $\mu > 1$ while the curve $\mu = 0$ is typical for the family $0 \leq \mu < 1$. Accounting for this fact, we may draw the following conclusions.

Firstly, a unique vibration field is obtained for each σ_p.

Secondly, for $\mu > 1$ increasing σ_p leads to an increase of the heterogeneity degree, whereas for $\mu < 1$ this results in its decrease.

What remains is to explain the relevance of this analysis to the problem of vibration in elastoplastic bodies.

First of all, the rheological equation (7.230) reduces to the equation for rigid-plastic material for $\mu = 0$.

Secondly, by means of (5.61) it is easy to show that for a narrow band process the complex rigidity of the Ishlinsky material is given by

$$c = k\left(1 - \delta\sigma_1^{\alpha} + i\beta\sigma_1^{\alpha}\right), \; k = EF, \tag{7.251}$$

where δ and β are some constants which depend upon the material properties only and F is the cross-sectional area. Comparing (7.251) with the second equation (7.214), and accounting for the first equation in (7.236) shows that in this case

$$\mu = 1 + \alpha, \; \gamma = \frac{k\beta}{\Omega^{\alpha+1}}, \tag{7.252}$$

$$h_1 = k\left(1 - \delta\sigma_1^{\alpha}\right). \tag{7.253}$$

Since the effect of plastic strains is usually small one can neglect the second term in the parentheses in (7.253) and take

$$h \approx k. \tag{7.254}$$

Then, inserting (7.252) and (7.254) into the equation of the third part of this Section, we obtain the solution of the problem of propagation of narrow band vibration in an elastoplastic rod. The value $\mu = 2$ corresponds to the case $\alpha = 1$, which is typical for the theory of internal friction.

8

Propagation of vibration in media with complex structure

8.1 On a model of the medium with complex structure

The laws of vibration decay with distance from the source of vibration were investigated in the previous Chapter. The character of the decay turned out to be essentially dependent on the rheological properties of the material. There exists, however, another important factor which influences the decay character, which is the dynamical structure of the medium. The analysis of media with complex structure has attracted much attention. The simplest among the media with complex structure is the Cosserat medium, its modern description being given in [91] and [117]. The Mindlin medium with a microstructure, see [106], is more sophisticated. Of extreme complexity is the multipolar mechanics developed by Green and Rivlin [45].

An essential feature of the above mentioned theories is the concept of a material point. Classical continuum mechanics deals with media of ordinary structure, that is a material point possesses three translational degrees of freedom. In contrast to this, a point in the Cosserat theory has all of the degrees of freedom of a rigid body, i.e. six degrees of freedom. In the Mindlin theory of a medium with microstructure each point possesses the degrees of freedom of a classical solid with a uniform strain, i.e. 12 degrees of freedom. In the multipolar mechanics a point is governed by n kinematic parameters, where n is finite but can be arbitrarily large. In what follows we consider one of the simplest media of such sort.

We postulate a carrier medium and assume that it is governed by the Lamé equation of the classical theory of elasticity

$$(\lambda + \mu) \nabla\nabla \cdot \mathbf{u} + \mu\Delta\mathbf{u} - \rho\ddot{\mathbf{u}} + \mathbf{K} = 0, \tag{8.1}$$

where ρ is the mass density, λ and μ are the Lamé moduli of elasticity, $\mathbf{u}$ is the vector of displacement of points in the carrier medium, and $\mathbf{K}$ is the intensity of external volumetric forces.

A set consisting of an infinite number of isotropic non-interacting oscillators with continuously distributed eigenfrequencies is attached to each point of the carrier medium. The equation of motion of a generic oscillator with an eigenfrequency k is as follows

$$m(k)\ddot{\mathbf{v}}_k + c(k)\left[1 + R_k(\partial/\partial t)\right](\mathbf{v}_k - \mathbf{u}) = \mathbf{Q}_k. \tag{8.2}$$

Here $\mathbf{v}_k$ is the vector of absolute displacement of the oscillator mass and $\mathbf{Q}_k$ is an external force acting on the oscillator mass. The value of $m(k)\,dk$ is equal to the mass of the oscillators having the eigenfrequencies in the interval $(k, k + dk)$ per unit volume. Thus, the total mass density of the oscillators attached to the carrier structure is given by

$$m = \int_0^\infty m(k)\,dk. \tag{8.3}$$

The static rigidity of an oscillator suspension is denoted in eq. (8.2) by $c(k)$ which is given by

$$c(k) = k^2 m(k). \tag{8.4}$$

The term with $R_k(\partial/\partial t)$ is introduced in eq. (8.2) in order to take into account the energy dissipation in the oscillator suspension. It will be shown below that this account of damping is absolutely necessary for obtaining physically meaningful results. Account of the damping will be achieved by means of two rheological models, namely a viscoelastic model and a model of an elastoplastic material. In the first case, $R_k(\partial/\partial t)$ denotes an operator of viscoelasticity while in the second case it is a hysteretic operator. The term with $R_k(\partial/\partial t)$ is assumed to be small for any oscillator motion in both cases.

It is now necessary to take into account the effect of the oscillators' suspension on the carrier medium. The force of this interaction per unit volume is given by

$$\mathbf{F} = \int_0^\infty c(k)\left[1 + R_k(\partial/\partial t)\right](\mathbf{v}_k - \mathbf{u})\,dk. \tag{8.5}$$

As this force is a volumetric force it must be added to eq. (8.1). Thus, the system of equations which governs the dynamics of the medium takes the following form

$$
\begin{aligned}
&(\lambda+\mu)\nabla\nabla\cdot\mathbf{u}+\mu\Delta\mathbf{u}-\rho\ddot{\mathbf{u}}-\int\limits_0^\infty m(k)\,\ddot{\mathbf{v}}_k dk+\mathbf{K}+\mathbf{Q}=0,\\
&m(k)\,\ddot{\mathbf{v}}_k+c(k)\left[1+R_k(\partial/\partial t)\right](\mathbf{v}_k-\mathbf{u})=\mathbf{Q}_k.
\end{aligned}
\tag{8.6}
$$

The vector $\mathbf{Q}$ is equal to the external volumetric force acting on the oscillators

$$
\mathbf{Q}=\int\limits_0^\infty \mathbf{Q}_k(k)\,dk. \tag{8.7}
$$

The boundary conditions for the introduced medium coincide with those of the classical theory of elasticity because they are prescribed only for the carrier medium.

The proposed model is useful for describing the vibration propagation in such mechanical structures as industrial buildings, vessels, airplanes, spacecraft etc. because all of these structures possess a primary structure and secondary systems attached to the primary structure. In the framework of the method proposed, the properties of the primary structure are described by the equations of elasticity theory. The suspended oscillators reflect the frequency properties of the secondary systems attached to the primary structure. This model allows us to determine the global properties of the vibration field of complex structures without unnecessary details of the structure and its vibrational field.

A one-dimensional variant of the theory, i.e. a rod with complex structure is of considerable interest for practical applications. Assuming that a generalised uniaxial stress state is realised in the rod, we obtain the dynamical equations by means of the conventional methods of elasticity theory

$$
\begin{aligned}
&Eu''-\rho\ddot{u}-\int\limits_0^\infty m(k)\,\ddot{v}_k dk+K+Q=0,\\
&m(k)\,\ddot{v}_k+c(k)\left[1+R_k(\partial/\partial t)\right](v_k-u)=Q_k,
\end{aligned}
\tag{8.8}
$$

where E is the Young's modulus of the carrier structure, and u is an axial displacement of the points of the carrier structure.

Equations (8.8) without volumetric forces K, Q_k and damping have been first derived by Slepyan [168]. Wave propagation in an anisotropic medium with complex structure was studied in [7], while the governing equations for a Cosserat rod of complex structure were reported in [8]. The latter model allows one to simulate longitudinal, bending and torsional vibration in extended complex structures in the framework of the same approach.

8.2 One-dimensional vibration in the medium with complex structure

Consider a rod with complex structure. Let its length be L and assume that the oscillators' suspension is viscoelastic, i.e. $R_k(\partial/\partial t)$ is an operator of viscoelasticity. Assume that the end $x = 0$ is free, while the second end $x = L$ is subject to a harmonic force of frequency ω and unit amplitude. The boundary conditions are therefore as follows

$$x = 0,\ u' = 0;\ x = L,\ Eu' = e^{i\omega t}. \tag{8.9}$$

It is not difficult to find a steady-state solution of eqs. (8.8) for $K = 0$ and $Q_k = 0$ satisfying the boundary conditions (8.9). In particular, the accelerations of the carrier medium and the oscillators are

$$\ddot{u} = \Phi(\omega, x)\, e^{i\omega t},\ \ddot{v}_k = \Psi_k(\omega, x)\, e^{i\omega t}, \tag{8.10}$$

where the frequency characteristics Φ and Ψ_k are given by

$$\Phi = \frac{\omega^2 \cos\lambda x}{E\lambda \sin\lambda L},\ \Psi_k = \Phi\left[1 - \frac{\omega^2}{k^2(1 + i\psi_k)}\right]^{-1}. \tag{8.11}$$

The parameters which appear in the latter equations are

$$\begin{aligned} \lambda^2 &= \frac{\omega^2}{E}\left[\rho + \int_0^\infty \frac{c(k)\,dk}{k^2 - \omega^2/(1 + i\psi_k)}\right], \\ \psi_k &= -iR_k(i\omega), \end{aligned} \tag{8.12}$$

with ψ_k being a complex function of frequency.

The form of the distribution of vibration amplitude along the rod is of paramount interest. The square of the vibration amplitude of the carrier structure A^2 and oscillators b_k^2 have the following expressions

$$A^2 = |\Phi(\omega, x)|^2,\ b_k^2 = A^2\left|1 - \frac{\omega^2}{k^2(1 + i\psi_k)}\right|^{-2}. \tag{8.13}$$

Introducing the real and imaginary parts of λ

$$\lambda = \mu - i\kappa, \tag{8.14}$$

we can rewrite A^2 in the following form

$$A^2 = \frac{\omega^4}{E^2|\lambda|^2}\frac{\cosh(2\kappa x) + \cos(2\mu x)}{\cosh(2\kappa L) - \cos(2\mu L)}. \tag{8.15}$$

Analysis of the numerator in this equation leads to the conclusion that parameter κ determines the vibration decay with distance from the vibration source, whereas parameter μ determines the number of wave lengths along the rod. Analysis of the denominator shows that the lower the value of κ, the higher are the resonance peaks. The density of the resonance peaks of the amplitude-frequency response function is determined by the value of parameter μ.

Let us next prove that the vibration decay does not depend primarily on the oscillators' damping and that it remains finite even for vanishingly small damping. To this end, it is sufficient to show that the expression for λ^2 remains complex when $\psi_k \to 0$. For the sake of simplicity we assume that the damping properties of all oscillators coincide, i.e. $\psi_k = \psi$ does not depend upon parameter k but can depend on the excitation frequency ω. Then, in order to ensure decaying free vibrations of each oscillator when the carrier medium does not vibrate it is necessary to require that $\operatorname{Re}\psi > 0$ for $\omega > 0$. Due to the oddness of the function $\operatorname{Re}\psi$ this means that $\operatorname{Re}\psi < 0$ for $\omega < 0$.

Let us introduce a complex function

$$z = \omega / \left(1 + i\psi\right)^{1/2} . \tag{8.16}$$

Let us agree that we take that branch of the radical in (8.16) which takes on the value of $+1$ for $\psi = 0$. The imaginary part of z is then negative for any frequency ω except $\omega = 0$. Hence, as $\psi \to 0$ the complex function z approaches the real axis from the lower half-plane of the complex variable z. By using the newly introduced variable z in (8.16) we rewrite eq. (8.12) as follows

$$\lambda^2 = \frac{\omega^2}{E}\left[\rho + \frac{1}{2z}\int_{-\infty}^{\infty}\frac{c\left(|k|\right)dk}{k - z}\right] . \tag{8.17}$$

In the Cauchy integral z always lies in the lower half-plane. As $\psi \to 0$ the complex variable z approaches the real axis, along which the integration (8.17) proceeds. In accordance with the Sokhotsky-Plemelj formula [110] and [169], we obtain the following limit in eq. (8.17) as $\psi \to 0$

$$\lambda^2 = \frac{\omega^2}{E}\left[\rho + \frac{1}{2\omega}\left(-\pi i c\left(|\omega|\right) + \text{v.p.}\int_{-\infty}^{\infty}\frac{c\left(|k|\right)dk}{k - \omega}\right)\right] . \tag{8.18}$$

Here the principal value of the integral is to be taken.

As the imaginary part in (8.18) does not vanish, the coefficient of the spatial decay has a finite value even for vanishingly small damping in oscillators, and is determined by the dependence of the oscillators' suspension on their eigenfrequencies. This effect is typical only for the model which accounts for the complex structure of the medium, i.e. existence of the suspended oscillators. From a physical point of view, the effect can be explained by the fact that the suspended oscillators act as dynamic absorbers.

Note that the account of damping in the oscillators is needed to determine their vibrational amplitudes. The second formula in (8.13) indicates that the amplitude of the oscillators b_k becomes unbounded at $\psi = 0$.

Another sophisticated analysis shows that the result (8.18) remains valid as $\psi_k \to 0$, with ψ_k depending not only on the frequency ω but also the eigenfrequency k.

Let us illustrate our analysis by a particular example. Let

$$m(k) = A\left(\beta^2 + k^2\right)^{-1}, \psi_k = \psi, \tag{8.19}$$

where A and β are positive parameters and $\psi_k = \psi$ does not depend upon parameter k. Substitution of eq. (8.19) into (8.3) and evaluation of the integral yields

$$m = A\pi\left(2\beta\right)^{-1}. \tag{8.20}$$

This result allows us to remove the parameter A from the forthcoming equations, retaining only m and β. Inserting now eq. (8.19) into (8.4) and (8.12) gives

$$\lambda^2 = \frac{\omega^2}{E}\left[\rho + \int\limits_0^\infty \frac{Ak^2dk}{\left(\beta^2 + k^2\right)\left[k^2 - \omega^2/\left(1 + i\psi\right)\right]}\right]. \tag{8.21}$$

Evaluating the integral by means of residue calculus we arrive at the following final result

$$\lambda^2 = \frac{\omega^2}{E}\left[\rho + m\left(1 + \frac{i\omega}{\beta\sqrt{1 + i\psi}}\right)^{-1}\right]. \tag{8.22}$$

The structure of this formula completely confirms the analysis of this Section. Further, it shows a weak dependence of λ on ψ for realistic, i.e. not very considerable, values of ψ.

This example is remarkable for another reason. If one puts $\rho = 0$ and $\psi = 0$ in eq. (8.22), then

$$\lambda^2 = \frac{\omega^2}{E}\frac{m}{1 + i\omega/\beta}. \tag{8.23}$$

Exactly the same expression for the wavenumber can be obtained for the problem of linear longitudinal vibrations in a rod of the Kelvin-Voigt material, cf. [14], i.e. a rod with a finite value of damping. Therefore, the medium under consideration possesses the following interesting peculiarity. Though the medium is "built" from high-quality elements, such as an ideally elastic carrier medium and slightly damped oscillators, it behaves like a medium of ordinary structure with considerable damping.

At first glance, this conclusion seems to be paradoxical. However, a careful analysis shows that the energy dissipation has a finite value even for small values of damping in oscillators which is explained by considerable amplitudes of vibrations for the resonating oscillators.

The problem of vibration in a rod subject to a random load is important for practical reasons. Its solution is easy to obtain by the methods of Section 7.1.

In particular, the dispersion of acceleration of the carrier medium and the dispersion of acceleration of the oscillators are as follows

$$D_{\ddot{u}} = \int_{-\infty}^{\infty} |\Phi(\omega, x)|^2 S(\omega) \, d\omega, \tag{8.24}$$

$$D_{\ddot{v}_k} = \int_{-\infty}^{\infty} |\Psi_k(\omega, x)|^2 S(\omega) \, d\omega, \tag{8.25}$$

where $S(\omega)$ denotes the spectral density of the stress in the cross-section $x = L$, while Φ and Ψ_k are given by eqs. (8.11).

Special attention should be given to the limiting case of asymptotically small damping. As explained above, the imaginary part of λ does not vanish as $\psi_k \to 0$, thus function $\Phi(\omega, x)$ remains bounded at all frequencies. Therefore, an asymptotic expression for integral (8.24) is

$$D_{\ddot{u}} = \int_{-\infty}^{\infty} |\Phi(\omega, x)|^2_{\psi_k \to 0} S(\omega) \, d\omega. \tag{8.26}$$

The subindex in the integrand means that the limiting expression of λ due to (8.18) as $\psi_k \to 0$ is to be used, while evaluating Φ from eq. (8.11). It is important to note that the asymptotic expression (8.26) does not depend upon the friction law in the oscillators' suspension at all. Nonetheless, equation (8.26) predicts an intensive decay of vibration with distance from its source.

Let us study an asymptotic representation of integral (8.25) for small values of damping. Inserting Ψ_k, due to (8.11), we obtain

$$D_{\ddot{v}_k} = 2 \int_0^{\infty} \frac{|\Phi(\omega, x)|^2 S(\omega) \left|k^2 [1 + i\psi_k(\omega)]\right|^2}{\left|k^2 [1 + i\psi_k(\omega)] - \omega^2\right|^2} d\omega. \tag{8.27}$$

The principal contribution to this integral as $\psi_k \to 0$ is seen to be due to the frequency $\omega = +k$. On this account we have asymptotically

$$D_{\ddot{v}_k} = 2k^4 S(k) |\Phi(k, x)|^2_{\psi_k \to 0} \int_0^{\infty} \frac{d\omega}{\left|k^2 [1 + i\psi_k(\omega)] - \omega^2\right|^2}. \tag{8.28}$$

The integral in this equation has already been evaluated. By means of (6.23) and (6.24) we find

$$2\int_{0}^{\infty} \frac{d\omega}{\left|k^2\left[1+i\psi_k(\omega)\right]-\omega^2\right|^2} = \frac{\pi}{k^3 \operatorname{Re}\psi_k(k)}. \tag{8.29}$$

Thus, the result is

$$D_{\ddot{v}_k} = \frac{\pi k S(k)}{\operatorname{Re}\psi_k(k)} \left|\Phi(k, x)\right|^2_{\psi_k \to 0}. \tag{8.30}$$

It is worth noting the physical interpretation of this asymptotic expression which is that the vibration of an oscillator with an eigenfrequency k depends only on its own damping at frequency k and does not depend on the damping of the other oscillators.

Let us briefly consider the following important simplification of eq. (8.26). If the spectral density of the load is a smooth function of frequency, then to simplify the calculations it is necessary to perform, in eq. (8.26), an averaging $|\Phi|^2$ by analogy with Section 7.1. The result of this averaging is

$$\left|\Phi(\omega, x)\right|^2_{av} = \frac{\omega^4}{E^2|\lambda|^2} \frac{\cosh(2\kappa x)}{\sinh(2\kappa L)},$$

and eq. (8.26) is replaced by the simpler expression

$$D_{\ddot{u}} = \int_{-\infty}^{\infty} \left|\Phi(\omega, x)\right|^2_{av,\ \psi_k \to 0} S(\omega)\, d\omega.$$

In the particular case of a load with a frequency band covering 2 or 3 resonance peaks we obtain approximately by analogy with Section 7.1 from the latter two equations

$$\sigma_{\ddot{u}} = \frac{\omega^2}{E|\lambda|} \sqrt{\frac{\cosh(2\kappa x)}{\sinh(2\kappa L)}}\,\sigma. \tag{8.31}$$

Here σ is the root-mean-square of the stress at the end $x = L$, $\sigma_{\ddot{u}}$ is the root-mean-square of acceleration in a cross-section x, and ω is the central frequency of the load.

Let us proceed to considering an elastoplastic suspension. The structure of expressions (8.26) and (8.30) essentially simplifies this transition. It seems evident that eq. (8.26) is immediately applicable since it does not depend on damping at all. A formal analysis, based on the method of statistical linearisation, leads to the same conclusion.

Let us study the vibration of oscillators. First, it is necessary to consider the equations which govern the vibration of oscillators with an elastoplastic suspension. Restricting our consideration to a one-dimensional variant of the model of a medium with complex structure we assume that each oscillator consists of a massless rod with a lumped mass at its end. Assuming that the mass of the oscillator moves in the longitudinal direction only, and using the general complex exponential form of Section 5.2, we write the governing equation for the oscillator in the following form

$$m(k)\,\ddot{v}_k + c(k)\left[1 + R_k\right](v_k - u) = 0, \tag{8.32}$$

where

$$R_k = -\frac{2(1+\nu)}{3}\int_0^1 \frac{p_k\left(\frac{\Gamma_k}{n}\eta\right)\frac{\Gamma_k}{n}\,d\eta}{1 + i\omega\varphi_k(\eta)}. \tag{8.33}$$

Here Γ_k is the root-mean-square of intensity of shear strain in the suspension rod, while φ_k is an auxiliary function which is the solution of eq. (5.49)

$$1 - \eta^2 = \int_{-\infty}^{\infty} \frac{S_k(\omega)}{\Gamma_k^2}\frac{d\omega}{1 + (\omega\varphi_k)^2}. \tag{8.34}$$

Here $S_k(\omega)$ is the spectral density of strain in the suspension of the oscillator with eigenfrequency k. Equation (8.32) has the same form as eq. (8.8) provided that we use the complex exponential form and put $Q_k = 0$ in the latter equation. Hence, to determine the vibration level we may use eq. (8.30) in which

$$\psi_k = -iR_k. \tag{8.35}$$

It should, however, be kept in mind that, in this particular case, ψ_k depends upon an unknown variable Γ_k as indicated by eq. (8.33). Before we proceed to determining Γ_k let us note that Γ_k differs from the root-mean-square σ_k of the elongation $v_k - u$ only by a known factor. For this reason

$$\Gamma_k = \lambda_k\sigma_k,\ \lambda_k = \frac{1+\nu}{\sqrt{3}L_k}, \tag{8.36}$$

where L_k is a length of the oscillator suspension.

The spectral density of shear is easy to find by means of eq. (8.32) and is given by

$$S_k(\omega) = \frac{\lambda_k^2}{k^4\left|1 + i\psi_k(\omega, \Gamma_k)\right|^2} S_{\ddot{v}_k},$$

or by virtue of (8.25)

$$S_k(\omega) = \frac{\lambda_k^2\left|\Psi_k(\omega, x)\right|^2 S}{k^4\left|1 + i\psi_k(\omega, \Gamma_k)\right|^2}.$$

Inserting these values into the expression for Γ_k and eq. (8.34), and evaluating the asymptotic representations of the corresponding integrals for small ψ_k, gives

$$\Gamma_k^2 = \frac{\pi \lambda_k^2 S(k)}{k^3 \operatorname{Re} \psi_k(k, \Gamma_k)} \left|\Phi(k, x)\right|^2_{\psi_k \to 0}, \tag{8.37}$$

$$1 - \eta^2 = \frac{1}{1 + (k\varphi_k)^2}. \tag{8.38}$$

Equation (8.37) enables the unknown parameter Γ_k to be determined. In particular, one can see that Γ_k depends on x, this dependence being determined by function $|\Phi|^2$. The latter characterizes the distribution of the vibration amplitude of the carrier medium driven harmonically at the eigenfrequency of the oscillator.

An important observation should be noted. Equation (8.38) looks like an equation for the harmonic vibration at frequency k. Then, the asymptotic representations (8.37) and (8.30) contain the damping at frequency k only. Substitution of solution (8.38) into eq. (8.33) at $\omega = k$ yields the following result

$$R_k = -\frac{2(1+\nu)}{3} \int_0^1 \left(1 - \eta^2 - i\eta\sqrt{1-\eta^2}\right) p_k \left(\frac{\Gamma_k}{n}\eta\right) \frac{\Gamma_k}{n} d\eta. \tag{8.39}$$

This value characterizes the damping in the suspension of an oscillator under harmonic vibration. There is nothing surprising in this result as it is known that an oscillator with asymptotically small damping subject to a broad band load experiences a narrow band vibration with a central frequency coinciding with its eigenfrequency. In this case, the problem of determining Γ_k from eq. (8.37) offers no great difficulty, and can be solved as has been done in Section 6.2.

8.3 On application of the theory of media with complex structure

It has been already pointed out that the equations of the theory of media with complex structure may be used for the analysis of vibration in such complex dynamical systems as vessels, spacecraft, submarines etc. In many cases a one-dimensional model turns out to be sufficient as the above structures are extended structures, and therefore the structure can be modelled in the form of a homogeneous rod vibrating longitudinally. However, there is a problem which impedes practical applications, namely it is not clear how to prescribe the parameters of a hypothetical rod with complex structure in order that this rod can simulate the vibrational behaviour of the real structure. Some parameters are easy to prescribe. It is natural to accept that the length L and the cross-sectional area F of a real structure coincide with those of the model. It is reasonable to require that the axial rigidity of the real structure must coincide with that of the model. This leads to the following equality

$$c = EF, \tag{8.40}$$

which gives the Young's modulus of the model.

In addition, the average mass per unit length of the real structure and the model coincide, i.e.

$$M = F\left(\rho + m\right). \tag{8.41}$$

The total mass density of the model $\rho + m = \rho_0$ is easily determined from this equation.

However, considerable difficulties arise when prescribing functions $c\left(k\right)$ and $m\left(k\right)$ which characterize the spectral properties of the secondary systems of the real structure. Since this problem cannot be solved theoretically, it is necessary to consider some experimental data of the vibrational field in the real structure. The question of what kind of experimental data is needed for this aim is very difficult and is not discussed here in detail. Only the practically important case of a relatively rigid suspension is considered here. The following modification of formula (8.12) appears to be useful

$$\lambda^2 = \frac{\omega^2}{E}\left[\rho_0 + \int\limits_0^\infty \frac{\omega^2}{1+i\psi_k}\frac{m\left(k\right)dk}{k^2 - \omega^2/\left(1+i\psi_k\right)}\right], \tag{8.42}$$

where ρ_0 is a total mass density

$$\rho_0 = \rho + m. \tag{8.43}$$

If the suspension of each oscillator were absolutely rigid, the second term in the square brackets would be equal to zero. This term is small compared to the first term provided that the frequency ω is not very high and the rigidity of the oscillator suspension is high, but finite. This leads to the following asymptotic expression for λ

$$\lambda = \frac{\omega}{a}\left[1 + \frac{\omega^2}{2\rho_0}\int\limits_0^\infty \frac{m\left(k\right)dk}{k^2\left(1+i\psi_k\right) - \omega^2}\right], \tag{8.44}$$

where a is the velocity of elastic wave propagation

$$a = \sqrt{E/\rho_0}. \tag{8.45}$$

In the limiting case of vanishing damping $\psi_k \to 0$ this expression takes the form

$$\lambda = \frac{\omega}{a}\left\{1 + \frac{\omega^2}{2\rho_0}\left[-\frac{\pi i m(|\omega|)}{2\omega} + \text{v.p.}\int\limits_0^\infty \frac{m(k)\,dk}{k^2-\omega^2}\right]\right\}. \tag{8.46}$$

This allows us to obtain the following asymptotic equations for the real and imaginary parts of λ and its absolute value

$$\mu = \frac{\omega}{a},\ \kappa = \frac{\omega}{a}\eta,\ |\lambda| = \frac{\omega}{a}, \tag{8.47}$$

where η is a non-dimensional factor

$$\eta = \frac{\pi\omega m(|\omega|)}{4\rho_0}. \tag{8.48}$$

Therefore, two of three parameters in eq. (8.15) are obtained explicitly, while the last one η (or κ) requires experimental data for its identification. The most suitable experiment to this aim is a test of a broad band excitation in the frequency domain which covers several eigenfrequencies of the longitudinal vibration. The results of such tests are usually available in the form of diagrams of the vibration level versus axial coordinate of the structure. Equation (8.31) is the most suitable one for approximation of these diagrams. By means of this equation we find a ratio of the vibration levels in a current cross-section x and in the loaded one $x = L$, and then remove κ by means of (8.47). The result is

$$\frac{\sigma_{\ddot{u}}(x)}{\sigma_{\ddot{u}}(L)} = \sqrt{\frac{\cosh\left(\frac{2L}{a}\omega\eta\frac{x}{L}\right)}{\cosh\left(\frac{2L}{a}\omega\eta\right)}}. \tag{8.49}$$

Comparing the latter equation for various values of η with the corresponding experimental curves one gets a possibility to find η at each frequency. Typical for complex structures values of η lie in the interval $\eta = 0.15 \div 0.3$.

After all parameters of the model have been identified the model can be used in technical calculations. Let us point out a few possible applications of the model.

1. Prediction of vibration in a structure due to known characteristics of pulsation of its engine or its propellant.

2. Choice of a shaker force which is required to excite a given field of vibration in the structure.

3. Determining the parameters of vibration and external excitation while testing not the entire structure but its large structural components.

8.4 Theory of vibroconductivity

Equations (8.31) and (8.30) show that an averaged spectral density of acceleration in the carrier medium is given by

$$S = \frac{\omega^4}{E^2 |\lambda|^2} \frac{\cosh 2\kappa x}{\sinh 2\kappa L} S_p, \tag{8.50}$$

where S_p is the spectral density of the load acting at the end $x = L$ of the rod with complex structure.

It has been proved above that eq. (8.50) is not sensitive to the character of the damping in the oscillator suspension.

It is easy to prove that function (8.50) is a solution of the following boundary-value problem

$$0 < x < L, \quad \frac{d^2 S}{dx^2} - (2\kappa)^2 S = 0, \tag{8.51}$$

$$x = 0, \quad \frac{dS}{dx} = 0, \tag{8.52}$$

$$x = L, \quad \frac{dS}{dx} = \frac{2\kappa\omega^4}{E^2 |\lambda|^2} S_p. \tag{8.53}$$

This boundary-value problem coincides with the problem of thermal conductivity in a one-dimensional object with a distributed heat sink, one end of the structure ($x = 0$) being thermally insulated, while the other ($x = L$) being subject to a thermal flux into the object. This leads to the idea that the found thermal analogy can be used to describe the propagation of high frequency vibrations in complex structures. This analogy is referred to as the theory of vibroconductivity. From a theoretical point of view, the very possibility of such a description cannot cause any objection since there is an analogy between thermal motion and high frequency vibration. From a practical point of view, this approach is useful as it leads to very simple solutions of particular problems.

An analogy between the vibrational and thermal problems has been pointed out in the papers of American authors [29], [30], [34], [39], [98], [102], [103], [111], [160], [161] and [186] in which it was shown that the problem of random vibration propagation in a rather broad class of mechanical systems reduces to a scheme of the heat transfer problem. However, the vibroconductivity equation was never derived in these papers. It was first derived for an isotropic object in [131], and was generalised to an anisotropic object in [10].

The boundary-value problem for the theory of vibroconductivity can easily be stated, cf. [131] and [10]. In the volume V, the following equation

$$\mathbf{r} \in V, \quad \nabla \cdot (\mathbf{K} \cdot \nabla S) - \alpha S = \gamma \frac{\partial S}{\partial t} \tag{8.54}$$

must hold. On the boundary of the object we pose the Neumann boundary condition

$$\mathbf{r} \in O, \quad \mathbf{N} \cdot (\mathbf{K} \cdot \nabla S) = F, \tag{8.55}$$

where F is an external flux of vibration and $\mathbf{N}$ is an exterior unit normal.

Assume the following representation for the vibroconductivity tensor $\mathbf{K}$

$$\mathbf{K} = \sum_{n=1}^{3} \mathbf{i}_n \mathbf{i}_n K_n, \tag{8.56}$$

where $\mathbf{i}_n$ are the unit vectors of the principal axes of the vibroconductivity tensor and K_n are the vibroconductivity factors of the principal axes of the vibroconductivity tensor. Factor γ should be termed a vibrocapacity, and, finally, S should be referred to as a quasi-temperature. Equation (8.54) differs from a classical equation of thermal conductivity in the term proportional to the quasi-temperature. This term plays the part of a vibration sink. In the original mechanical problem, this terms characterizes the spatial decay of vibration with distance from a vibration source.

The governing equation and the boundary conditions have been written down. However, it is necessary to have knowledge of the vibroconductivity factors K_n, the vibrocapacity γ and the factor of spatial decay α. Generally speaking this should be achieved under some special test conditions. The most suitable tests to this aim are those in which the quasi-temperature varies along one of the anisotropy directions, e.g. along axis x_n. Assume that cross-section $x_n = 0$ is insulated, while a vibration flux F_n is prescribed in the cross-section $x_n = L$. We consider first the stationary problem. The boundary-value problem (8.54) and (8.55) takes the form

$$\begin{aligned} 0 &< x_n < L, \quad \frac{d}{dx_n}\left(K_n \frac{dS}{dx_n}\right) - \alpha S = 0, \\ x_n &= 0, \quad \frac{dS}{dx_n} = 0, \\ x_n &= L, \quad \frac{dS}{dx_n} = F_n. \end{aligned}$$

In the case of a homogeneous structure the solution is as follows

$$S = \frac{F_n}{K_n \sqrt{\alpha/K_n}} \frac{\cosh \sqrt{\alpha/K_n} x_n}{\sinh \sqrt{\alpha/K_n} L} S_p. \tag{8.57}$$

This theoretical dependence predicts an essential reduction of the quasi-temperature S as x_n decreases. Experimental data of vibrations have exactly this character. In this sense, the theory of vibroconductivity agrees well with the observed vibrational phenomena. Comparing the theoretical dependence (8.57) and the corresponding experimental curve we can find α/K_n. If F_n can be experimentally measured then parameter K_n can be found. By performing this one-dimensional experiment for all orthotropic axes and processing the data using the above strategy, we obtain the values of K_n and α. The problem of identifying these parameters can now be considered to be solved.

The question still remains as to what are S and F_n? These are to be measured experimentally, but nothing has been said about them so far. The posed question is the main problem of the theory of vibroconductivity. An answer can be given with the help of the results of an "imaginary experiment", i.e. results of a theoretical analysis of a particular complex dynamical structure, provided that the results are obtained by means of an available method of vibration theory. Comparison of these results with the result of the theory of vibroconductivity allows us to answer the question on the mechanical meaning of the quasi-temperature S and the load F, as well as give the theoretical formulae for parameters α and F_n.

Let us consider eq. (8.50) as an appropriate solution from the vibration theory, however, we will write it for an arbitrary axis of anisotropy

$$S_n = \frac{\omega^4}{E_n^2 \left|\lambda_n\right|^2} \frac{\cosh 2\kappa_n x}{\sinh 2\kappa_n L} S_{pn}. \tag{8.58}$$

Here E_n is the Young's modulus for the direction $\mathbf{i}_n$, S_n is the averaged spectral density of the vibration acceleration, and S_{pn} is the spectral density of the load at $x_n = L$.

Comparing (8.57) and (8.58) we conclude that these equations have the same character of functional dependence. Quantitatively coinciding results are obtained provided that

$$\sqrt{\alpha / K_n} = 2\kappa_n, \tag{8.59}$$

$$\frac{S K_n}{F_n} \sqrt{\frac{\alpha}{K_n}} = \frac{S_n E_n^2 \left|\lambda_n\right|^2}{\omega^4 S_{pn}}. \tag{8.60}$$

Combining them we obtain

$$K_n = \frac{\alpha}{\left(2\kappa_n\right)^2}, \tag{8.61}$$

$$S = S_n \frac{2\kappa_n E_n^2 \left|\lambda_n\right|^2}{\alpha\omega^2} \frac{F_n}{\omega^2 S_{pn}}. \tag{8.62}$$

Recall the following expressions for λ_n and κ_n due to eqs. (8.12) and (8.14)

$$\lambda_n^2 = \frac{\omega^2}{E_n} \left[\rho + \int_0^\infty \frac{c(k)\, dk}{k^2 - \omega^2 / (1 + i\psi_k)} \right], \tag{8.63}$$

$$\lambda_n = \mu_n - i\kappa_n. \tag{8.64}$$

We introduce the following non-dimensional parameters

$$\bar{\lambda}^2 = \frac{1}{\rho_0}\left[\rho + \int_0^\infty \frac{c(k)\,dk}{k^2 - \omega^2/(1+i\psi_k)}\right], \tag{8.65}$$

$$\bar{\lambda} = \bar{\mu} - i\bar{\kappa}, \tag{8.66}$$

and the velocity of the elastic waves along the axis x_n

$$a_n = \sqrt{\frac{E_n}{\rho_0}}, \tag{8.67}$$

where ρ_0 is the total mass per unit length of the object.

By means of these denotations we obtain

$$S = S_n a_n \frac{2\omega\bar{\kappa}\rho_0^2\left|\bar{\lambda}\right|^2}{\alpha}\frac{F_n}{\omega^2 S_{pn}}. \tag{8.68}$$

As the ratio F_n/S_{pn} may depend only on the parameters of external excitation and frequency ω we take

$$F_n = \omega^2 S_{pn}. \tag{8.69}$$

Let us analyse the first quotient in eq. (8.68) which is seen to be independent on n. Let us equate it to unit, then

$$S = S_n a_n, \tag{8.70}$$

$$\alpha = 2\omega\bar{\kappa}\rho_0^2\left|\bar{\lambda}\right|^2 \tag{8.71}$$

and, then due to eq. (8.61),

$$K_n = a_n^2\frac{\rho_0^2\left|\bar{\lambda}\right|^2}{2\omega\bar{\kappa}}. \tag{8.72}$$

Thus, eqs. (8.71) and (8.72) determine the factor of spatial decay and vibroconductivity factors K_n, respectively.

Let us proceed to eq. (8.70) and answer the following question: what is an expression for S in the three-dimensional case which reduces to eq. (8.70) in the one-dimensional case? Assuming that it is a linear function of S_n we easily find

$$S = \sum_{n=1}^{3} a_n S_n.$$

Therefore, the part of the quasi-temperature is played by a weighted frequency-averaged value of spectral densities of vibrational accelerations along the orthotropic axes, the coefficients being the corresponding velocities of sound. Hence, each frequency component of vibration requires its own "temperature", meaning that there is a difference between vibrational and thermal problems.

The spectral density of the stress which is normal to the border $x = x_n$ appears in eq. (8.69). Hence, in the general case, eq. (8.69) must be replaced by the following one

$$F = \omega^2 S_{pN},$$

where S_{pN} denotes the spectral density of the external normal stress (we assume that any tangential load is absent).

Therefore, an expression for the external load in the theory of vibroconductivity is obtained. The vibrocapacity γ remains undetermined. In order to find it we consider a non-stationary problem of vibroconductivity for one of the orthotropic directions. The vibroconductivity equation, the boundary and initial conditions for a semi-infinite object $x_n \geq 0$ driven at $x_n = 0$ is given by

$$x_n > 0, \qquad t > 0 \qquad K_n \frac{d^2 S}{dx_n^2} - \alpha S = \gamma \frac{\partial S}{\partial t},$$

$$x_n = 0, \qquad t > 0 \qquad -K_n \frac{dS}{dx_n} = F_n,$$

$$x_n \to \infty, \qquad t > 0 \qquad S \to 0,$$

$$x_n \geq 0, \qquad t = 0, \qquad S = 0,$$

where the load F_n is assumed to be time-independent.

The formal solution of this problem is taken from [63] or [10]

$$\begin{aligned} S \;=\; & \frac{F_n}{2K_n} \int_{x_n}^{\infty} \left\{ \exp\left(-\sqrt{\frac{\alpha_n}{K_n}} x_n\right) \left[1 - \Phi\left(\frac{x_n - 2t\sqrt{\alpha K_n}/\gamma}{2\sqrt{tK_n/\gamma}}\right)\right] + \right. \\ & \left. + \exp\left(\sqrt{\frac{\alpha_n}{K_n}} x_n\right) \left[1 - \Phi\left(\frac{x_n + 2t\sqrt{\alpha K_n}/\gamma}{2\sqrt{tK_n/\gamma}}\right)\right] \right\} dx_n, \end{aligned} \tag{8.73}$$

with $\Phi(z)$ being the probability integral.

Recall an asymptotic property of the probability integral for large values of the argument $|z| \gg 1$

$$\Phi(z) \approx \operatorname{sgn} z \tag{8.74}$$

or more exactly

$$|\operatorname{sgn} z - \Phi(z)| \ll \exp(-z^2). \tag{8.75}$$

Analysis of eq. (8.73), together with the asymptotic representations (8.74) and (8.75), shows that the value of the integral is asymptotically small if

$$x_n \gg 2t\sqrt{\alpha K_n}/\gamma.$$

On the other hand, the integral has a finite value if

$$x_n \ll 2t\sqrt{\alpha K_n/\gamma}.$$

Hence, a zone of rapid increase of the quasi-temperature is localised near the cross-section

$$x_n = 2t\sqrt{\alpha K_n/\gamma},$$

which means that this zone progresses with the following velocity

$$v = 2\sqrt{\alpha K_n/\gamma}. \tag{8.76}$$

In the analogous mechanical problem, the wave travels with velocity $a_n/\bar{\mu}$ where $\bar{\mu}$ is given by eq. (8.64). Equating these velocities yields an equation for γ

$$2\sqrt{\alpha K_n/\gamma} = \frac{a_n}{\bar{\mu}}.$$

From this equation we obtain due to eqs. (8.72) and (8.71)

$$\gamma = 2\bar{\mu}\rho_0^2 \left|\bar{\lambda}\right|^2. \tag{8.77}$$

Therefore, the vibrocapacity is determined. It does not depend upon n, which conforms to the general idea of the approach.

References

[1] Afanasiev, N.N., Statistical theory of fatigue strength (in Russian). Publishers of the Academy of Science of the Ukrainian SSR, 1953.

[2] Arutyunyan, N.Kh., Abramyan, B.L., Torsion of elastic bodies (in Russian). Fizmatgiz, 1963.

[3] Arutyunyan, R.A., Vakulenko, A.A., On repeated loading of an elastoplastic medium (in Russian), MTT, No. 4, 1965.

[4] Astafiev, V.I., Meshkov, S.I. Forced vibrations of a semi-infinite rod of nonlinear hereditary-elastic material (in Russian). MTT, No. 4, 1970.

[5] Bashta, O.T., Matveev, V.V., On influence of the cross-sectional form and sort of deformation of rods on their vibration damping (in Russian). Problem of Strength No. 10, 1971.

[6] Bellman, R., Introduction to matrix analysis, 2nd ed.. McGraw-Hill, New York, 1970.

[7] Belyaev, A.K., Propagation of plane waves in anisotropic medium having a complex structure. Soviet Applied Mechanics (USA), vol. 14, No. 5, pp. 490-494, 1978.

[8] Belyaev, A.K., Dynamical simulation of high-frequency vibration of extended complex structures. Int. Journ. Mechanics of Structures and Machines, vol. 20, No. 2, pp. 155-168, 1992.

[9] Belyaev, A.K., Irschik, H., Non-linear waves in complex structures modelled by elastic-viscoplastic stochastic media. International Journal of Non-Linear Mechanics, vol. 31, No. 5, pp. 771-777, 1996.

[10] Belyaev, A.K., Palmov, V.A., Integral theories of random vibration of complex structures. In: Random Vibration - Status and Recent Developments, Eds. I. Elishakoff and R.H. Lyon, Elsevier, Amsterdam, pp. 19-38, 1986.

[11] Berg, G.V., Da Peppo, D.A., Dynamic analysis of elastoplastic structures. Proc. ASCE, vol. 86, EM 2, 1960.

[12] Besseling, J.F., A theory of elastic, plastic and creep deformations of an initially isotropic material, showing anisotropic strain-hardening creep recovery and secondary creep. Journ. Applied Mechanics, vol. 27, No. 4, 1960.

[13] Besseling, J.F., van der Giessen, E., Mathematical modelling of inelastic deformation. Chapman and Hall, London, 1994.

[14] Bland, D. The theory of linear viscoelasticity. Pergamon Press, Oxford, 1966.

[15] Blombergen, N., Nonlinear optics (in Russian). Mir, 1966.

[16] Bogolyubov, N.N., Mitropolsky, Yu.A., Asymptotic methods in theory of nonlinear vibrations (in Russian). Fizmatgiz, 1958.

[17] Bolotin, V.V., Statistical methods in structural dynamics (in Russian). Gosstroiizdat, 1965.

[18] Bolotin, V.V., On elastic vibrations caused by random forces with broad band spectra (in Russian). Izv. Vuzov, Mashinostroenie, No. 4, 1963.

[19] Bolotin, V.V., Application of the methods of probability theory in the theory of plates and shells (in Russian). In: Theory of shells and plates. Proceedings of the 4th USSR Conference on theory of shells and plates. Publishers of the Academy of Science of Armenian SSR, 1964.

[20] Bolotin, V.V., Fringe effect in vibrating elastic shells (in Russian). PMM, No.5, 1960.

[21] Bolotin, V.V., Dynamic fringe effect in the elastic vibration of plates (in Russian). Inzhenernyi Zbornik, vol. 31, 1960.

[22] Bolotin, V.V., Makarov, B.P., Mishenkov, G.V., Shveiko, Yu.Yu., Asymptotic method for the investigation of the spectrum of eigenfrequencies of elastic plates (in Russian). In: "Raschety na prochnost" No. 6, Mashgiz, 1960.

[23] Book, R., Random excitation of a system with hysteresis (in Russian). In: "Mechanika", collection of translations, Mir, No. 6, 1968.

[24] Coughay, T.K., Sinusoidal excitation of a system with bilinear hysteresis. Journ. Appl. Mech. vol. 27, No. 4, 1960.

[25] Coughay, T.K., Forced vibration of a semi-infinite rod exhibiting weak bilinear hysteresis. Journ. Appl. Mech. vol. 27, No. 4, 1960.

[26] Coughay, T.K., Random excitation of a system with bilinear hysteresis. Journ. Appl. Mech. vol. 27, No. 4, 1960.

[27] Coughay, T.K., Equivalent linearisation technique. Journ. Acoust. Soc. Am. vol. 35, No. 11, 1963.

[28] Davidenkov, N.N., On energy dissipation under vibrations (in Russian). ZhTF, vol. 8, No. 6, 1938.

[29] Davies, H., Power flow between two coupled beams. Journ. Acoust. Soc. Am. vol. 51, pp. 393-401, 1972.

[30] Davies, H., Random vibration of distributed systems strongly coupled at discrete points. Journ. Acoust. Soc. Am. vol. 54, pp. 507-515, 1973.

[31] Daer, I., Vibration of a spacecraft body caused by the noise of the rocket engine. In: "Random Vibration", Ed. S.H. Crandall, Cambridge, Mass., Technology Press of the Massachusetts Institute of Technology, 1958.

[32] Dobroslavsky, V.L., On models and mathematical description of elastic links with hysteresis (in Russian). In: Energy dissipation in vibrating elastic systems, Naukova dumka, 1968.

[33] Duszek, M.K., Perzyna, P., Stein, E., Adiabatic shear band localization in elastic-plastic damaged solids. Int. J. Plasticity, vol. 8, pp. 361-384, 1992.

[34] Eichler, E., Thermal circuit approach to vibrations in coupled systems and noise reduction of a rectangular box. Journ. Acoust. Soc. Am. vol. 37, pp. 995-1007, 1965.

[35] Esin, A., Jones, W.J., A statistical approach to microplastic strains in metals. Journ. Strain. Analysis, vol. 1, No. 5, 1966.

[36] Evgrafov, M.A., Asymptotic estimations and entire functions (in Russian). Fizmatgiz, 1962.

[37] Fung, Y.C., Foundations of solid mechanics. New Jersey, Prentice-Hall, Englewood Cliffs, 1965.

[38] Gantmakher, F.R., Lectures on analytical mechanics (in Russian). Nauka, 1966.

[39] Gersch, W., Average power and power exchange in oscillators. Journ. Acoust. Soc. Am. vol. 46, pp. 1180-1185, 1969.

[40] Godunov, S.K., Ramensky, E.I., Non-stationary equations of nonlinear theory of elasticity in Eulerian coordinates (in Russian). PMTF, No. 6, 1972.

[41] Goldenweiser, A.L., Theory of elastic thin shells (in Russian). Gostekhizdat, 1953.

[42] Goldenweiser, A.L., On density of eigenfrequencies of a thin shell (in Russian). PMM, No. 5, 1970.

[43] Goncharenko, V.M., On the influence of energy dissipation of the random vibration in elastic elements (in Russian). In: Energy dissipation in vibrating elastic systems, Naukova dumka, 1968.

[44] Gradshtein, N.S., Ryzhik, I.M., Tables of integrals, sums, series and products (in Russian). Fizmatgiz, 1962.

[45] Green, A.E., Rivlin, R.S., Multipolar continual mechanics, Arch. Rat. Mech. Anal. vol. 17, 1964.

[46] Green, A.E., Zerna, W., Theoretical elasticity. Oxford, Clarendon Press, 1954.

[47] Gusenkov, A.P., Properties of diagrams of cyclic deformations at normal temperatures (in Russian). In: Resistance to deformation and failure at small numbers of cycles. Nauka, 1967.

[48] Heading, J., An introduction to phase-integral method. New York, Wiley, 1962.

[49] Hill, R., The mathematical theory of plasticity. Clarendon Press, Oxford, 1950.

[50] Hill, R., Macroscopic deformation measures and works on plastic deformations of microheterogeneous medium (in Russian). PMM, No. 1, 1971.

[51] Holzapfel, G.A., Simo, J.C., A new viscoelastic constitutive model for continuous media at finite thermomechanical changes. Int. J. Solids and Structures, vol. 33, No. 20-22, pp. 3035-3056, 1996.

[52] Ilyushin, A.A., Plasticity (in Russian). Publishers of the Academy of Science of the USSR, 1963.

[53] Irschik, H., Ziegler, F., Dynamic processes in structural thermo-viscoplasticity. Appl. Mech. Rev., vol. 48, No. 6, pp. 301-315, 1995.

[54] Ishlinsky, A.Yu., Some applications of statistics to description of laws of body deformation (in Russian). Reports of the Academy of Science of the USSR, ONT, No. 9, 1944.

[55] Ishlinsky, A.Yu., On equations of spatial deformation of not completely elastic and elastoplastic bodies (in Russian). Reports of the Academy of Science of the USSR, ONT, No. 3, 1945.

[56] Ishlinsky, A.Yu., General theory of plasticity with linear hardening (in Russian). Ukrainian Math. Journ. vol. 6, No. 3, 1954.

[57] Iwan, W.D., Amplitude-frequency characteristics of a system with two degrees of freedom and a rectilinear hysteresis loop (in Russian). Applied Mechanics, Mir, No. 1, 1965.

[58] Iwan, W.D., A distributed element model for hysteresis and its steady state dynamic response. Trans. ASME, ser. E, vol. 88, 1966.

[59] Iwan, W.D., On a class of models for the yielding behaviour of continuous and composite systems. Trans. ASME, ser. E, vol. 89, 1967.

[60] Iwan, W.D., Response of multi-degree of freedom yielding systems. Proc. ASCE, vol. 94, EM2, 1968.

[61] Iwan, W.D., On the steady state response of a one-dimensional yielding continuum. Trans. ASME, ser. E, vol. 92, 1970.

[62] Iwan, W.D., Lutes, L.D., Response of the bilinear hysteretic system to stationary random excitation. Journ. Acoust. Soc. Am. vol. 43, No. 3, 1968.

[63] Jeffreys, H., Swirles, B., Methods of mathematical physics, 3rd edition. Cambridge University Press, Cambridge, 1966.

[64] Jennings, P.G., Periodic response of a general yielding structure. Proc. ASCE, vol. 90, EM2, 1964.

[65] Kachanov, L.M., Foundations of plasticity theory (in Russian). Fizmatgiz, 1969.

[66] Kadashevich, Yu.I., A generalised theory of plastic flow (in Russian). In: "Investigations on elasticity and plasticity", Publishers of the Leningrad University, 1967.

[67] Kadashevich, Yu.I., On cyclic deformation of metals (in Russian). In: Strength at small numbers of cycles. Nauka, 1969.

[68] Kadashevich, Yu.I., Novozhilov, V.V., Theory of plasticity accounting for the Bauschinger effect (in Russian). Reports of the Academy of Science of the USSR, vol. 116, No. 4, 1957.

[69] Kadashevich, Yu.I., Novozhilov, V.V., On account of microstresses in plasticity theory (in Russian). MTT, No. 3, 1968.

[70] Kadashevich, Yu.I., Novozhilov, V.V., On the influence of initial microstresses on macroscopic deformation in polycristals (in Russian). PMM, No. 5, 1968.

[71] Kafka, V., Der Einfluß der mikroplastischen Homogenität auf die elastisch-plastischen Verformungsgesetze. ZAMM, Bd. 46, No. 8, 1966.

[72] Kaliski, S., Wlodarczyk, E., The problem of resonance for longitudinal elasto-plastic waves in a finite bar. Proc. of Vibration Problem, vol. 8, No. 1, 1967.

[73] Kaliski, S., Wlodarczyk, E., Resonance of a longitudinal shock wave in an elastic-plastic bar. Proc. of Vibration Problem, vol. 8, No. 2, 1967.

[74] Kaliski, S., Wlodarczyk, E., Resonance of longitudinal elastoviscoplastic waves in a finite bar. Proc. of Vibration Problem, vol. 8, No. 2, 1967.

[75] Kaliski, S., Wlodarczyk, E., On certain closed-form solutions of the problem of propagation and reflections of elastoviscoplastic waves in bars. Archiwum Mech. Stosowanje, vol. 19, No. 3, 1967.

[76] Kantorovich, L.V., Krylov, V.I., Methods of approximate solution of partial differential equations (in Russian). ONTI, 1936.

[77] Karnopp, D., Scharton, T.D., Plastic deformation in random vibration. Journ. Acoust. Soc. Am. vol. 39, No. 6, 1966.

[78] Kazakov, I.E., Dostupov, B.G., Statistical dynamics of nonlinear control systems (in Russian). Fizmatgiz, 1962.

[79] Khan, A.S., Huang, S., Continuum theory of plasticity. Willey & Sons, 1995.

[80] Khilchevsky, V.V., On the influence of the stress state type on energy dissipation in materials (in Russian). In: Proceedings of the scientific-technological seminar on damping of vibrations, Publishers of Academy of Science of the Ukrainian SSR, 1960.

[81] Khilchevsky, V.V., On energy dissipation under bending-torsional vibrations (in Russian). Vestnik KPI, "Mashinostroenie", No. 6, 1969.

[82] Khilchevsky, V.V., On the influence of the asymmetry of the cycle on the decay of torsional vibrations (in Russian). Problems of Strength, No. 2, 1969.

[83] Khilchevsky, V.V., Investigation of decay of independent bending-torsional vibrations (in Russian). Problems of Strength, No. 3, 1970.

[84] Khilchevsky, V.V., Petushkov, V.G., Fedorov, V.A., Method of investigation of the energy dissipation in a special case of plane stress state (in Russian). Zavodskaya laboratoriya, vol. 35, No. 7, 1969.

[85] Koiter, W.T., Stress-strain relations, variational theorems and uniqueness theorem for elastoplastic materials with a singular yield surface (in Russian). In: "Mekhanika", collection of translations, Mir, No. 2, 1960.

[86] Kolovsky, M.Z., Nonlinear theory of systems of vibration protection (in Russian). Nauka, 1966.

[87] Kolovsky, M.Z., Pervozvansky, A.A., On linearisation by means of the distribution function in problems of the theory of nonlinear vibration (in Russian). Reports of the Academy of Science of the USSR, OTN, Mechanics and Mechanical Engineering, No. 5, 1962.

[88] Kolovsky, M.Z., Pervozvansky, A.A., Application of the method of distribution functions for determining polyharmonic solutions of nonlinear equations (in Russian). Izv. Vuzov "Mashinostroenie" No. 4, 1963.

[89] Kröner, E., Zur plastischen Verformung des Vielkristals. Acta Metallurgica, vol. 9, No. 2, 1961.

[90] Kröner, E., A new concept in the continuum theory of plasticity. Journ. Math. and Phys. vol. 42, No. 1, 1963.

[91] Kuvshinsky, E.V., Aero, E.L., Continual theory of asymmetric elasticity, account of internal rotation (in Russian). Physics of Solids, vol. 5, No. 9, 1963.

[92] Kulemin, A.B., Mitskevich, A.M., On energy losses in metals at low ultrasound frequencies (in Russian). Akustichesky Zhurnal, vol. 16, No. 2, 1970.

[93] Kuznetsov, D.S., Special functions (in Russian). Vysshaya shkola, 1965.

[94] Leibenson, L.S., Course of elasticity theory (in Russian). Gostekhizdat, 1947.

[95] Lensky, V.S., On a problem of elastoplastic wave propagation (in Russian). Vestnik MGU, Phys.-Math. Series, No. 3, 1949.

[96] Leonov, A.I., On the description of the rheological behaviour of elastoviscous media under large elastic deformations (in Russian). Institute for Problems in Mechanics, Academy of Science of the USSR, preprint No. 34, 1973.

[97] Lighthill, J., Waves in fluids. Univ. Press, Cambridge, 1978.

[98] Lotz, R., Crandall, S., Prediction and measurement of the proportionality constant in statistical energy analysis of structures. Journ. Acoust. Soc. Am. vol. 54, pp. 516-524, 1973.

[99] Lurie, A.I., Theory of elasticity (in Russian), Nauka, 1970.

[100] Lurie, A.I., Nonlinear theory of elasticity. North-Holland, Amsterdam, 1990.

[101] Lutes, L.D., Approximate technique for treating random vibration in hysteretic systems. Journ. Acoust. Soc. Am. vol. 48, No. 1, 1970.

[102] Lyon, R., Maidanik, G., Power flow between linearly coupled oscillators. Journ. Acoust. Soc. Am. vol. 34, pp. 623-639, 1962.

[103] Lyon, R., Scharton, T., Vibration energy transmission in a three element structures. Journ. Acoust. Soc. Am. vol. 38, pp. 253-261, 1965.

[104] Mason, W., Physical Acoustics and the properties of solids. Van Nostrand, Princeton N.J.. 1958.

[105] Matveev, V.V., Yakovlev, A.P., Risavin, L.I., Chaikovsky, B.S., On the influence of statical tension in transversely vibrating material (in Russian). In: Energy dissipation in vibrating elastic systems, Naukova dumka, 1968.

[106] Mindlin, R.D., Microstructure in linear elasticity (in Russian). In "Mekhanika", collection of translations, No. 4, Mir, 1964.

[107] Mironov, M.V., On propagation of longitudinal vibrations with slowly-varying parameters in rods (in Russian). MTT, No. 4, 1969.

[108] Mitropolsky, Yu.A., Moseenkov, V.I., Lectures on the application of asymptotic methods to solving partial differential equations (in Russian). Publishers of the Academy of Science of the Ukrainian SSR, 1968.

[109] Moskvitin, V.V., Plasticity under variable loadings (in Russian). Publishers of MGU, 1965.

[110] Muskhelishvili, N.I., Some basic problems of the theory of elasticity. Noordhoff, Groningen, 1953.

[111] Newland, D., Power flow between a class of coupled oscillators. Journ. Acoust. Soc. Am. vol. 43, pp. 553-559, 1968.

[112] Novikov, N.V., Influence of the stress state on energy dissipation in vibrating materials (in Russian). In: Proceedings of the scientific-technological seminar on damping of vibrations, Publishers of Academy of Science of the Ukrainian SSR, 1960.

[113] Novikov, N.V., Characteristics of energy dissipations under longitudinal and torsional vibrations (in Russian). In: Energy dissipation in vibrating elastic systems, Publishers of Academy of Science of the Ukrainian SSR, 1963.

[114] Novozhilov, V.V., On complex loading and perspectives of the phenomenological approach to analysis of microstresses (in Russian). PMM, No. 3, 1964.

[115] Osetinsky, Yu.V., On methods of accounting for internal dissipation of vibrational energy (in Russian). Engineering Journal, No. 4, 1964.

[116] Osiecki, J., Propagation of plastic strain waves in a semi-infinite bar produced by a periodic load. Proc. Vibr. Prob., vol. 3, No. 2, 1962.

[117] Palmov, V.A., The main equations of the theory of non-symmetric elasticity (in Russian). PMM, No. 3, 1964.

[118] Palmov, V.A., Thin plates subject to broad-band random load (in Russian). Proceedings of LPI, No. 252, Mashgiz, 1965.

[119] Palmov, V.A., Thin shells subject to broad-band random load (in Russian). PMM, No. 4, 1965.

[120] Palmov, V.A., Correlational properties of the vibrational field in a plate subjected to broad band random load (in Russian). MTT, No. 5, 1967.

[121] Palmov, V.A., Propagation of vibration in a nonlinear dissipative medium (in Russian). PMM, No. 4, 1967.

[122] Palmov, V.A., Propagation of random vibration in a viscoelastic rod (in Russian). In: Reliability Problems in Structural Mechanics, Vilnus, 1968.

[123] Palmov, V.A., Propagation of vibration in a nonlinear inelastic rod (in Russian). Proceedings of LPI, No. 307, Mashgiz, 1969.

[124] Palmov, V.A., Propagation of random vibration in a rod with nonlinear properties (in Russian). PMM, No. 3, 1969.

[125] Palmov, V.A., On a model of a medium of complex structure (in Russian). PMM, No. 4, 1969.

[126] Palmov, V.A., Propagation of vibration in a nonlinear medium (in Russian). In: Proceedings of 5th Int. Conference on nonlinear vibrations, vol. 3, Publishers of Academy of Science of the Ukrainian SSR, 1970.

[127] Palmov, V.A., Propagation of vibration in an elastoplastic rod (in Russian). In: "Waves in inelastic media", Publishers of Academy of Science of the Moldavian SSR, 1970.

[128] Palmov, V.A., Random vibrations of elastoplastic plates (in Russian). In: Reliability Problems in Structural Mechanics, Vilnus, 1971.

[129] Palmov, V.A., Vibrations of elastoplastic bodies (in Russian). MTT, No. 4, 1971.

[130] Palmov, V.A., Random vibrations of elastoplastic bodies (in Russian). MTT, No. 5, 1971.

[131] Palmov, V.A., Description of high-frequency vibration of complex dynamical objects by the method of the theory of thermal conductivity (in Russian). In: Selected Methods of Applied Mechanics, dedicated to 60th anniversary of V.N. Chelomei, pp. 214-221, 1974.

[132] Palmov, V.A., Integral methods for the analysis of vibration of dynamic structures (in Russian). Advances in Mechanics (Warsaw), vol. 2, No. 4, pp. 3-24, 1979.

[133] Palmov, V.A., Rheological models in nonlinear mechanics of solids (in Russian). Advances in Mechanics (Warsaw), vol. 3, No. 3, pp. 75-115, 1980.

[134] Palmov, V.A., Rheological models in nonlinear thermodynamics (in Russian). ZAMM, vol. 64, pp. 481-484, 1984.

[135] Palmov, V.A., Verification of mathematical models of plasticity and viscoplasticity by means of nonlinear thermodynamics. In: Proc. 4th Int. Conf. on Computational Plasticity. Fundamentals and Applications (COMPLAS IV), eds. D.R.J. Owen and E. Oñate. Swansea, Pineridge Press, pp. 957-962, 1995.

[136] Palmov, V.A., Comparison of decomposition methods in nonlinear viscoelasticity and elastoplasticity. ZAMM, vol. 77, Supplement 2, pp. 643-644, 1997.

[137] Palmov, V.A., Large strains in viscoelastoplasticity. Acta Mechanica, vol. 125, No. 1-4, pp. 129-139, 1997.

[138] Palmov, V.A., Pupyrev, V.A., Forced vibrations of cylindrical shells containing an acoustical medium subject to broad band and high frequency random loading (in Russian). Proceedings of LPI, No. 279, Mashgiz, 1967.

[139] Palmov, V.A., Pupyrev, V.A., Vibration and sound radiation of a plate subject to a random load (in Russian). Akustichesky Zhurnal, No. 2, 1967.

[140] Panovko, Ya.G., Internal friction of vibrating elastic systems (in Russian). Fizmatgiz, 1960.

[141] Pervozvansky, A.A., Random processes in nonlinear control systems (in Russian). Fizmatgiz, 1962.

[142] Pisarenko, G.S., Vibration in elastic systems with account of energy dissipation (in Russian). Publishers of Academy of Science of the Ukrainian SSR, 1955.

[143] Pisarenko, G.S., Energy dissipation under mechanical oscillations (in Russian). Publishers of Academy of Science of the Ukrainian SSR, 1962.

[144] Pisarenko, G.S., Vibration in elastic systems with account of non-ideal material elasticity (in Russian). Publishers of Academy of Science of the Ukrainian SSR, 1970.

[145] Pisarenko, G.S., Boginich, O.E., Account of the energy dissipation of cyclic deformable materials in the plane stress state applied to transverse vibrations in plates (in Russian). Problems of Strength, No. 9, 1970.

[146] Pisarenko, G.S., Khilchevsky, V.V., Goncharov, T.I., Investigation of energy dissipation in material at bending in a field of statically normal stresses, In: Energy dissipation in vibrating elastic systems, Naukova dumka, 1968.

[147] Pisarenko, G.S., Vashchenko, K.I., Khilchevsky, V.V., Snezhko, A.A., Investigation of energy dissipation in magnesium cast iron with globular graphite for bending-torsional vibrations (in Russian). In: Energy dissipation in vibrating elastic systems, Naukova dumka, 1968.

[148] Pisarenko, G.S., Vasilenko, N.V., Yakovlev, A.P., Energy dissipation in rods for various kinds of vibration (in Russian). In: Energy dissipation in vibrating elastic systems, Gostekhizdat of the Ukrainian SSR, 1962.

[149] Popov, E.P., Paltov, I.P., Approximate methods of investigation of nonlinear control systems (in Russian). Fizmatgiz, 1960.

[150] Postnikov, V.S., Internal friction in metals (in Russian). Metallurgiya, 1969.

[151] Prager, W., A new method of analysing stresses and strains in work-hardening plastic solids. Journ. Appl. Mech. vol. 23, No. 4, 1956.

[152] Prager, W., Elementary analysis of determining of stress rates (in Russian). "Mekhanika", collection of translations, No. 3, Mir, 1960.

[153] Prager, W., Einführung in die Kontinuumsmechanik. Birkhauser, Basel & Stuttgart, 1961.

[154] Pabotnov, Yu.N., Stresses and strains under cyclic loading (in Russian). PMM, No. 1, 1952.

[155] Rakhmatullin, Kh.A., Shapiro, G.S., On propagation of plane elasto-plastic waves (in Russian). PMM, No. 4, 1948.

[156] Reiner, M., Rheology. Springer-Verlag, Göttingen Heidelberg, 1958.

[157] Reiner, M., Rheology (in Russian). Nauka, 1965.

[158] Roberts, J.B., Spanos, P.D., Random vibration and statistical linearization. Wiley & Sons, Chichester, 1990.

[159] Robertson, I.B., Yorgiadis, A.S., Internal friction in engineering materials. Journ. Appl. Mech. vol. 13, No. 3, 1946.

[160] Scharton, T.D., Frequency averaged power flow into a one-dimensional acoustic system. Journ. Acous. Soc. Am. vol. 50, pp. 373-381, 1971.

[161] Scharton, T.D., Lyon, R.H., Power flow and energy sharing in random vibration. Journ. Acous. Soc. Am. vol. 43, pp. 1332-1343, 1968.

[162] Sedov, L.I., Mechanics of continuous media (in Russian). vol. 1 and 2, Nauka, 1976.

[163] Sedov, L.I. , Mathematical methods of construction of models of continuous medium (in Russian). Uspekhi Mat. Nauk, vol. 20, No. 5, 1965.

[164] Shapiro, G.S., Longitudinal vibrations in a rod (in Russian). PMM, No. 5-6, 1946.

[165] Shatalov, K.T., Forced vibrations of a linear chain-like systems accounting for all internal and external frictions (general solution of the problem) (in Russian). Publishers of Academy of Science of the USSR, Institute for Problems in Mechanics, 1949.

[166] Simo, J.C., A framework for finite strain elastoplasticity based on maximum plastic dissipation and multiplicative decomposition: Part 1, Continuum formulation. Comp. Meth. Appl. Mech. Engineering, vol. 66, pp. 199-219, 1988.

[167] Simo, J.C., Miehe, C., Associate coupled thermoplasticity at finite strains: Formulation, numerical analysis and implementation. Comp. Meth. Appl. Mech. Engineering, vol. 98, pp. 41-104, 1992.

[168] Slepyan, L.I., Wave of deformation in a rod with flexible mounted masses (in Russian). MTT, No. 5, 1967.

[169] Sokolnikoff, I.S., Mathematical theory of elasticity. New York, McGraw-Hill, 1956.

[170] Sorokin, E.S., Method of consideration of inelastic resistance for vibrational analysis of structures (in Russian). Stroiizdat, 1951.

[171] Sorokin, E.S., On the theory of internal friction in vibrating elastic systems (in Russian). Gosstroiizdat, 1960.

[172] Spenser, A.I., Rivlin, R.S., The theory of matrix polynomials and its application to the mechanics of isotropic continua. Archive for Rational Mechanics and Analysis, vol. 2, No. 4, 1959.

[173] Steklov, V.A., On asymptotic behaviour of solutions of linear differential equations (in Russian). Publishers of the Kharkov University, 1956.

[174] Strutt, J.W., The theory of sound. MacMillan, London, 1926.

[175] Szemplinska-Stupnicka, W., On the asymptotic, averaging and Ritz method in the theory of steady-state vibrations of nonlinear systems with many degrees of freedom. Arch. Mech. Stosow. vol. 22, No. 2, 1970.

[176] Timoshenko, S.P., Woinowsky-Krieger, S., Theory of plates and shells. McGraw-Hill, New York, 1959.

[177] Timoshenko, S.P., Strength of materials. Van Nostrand, Princeton N.J., 1960.

[178] Truesdell, C., A first course in rational continuum mechanics (in Russian). Mir, 1975.

[179] Truesdell, C., A first course in rational continuum mechanics. Academic Press, New York, 1977.

[180] Truesdell, C., Rational thermodynamics. McGraw-Hill, 1969.

[181] Tsu, T.C., Interplanetary travel by solar sail. ARS Journal, vol. 26, No. 6, 1959.

[182] Vasilenko, N.V., Influence of the hysteresis loop form on characteristics of vibrational motion (in Russian). In: Energy dissipation in vibrating elastic systems, Gostekhizdat of the Ukrainian SSR, 1962.

[183] Vasilenko, N.V., Stress-strain relation in real isotropic bodies (in Russian). In: Energy dissipation in vibrating elastic systems, Naukova dumka, 1966.

[184] Vasilenko, N.V., Small forced elastoplastic vibration in systems with distributed parameters (in Russian). In: Problem of Strength, No. 9, 1970.

[185] Vashchenko, K.I., Khilchevsky, V.V., Snezhko, A.A. Cyclic ductility of magnesium cast iron (in Russian). In: Casting, No. 9, 1970.

[186] White, P., Sound transmission through a finite, closed, cylindrical shell. Journ. Acous. Soc. Am. vol. 40, pp. 1124-1130, 1966.

[187] Witeman, I.R., A mathematical model depicting the stress-strain diagram and the hysteresis loop. Journ. Appl. Mech. vol. 26, No. 3, 1959.

[188] Wittaker, E.T., Watson, G.N., A course of modern analysis, 4th ed.. Cambridge University Press, London, 1952.

[189] Wittaker, E.T., Analytical dynamics. Dower Publications, New York, 1944.

[190] Yaglom, A.M., Correlation theory of stationary and related random functions. Springer-Verlag, New York, Berlin, 1987.

[191] Zarembo, L.K., Krasilnikov, V.A., Introduction to nonlinear acoustics (in Russian). Nauka, 1966.

[192] Ziegler, F., Mechanics of solids and fluids. Springer-Verlag, New York Vienna, 1995.

Index

MIX
Papier aus verantwortungsvollen Quellen
Paper from responsible sources
FSC® C105338

If you have any concerns about our products,
you can contact us on
ProductSafety@springernature.com

In case Publisher is established outside the EU,
the EU authorized representative is:
Springer Nature Customer Service Center GmbH
Europaplatz 3, 69115 Heidelberg, Germany

Printed by Libri Plureos GmbH
in Hamburg, Germany